Transportation Engineering and Technology

Volume I

Transportation Engineering and Technology
Volume I

Edited by **Samuel Morgan**

CLANRYE INTERNATIONAL

New Jersey

Published by Clanrye International,
55 Van Reypen Street,
Jersey City, NJ 07306, USA
www.clanryeinternational.com

Transportation Engineering and Technology
Volume I
Edited by Samuel Morgan

International Standard Book Number: 978-1-63240-497-8 (Hardback)

Printed in the United States of America.

Contents

Preface

Transportation Technologies is a discipline which deals with creating innovative ways to run various modes of transport. Principles of civil engineering and mechanical engineering are integral components of the discipline of transportation technologies.

Conventional transportation technologies have always been premised on the use of fossil fuels for vehicle propulsion. Though the conventional modes continue, the gradual depletion of fossil resources has led to rising fuel prices. This has compelled an interest in alternative sources of fuel energy. In order to reduce dependence on fossil fuels and prevent the harmful environmental effects of fuel energy, today, automobile companies are working to develop more sustainable vehicles.

Numerous methods have been created and are in the process of development to reduce fossil fuel dependency. Common algae are being used for biofuel production. Solar powered electric hybrid vehicles have been developed as an alternative vehicle. In order to accommodate the demands created by increased air traffic, experiments are being conducted to develop quieter subsonic and supersonic commercial planes. This will help accommodate the demands created by increased air traffic and will consume less fuel and pollute less.

Thus, Transportation Technologies is an emerging discipline and there's a lot of interesting work happening in this discipline. I'd like to thank all the researchers who've shared their latest works with us in this book. I would personally like to thank the publishing house for giving me this opportunity, and my family for their constant support.

Editor

FPGA-Based Traffic Sign Recognition for Advanced Driver Assistance Systems

Sheldon Waite, Erdal Oruklu

Department of Electrical and Computer Engineering, Illinois Institute of Technology, Chicago, USA

ABSTRACT

This paper presents the implementation of an embedded automotive system that detects and recognizes traffic signs within a video stream. In addition, it discusses the recent advances in driver assistance technologies and highlights the safety motivations for smart in-car embedded systems. An algorithm is presented that processes RGB image data, extracts relevant pixels, filters the image, labels prospective traffic signs and evaluates them against template traffic sign images. A reconfigurable hardware system is described which uses the Virtex-5 Xilinx FPGA and hardware/software co-design tools in order to create an embedded processor and the necessary hardware IP peripherals. The implementation is shown to have robust performance results, both in terms of timing and accuracy.

Keywords: Traffic Sign Recognition; Advanced Driver Assistance Systems; Field Programmable Gate Array (FPGA)

1. Introduction

Advances in materials, engine design, embedded electronics, and production methods have made the personal vehicle one of the most transformative technologies of the past century [1,2]. With cars becoming almost ubiquitous in developed nations, there has also been a large rise in associated risks. According to the US Census Bureau in 2009 alone 10.8 million motor vehicle accidents occurred resulting in almost 36 thousand fatalities. As horrible as these number are, they are a significant improvement from the previous decade. In 1990 there were 11.5 million accidents resulting in 46.8 thousand deaths [3]. This marked reduction in accidents (6%) and fatalities (23%) despite the population increasing by nearly 10% [4] is a testament to the efforts being made to driver safety and accident avoidance. Technologies like airbags, antilock brakes, tire pressure monitors, and traction control have become very common if not standard. More recently a new level of intelligence and intervention has surfaced in the form of systems commonly referred to as Advanced Driver Assistance Systems (ADAS) such as lane departure warning systems, intelligent speed adaptation, and driver drowsiness detection [5]. These technologies have the capacity to greatly increase driver safety by monitoring the driver and their environment and providing information, warnings, or even taking action. *Traffic sign recognition* (*TSR*) is one of these technologies.

Over the years, as our road system has matured, road signs have become the de facto way of communicating information to the driver. These road signs communicate the local traffic laws, such as right of way and speed limits as well as information like city limits and distances to destinations. Road signs, however, are only useful if the driver notices them. Though this fact is inherently obvious, it nonetheless is worth noting because it highlights the fact that increasing a driver's ability to see road signs can increase road safety [6]. A driver may be distracted, tired, or simply overwhelmed driving in a new environment and miss an important road sign. A system that monitors the road ahead of a vehicle and detects road signs could be a great service to the driver. This information, summarizing the traffic sign topology of the area, could be displayed to the driver or used in conjunction with other vehicle information to take action such as slowing the vehicle as it approaches a stop sign.

This paper describes in detail a specific implementation of a traffic sign recognition system done in reconfigurable hardware. Following section discusses the challenges encountered for building traffic sign recognition systems and briefly describes existing research and product embodiments in this technology. The subsequent sections detail the algorithm design, the actual hardware implementation, and performance results.

A traffic sign recognition system monitors a complex and ever changing environment and must do so accurately and continuously. Here in lies the challenges for this type of system: complex environment, expectation of accuracy, and short response time. Essentially it must identify the road signs that are in view in real time. These efforts to determine the presence of a road sign in real time are complicated by the fact that the environment is continually changing. Road signs will appear significantly different to an artificial (*i.e.*, computer) vision system depending on the amount, direction, and type of light as well as weather conditions. Road signs may also be damaged or tilted confusing an automated system.

2. Background

Traffic sign recognition is an arena of active research. Although there are many different algorithms and approaches, some patterns do emerge as the existing body of work is examined. The following is a summary of some of the more recent and relevant work.

Lai *et al.* [7] present a sign recognition scheme aimed for intelligent vehicles and smart phones. Color detection is used and is performed in HSV color space. Template based shape recognition is done by using a similarity calculation. OCR is used on the pixels within the shape boarder to determine provide a match to actual sign. The description is purely algorithmic and implemented in software. Andrey *et al.* [8] use a very similar approach involving color segmentation and shape analysis. Histograms, however, are used as the shape classification method after connected regions are labeled. Actual sign recognition is done via template matching by using a weighted direct comparison of the interior portion of each shape to templates.

A different approach is used by Soendoro *et al.* [9]. Here, a binary image is created using color segmentation. Morphological filtering is used to improve the image. Shape estimation is done by an application of Ramer-Douglas-Peucker algorithm followed by classification using Support Vector Machines (SVM).

Liu *et al.* [10] limit their application to speed limit signs found in Europe, Asian, and Australia. This causes their target signs to all be outlined in a red circle. The Canny edge detection algorithm is applied followed by the Fast Radial Symmetry Transform to detect circular signs. A fuzzy template matching, a comparison of correlation coefficient is used to initially match the information within the circular sign. A character recognizer based on the local feature vector is used to make the final match selection. This algorithm is implemented in software running on a standard PC.

A novel approach is described by Kastner *et al.* [11]. They use the difference of Gaussian and Gabor filter kernels to model the characteristics of neural receptive fields measured in the mammal brain. They attempt to process an image in a similar way that human's brains do. This highlights areas of important information in a frame identifying regions of interest (ROIs). These ROIs have relevant information extracted in the form of week classifiers, essentially a probability value that corresponds to a certain traffic sign class. Their algorithm is implemented on a standard PC running software.

The bulk of the research materials on this topic have focused on algorithms that are implemented in software. A few have been targeted toward reconfigurable hardware, but even these often have all or a large portion of the work performed by an embedded processor running software. Souki *et al.* [12] describe an algorithm implemented on FPGA-Cyclone II 2C35 FPGA manufactured by Altera. They use color segmentation to create a binary image. Edge detection and Hugh Transforms are use to detect shapes. Classification uses template matching. This algorithm is implemented exclusively in software that runs on the Nios II embedded processor. Computation times exceed 17 seconds per frame.

Designed using SystemC, Muller *et al.* [13] present a system that uses a Virtex-4 LX100 FPGA with implementations of multiple embedded LEON CPU cores. Their algorithm of preprocessing, shape detection, segmentation, extraction, and classification, is initially implemented exclusively in software. To improve their performance they move the classification stage to a synthesized hardware block. With this improvement they are able to achieve a computation time of about 0.5 seconds.

Similarly, Irmak [14] uses an embedded processor approach with minimal support from portions of the algorithm implemented in hardware. Color segmentation, shape extraction, morphological operations and template matching are all performed in the Power PC processor that is part of the system. Only edge detection is implemented in hardware.

In addition to being a topic of active academic research, Traffic Sign Recognition is also a technology that is being researched and implemented in the industry. This technology is developed by many car manufacturers who are partnering with traditional automotive suppliers such as Continental Automotive, TRW, Bosch [15] and newer image recognition software product developers like Ayonix. Continental Automotive developed several products for traffic sign recognition. Its Multi Function Camera specification details its abilities for use in TSR [16]. This traffic sign recognition system [17] began production in 2010 on the BMW 5-series. In addition to BMW, many other carmakers have rolled out some version of this technology. Volkswagen has done so on Phaeton and the Audi A8. Mercedes-Benz E and S class both have an implementation of TSR. As well as the Saab 9-5, Opel, Insignia, and the European 2011 Ford Focus. Additionally, Google has developed technology that allows a vehicle to drive itself. Using a combination of data stored in its map database and data that it collects from its environment in real-time, the Google Car is able to safely navigate complex urban environments [18].

3. Algorithm Design

3.1. Algorithm Overview

The problem of identifying traffic signs within an image can be broken into the two sub-problems of detection and recognition. Detection presents the challenge of analyzing the image to identify portions of the image that could contain a traffic sign. Recognition is the challenge of determining if these candidates are indeed traffic signs and if so which one.

Figure 1 depicts the proposed algorithm used for detection and recognition of traffic signs. The following sections discuss each portion of the algorithm and present the specifics of the implementation of each part.

3.2. Hue Calculation and Detection

Traffic signs consist of solid color text, symbols, or shapes on a solid color background as seen in **Figure 2**. Scanning an image looking for this color signature will allow for the quick identification of possible traffic signs and the rejection of the remaining parts of the image.

Stop, *yield*, *do not enter*, *wrong way*, and prohibition signs such as *no left turn*, all contain red backgrounds with white text or white backgrounds with red text. Main distinguishing color for these signs is red. Similar groupings can be done for signs that are primarily green, yellow, blue, or black and white. The algorithm described here and in the sections following must be performed for each of these color groupings.

For each group of similarly colored signs, the algorithm begins by scanning the image to calculate the *hue* of each pixel. There are a variety of ways in which to express the color of a pixel. Perhaps the most common is by using the color's primary color components or the RGB value. Although this is very useful when displaying that color, it is not as helpful when trying to extract all the pixels of a specific hue. If, as in this case, the desire is to identify all the pixels that would be considered red, there are colors that contain a significant about of red as a primary color contributor, but are themselves not red. The color yellow is one such example. **Figure 3(a)** depicts the color representations using RGB.

To determine the hue of a pixel, or its color regardless of shade, a conversion must be made. Each RGB pixel is converted to a different triplet called HSV: Hue, Saturation, and Value. HSV represents the color spectrum by having a value for the color (hue), the amount of that color (saturation), and the brightness of that color (value) [19]. The hue parameter represents the angle where the pixel's color lies on the cylinder depicted in **Figure 3(b)**. Thresholds can be chosen to categorize any hue value found. Once this conversion is made, the hue values for each pixel can be scanned. Detection is the process of identifying the pixels whose hue value falls between the thresholds for the relevant color. This will split the image into two categories: pixels that have the hue of interest and those that do not. At this point the full color image being processed can be simplified into a binary image. Active pixels had the desired hue while inactive pixels do not. This step is called *Hue Detection*.

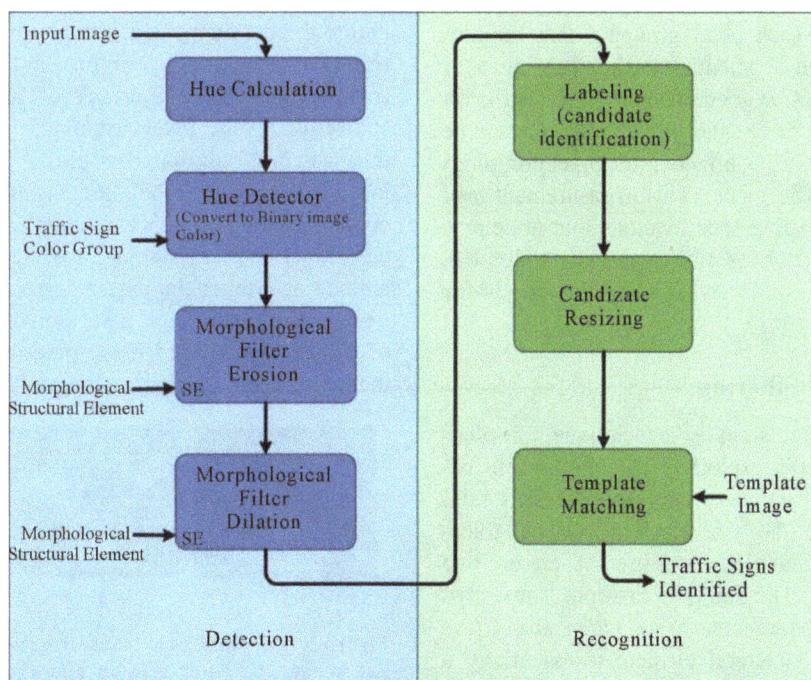

Figure 1. Algorithm flow chart.

Figure 2. Example traffic signs (red).

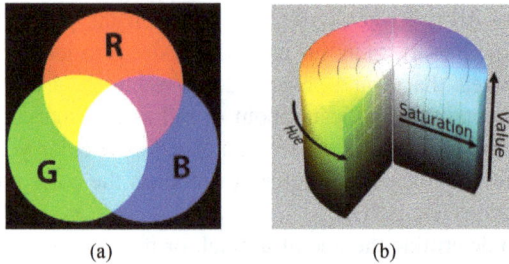

(a)   (b)

Figure 3. (a) RGB color space, (b) HSV color space [19].

Figure 4. Hue calculation and detection example.

Figure 4 shows the effect of Hue Calculation and Detection on an example image. The red pixels have been extracted and are now represented in the binary image as active white pixels. Pixels of all other colors are represented as inactive or black pixels. At this point the system has eliminated a significant amount of information to consider it its effort to identify traffic signs. However, in many images a peppering of red pixel groupings will be found throughout. If each pixel grouping that remains were to be considered a potential traffic sign, most of these would clearly not be good candidates. A step is required where many of these small pixel groupings can be eliminated without adversely affecting the pixel groupings that are actually traffic signs. This measurement and others are deliberate, using specifications that anticipate your paper as one part of the entire journals, and not as an independent document. Please do not revise any of the current designations.

3.3. Morphological Filtering

Morphological transformation is a technique for processing digital images by exploiting the relationship between each pixel and its neighboring pixels. This relationship is defined by the type of morphological transformation and the structural element used. There are two basic transformations: Dilation and Erosion. These two can also be combined to create the *Open* and *Close* transformations. The structural element is essentially a small geometric shape such as a square, disk, line, or cross. During a transformation operation, each pixel is compared to its neighbors as defined by the window specified by the structural element. During erosion, the pixel in question will be deactivated unless all of the other pixels in the structural element window are also active. During dilation, the pixel will be activated if any of the other pixels in the structural element window are active. As their names might suggest, erosion has the effect of trimming the edges of objects where dilation has the effect of puffing out the edges [20,21]. Open and close operations are achieved by combining erosion and dilation. An open is obtained by erosion followed by dilation and a close is a dilation followed by erosion. The results of these operations are less dramatic because the second step tends to temper the effects of the first. **Figure 5** shows example morphological transformations, erosion and dilation.

3.4. Labeling

After detecting the pixels in the image that match the desired hue and after filtering the resulting binary image to reduce the number of pixel "blobs", the resulting image is ready for the labeling step. Labeling is the process of scanning the image to detect pixel groupings. Pixels that share an edge or a vertex are considered to be members of the same pixel grouping or "blob". The remainder of this section will detail our proposed algorithm used to detect these *blobs*.

The goal with labeling is to identify a bounding box for each and every pixel grouping in the image. This will result in four parameters for every blob: Xmin, Ymin, Xmax, Ymax. These are the two diagonal corners that define the bounding box. Simply put, the labeling algorithm scans the image from bottom to top and left to right (raster order) keeping track of Xmin, Ymin, Xmax, Ymax for all the pixel groupings it encounters. During this scanning only one row plus 2 pixels are stored and labeled at a time. Labels are assigned and a table built containing the Xmin, Ymin, Xmax and Ymax values for each label. While scanning, a window of 5 pixels is examined at a time: the current pixel and its 4 previously visited neighbors. Each pixel initially starts with a label of 0 (not labeled) and if the pixel is a foreground pixel then a non-zero label is assigned. The value of that label

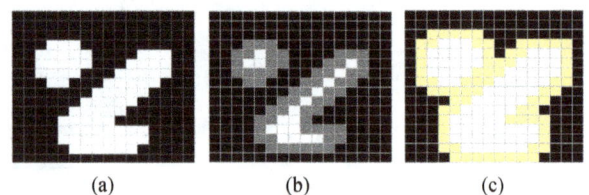

(a)   (b)   (c)

Figure 5. Morphological transformation example: (a) Original image; (b) After erosion operation with 3 × 3 square structural element (gray pixels are removed); (c) After dilation operation (yellow pixels added to the original).

depends on the labels of its four neighbors.

Figure 6 shows all the possible combinations for consideration when determining the label of the current pixel. Each of these diagrams represents the current pixel, X, its immediately previously scanned neighbor, P, and it's previously scanned neighbors 1, 2, and 3. Shaded pixels are active and labeled. The key to this labeling algorithm is in recognizing that these sixteen possible combinations can be grouped into just three categories. There are two special cases distinguished by the red and blue borders, and the remaining are all the normal cases. The first special case, marked in red, is the case in which none of the current pixel's neighbors have a label. In this case the current pixel receives a new label. The second special case, marked in blue, is the case when the current pixel has multiple neighbors that have labels, and these labels may not be the same. This may be the most interesting case because it happens when different labels meet at the current pixel. When this occurs, relabeling must happen so that a single bounding box is generated to encompass what had been two separate labels. The remaining case is the simplest: the current pixel has one or more previously labeled neighbors and they all have the same label. In this case the current pixel is simply labeled in kind. At the same time that pixel labels are being assigned, the bounding box for each label is being grown to encompass all the pixels that contain that label. This is done by maintaining a record table that holds Xmin, Ymin, Xmax, and Ymax for each label.

The second special case described above is the most challenging part of the labeling algorithm. When two labels meet, the algorithm is discovering for the first time that what was previously thought to be two distinct blobs is indeed one. The records that have been maintained thus far will need to be updated to include this information. One possible approach would be to rescan all or portions of the image. However, given that doubling back to reliable portions of the image can be a time consuming step, it was chosen to solve this problem by using a level of indirection: Each pixel is given a label. This label does not point to a record of Xmin, Xmax, Ymin

and Ymax, but rather to a record number. This record number points to these actual bounding box parameters. In this way, when two labels meet both labels can be made to point to the same record that will now encompass what was once two label groupings [22].

For example, consider the image in **Figure 7**. This is a simple 16×16 pixel image that contains two blobs that should be labeled by the labeling algorithm. As the algorithm begins as shown in **Figure 7(a)**, the first three pixels encountered are labeled with new labels as they have no labeled neighbors that have been encountered yet. The pixels that are marked with red indicate special case, and a new label is used. The Record Table to the right shows the values it would take at this point in the scan of the image. **Figures 7(b)** and **(c)** show what happens when labels 1 and 2 meet and when labels 2 and 3 meet. The yellow box shows the window the algorithm is considering at that point in time. When two labels meet, the label table is updated to point to the lower record entry. This entry is also updated so that the points it describes encompass all the pixels of both labels. In this way this information is carried along. When labels 2 and 3 meet, label table 3 is made to point to record 1 also and record 1 is updated to encompass label 3's pixels. This way the algorithm encompasses all the pixels of each pixel grouping. **Figure 7(d)** shows the fully labeled image with its completed record table.

3.5. Candidate Scaling

During the Labeling portion of the algorithm, contiguous pixel groupings are identified. These groupings are candidate traffic signs. Not all of these will be good candidates; many will simply be too small to contain enough information. These candidates are filtered to remove ones that do not contain enough resolution to match a traffic sign template. The remaining candidates are considered good candidates. Since the matching algorithm discussed in the next section requires that the template and the candidate images both be of the same resolution, these remaining candidates must next be scaled to match the template images resolution. Once the candidate is scaled to match the resolution of the traffic sign templates, they are subjected to the template matching portion of the algorithm.

3.6. Template Matching

Template Matching is the portion of the algorithm that takes candidates found in the previous steps and compares them with images of known good traffic signs. These traffic sign images will comprise the template library. Each potential traffic sign will be compared with each member of the template library and, if a close match is found, it is declared to be that traffic sign. This template

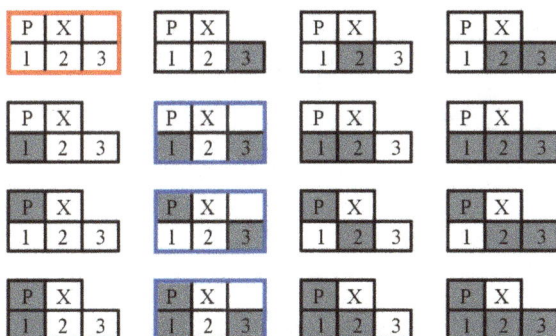

Figure 6. Possible pixel configurations.

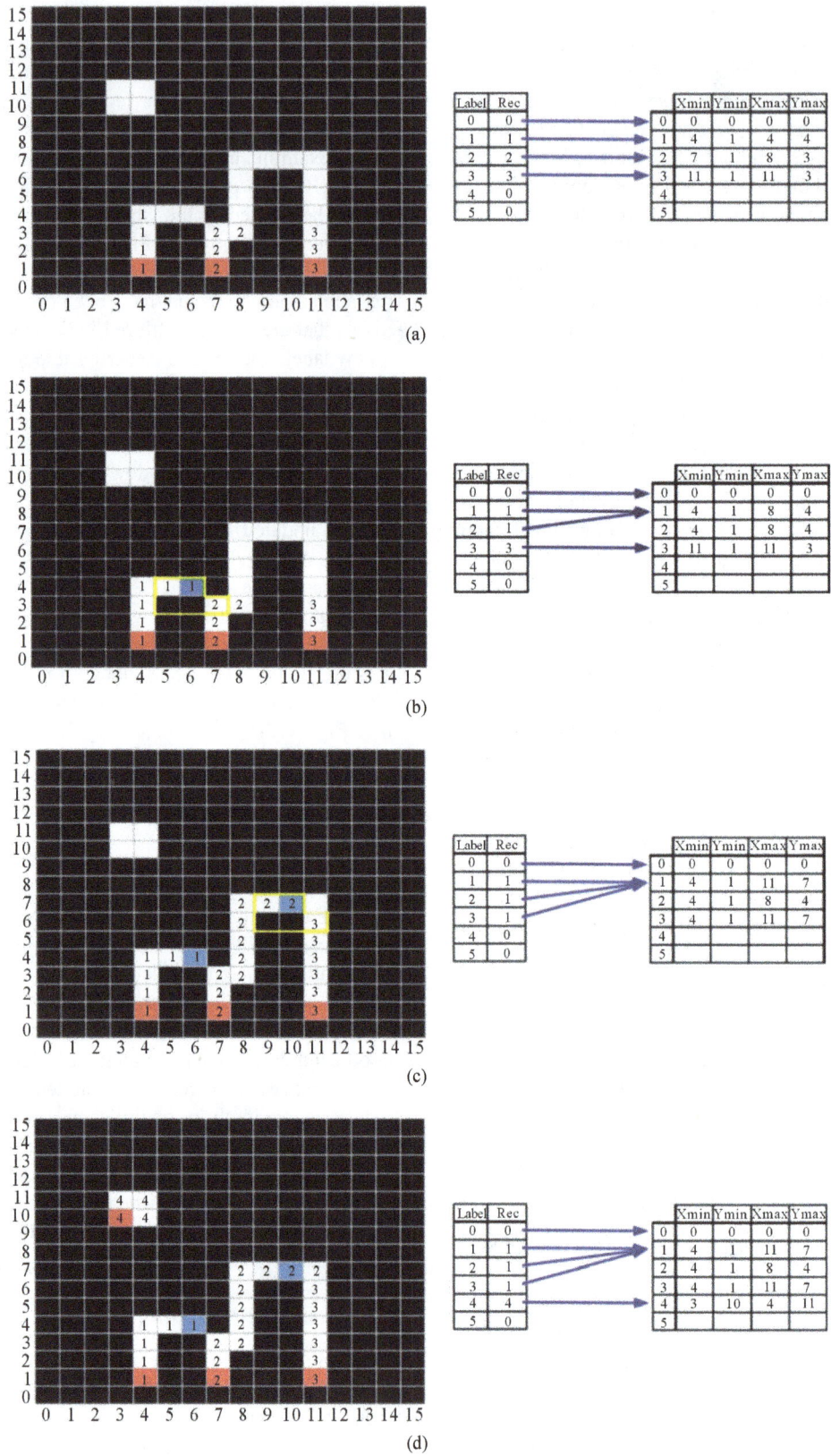

Figure 7. Labeling algorithm example. (a) Initial Labeling and Record table; (b) Labeling and Record table when Labels 1 and 2 meet; (c) Labeling and Record table when Labels 2 and 3 meet; (d) Completed Labeling (Labels 1-3 all point to the same segment).

matching is done by calculating the *Hausdorff distance* between the candidate and each template. The Hausdorff distance calculation is a method and metric used to compare two binary image shapes. A benefit over other comparison methods is that does not rely on explicit point correspondence measuring proximity rather than exact superposition [23]. The Hausdorff distance is defined as [24]:

Given two non-empty finite sets of points

$$F = \left\{ f_1, \cdots, f_{n_F} \right\} \text{ and } G = \left\{ g_1, \cdots, g_{n_G} \right\} \text{ of } \Re^2,$$

and an underlying distance d, the Hausdorff distance is given:

$$D_H \left(F, G \right) = \max \left\{ h \left(F, G \right), h \left(G, F \right) \right\} \qquad (1)$$

where

$$h \left(F, G \right) = \max_{f \in F} \left(\min_{g \in G} d \left(f, g \right) \right) \qquad (2)$$

If two binary images are considered to each be a set of points, then this definition essentially describes an algorithm that for each pixel of the first set will calculate the minimum distance between that pixel and every pixel in the second set. A minimum distance is also calculated for each pixel of the second image with relation to every pixel of the first image. The maximum of these minimums is the Hausdorff distance. In other words, *it is a measurement of the most out-of-place pixel*. This distance calculation technique has been successfully applied in other image recognition applications such as face recognition and object matching [25].

There are other approaches that can be used for this step in the overall algorithm. Several have shown good success in a variety of applications especially in contexts where the candidates may be shifts or tilts of the template. Neural Networks can be used to create a cascade of classifiers that will allow for very fast sign detection [26]. However, the process of teaching a neural network is a long one; on the order of days. Therefore, building classifiers for multiple signs become a time prohibitive approach [27]. Other possible methods for traffic sign identification use interest point correspondence to find matches. One such approach is SURF or Speeded Up Robust Features [28]. SURF makes use of 2D Haar wavelet responses to find these points of interest [29]. Various applications of SURF have been shown to be fast and robust to image transformations. Since the main objective is a practical hardware implementation, Hausdorff distance calculation was used in this work.

4. Hardware Implementation

4.1. Environment

To implement this design, a Xilinx FPGA platform and toolkit was chosen. The hardware selected was Digilent's

Virtex-5 OpenSPARC Evaluation Platform, Xilinx XUPV5-LX110T, which is a version of the ML505 Xilinx evaluation board with a larger version of the Virtex 5 FPGA. This is a versatile board with many on-board peripherals that allow for a diverse number of applications. To develop for this platform, Xilinx's Embedded Development Kit (EDK) was used [29]. EDK enables the quick creation of an on-chip embedded processor (MicroBlaze) and user specific logic (peripherals) on a Field Programmable Gate Array (FPGA). EDK allows for strategic decisions regarding the partitioning of hardware and software implementation. Although software run time is often slower than hardware implementation, there are tasks that may be simpler to accomplish in software.

4.2. System Overview

The testing system included a Intel PC and the Xilinx XUPV5-LX110T which implements a MicroBlaze processor that uses external SDRAM memory and the image processing IP peripheral. The PC ran a custom application that opened BMP images and transferred the RGB data to the MicroBlaze over a serial UART connection. The MicroBlaze received this image data, stored it to RAM and transferred it in turn to the hardware peripheral. The hardware IP peripheral processed the image and returned it to the MicroBlaze. After highlighting any traffic signs that were found, the original image was transferred back to the PC. The MicroBlaze also used a second UART connection used for debugging and reporting.

Figure 8 shows a flowchart for the design. Here, the operational flow of the system is described in its entirety. It can be noted that each subsystem (PC, MicroBlaze, and IP peripheral) fulfills key portions of the work required. The application running on the PC is concerned with preparing the image RGB data, sending it to the embedded system, receiving it back after it has been processed, and writing it back to the file system. The MicroBlaze, in turn, receives this RGB image data, stores it, transfers it to the IP peripheral, receives it from the IP peripheral after processing, and finishes processing it before returning an updated image back to the PC. The IP peripheral receives the image data and processes it according to the algorithm steps and returns it to the MicroBlaze.

An image resolution of QVGA or quarter VGA of 320 × 240 pixels was chosen for implementation. This choice was made because it is large enough for images to contain enough information to process, but small enough to not become a limiting factor during development. The following sections describe the implementation of the key five algorithm steps as they are shown in **Figure 9**.

4.3. Hue Calculation and Detection

Hue calculation involves taking the stream of pixels in

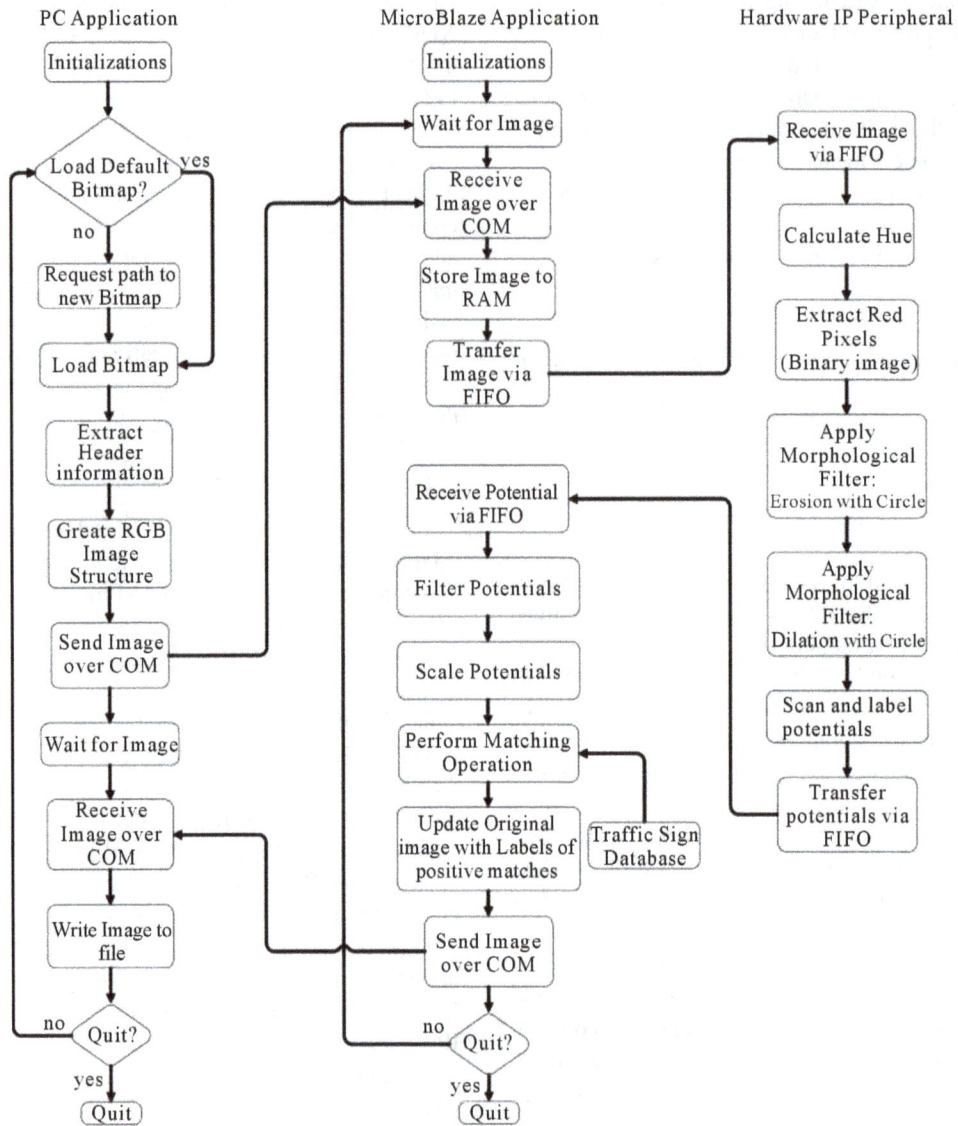

Figure 8. System flow chart.

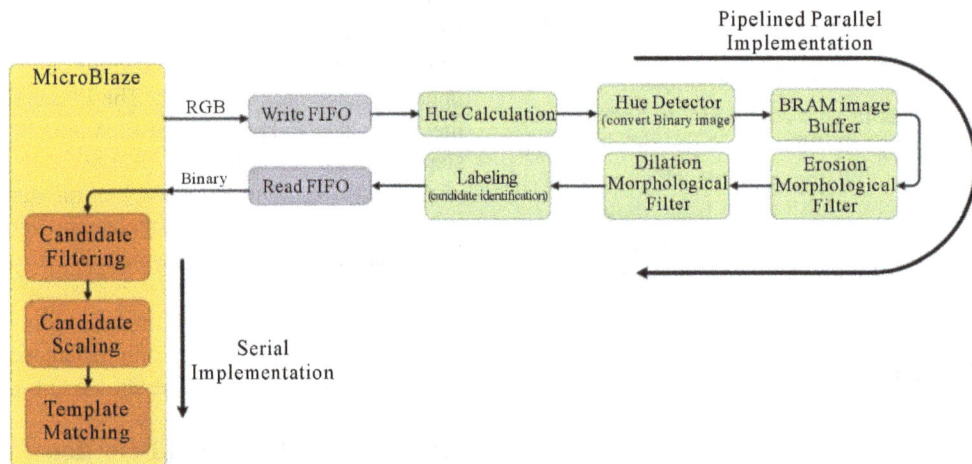

Figure 9. Image processing algorithm implementation.

RGB format and determining which pixels are red. As it was described in the previous chapter, this is done by converting the RGB triplet into a different representation of the color spectrum called HSV. However, it should be noted that one byte of this new triplet contains all the information we need.

Hue indicates that the pixel is red, the Saturation indicates where it lies between the lightest pink and solid red, and Value indicates where it lies between the darkest red and solid red. Therefore, the Saturation and Value components do not need to be calculated. Given this simplification, **Figure 10** shows the implementation in hardware. This circuit accepts the RGB values and calculates the representative hue value. Most of the circuit is combinatorial. The division, however, introduces the need to add some pipelining. The division requires 17 clock cycles to fill the pipeline, but afterwards will calculate a new quotient every clock cycle. This fact required the introduction of a delay register that would allow the control signals, offsets, and quotient to align. The Hue value is taken as an angle, meaning that it takes a value between 0 and 359. This 360-degree range is normally divided into sextants representing Magenta, Red, Yellow, Green, Cyan, and Blue. These sextants then, cover a range of 60 degrees each. To simplify calculation, this implementation used sextants with a range of 0 to 63.

This resulted in a color wheel with an angle range between 0 and 383 [22]. In addition, the normal color wheel

has red at the 0 angle. To avoid modulo calculation for red pixels, the color wheel was rotated in this implementation. Magenta, a much less likely color to find in nature, was placed at the origin.

The Hue detection portion of the algorithm involves taking the stream of hue values and categorizing them as red or not-red. Given the adjusted color wheel described above, red is centered at value 64. Therefore the range within which each pixel, P, must fall is ($48 \leq P \leq 80$). As pixels are categorized, the resulting stream is binary. **Figure 11(a)** shows an example image and **Figure 11(b)** shows how this image has been converted to a binary image. This second image contains only pixels that were detected to be of a red hue.

4.4. Morphological Filtering

To implement morphological filtering, a pixel must be compared to its neighbors. Given that the image data arrives one pixel at a time, structures must be created to buffer enough pixels that will allow for these comparisons. These buffers will create a window. The center of the window is the pixel under evaluation and the other pixels in the window will contribute to whether this pixel is output as or active inactive. In this design, a 5 × 5 window was chosen. In addition to this window, a BRAM (Block RAM) buffer was used to guarantee the arrival to all pixels to be at consecutive clock cycles. The FIFO

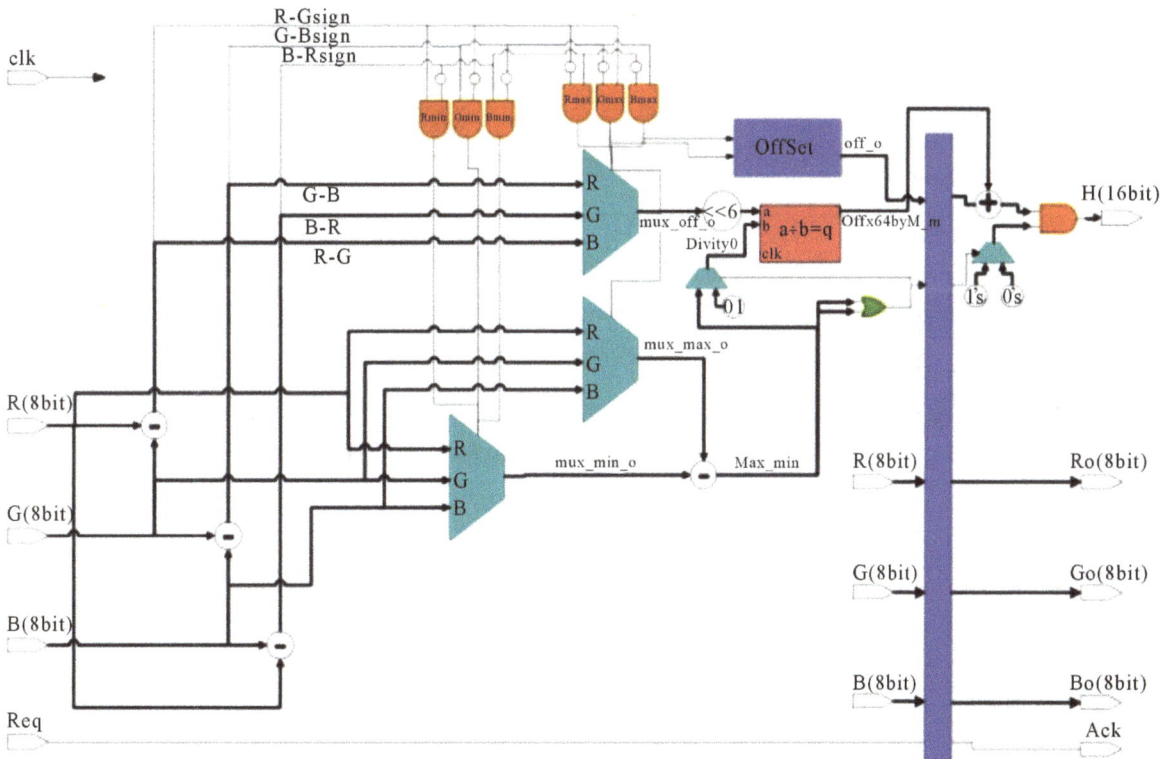

Figure 10. Hue calculator.

(a)

(b)

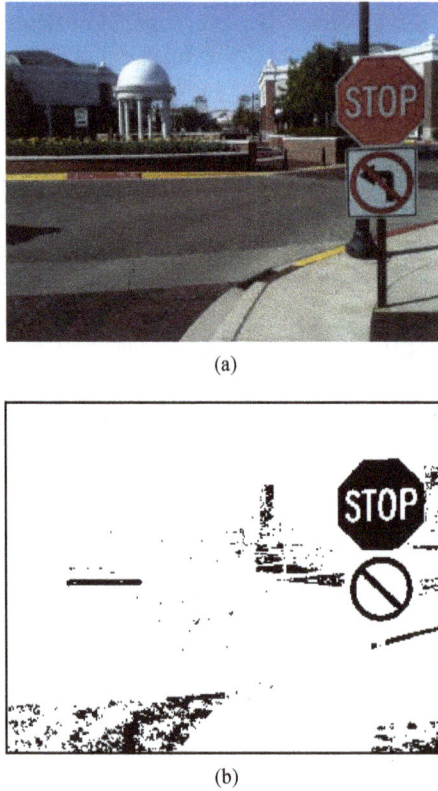

Figure 11. Results of Hue detection (red). (a) Original test image; (b) Image after detecting Red Hue.

structures linking the MicroBlaze and the IP peripheral do not send a solid stream of pixels and these breaks will affect the outcome of the morphological filtering.

Using a 5 × 5 window means that the algorithm is considering pixels from 5 different rows of the image. To allow for this, row buffers are used. A row buffer holds all the pixels in a given row to make them available when they are needed. Single bit registers were used to hold the pixels within the window. Combinational logic compares the structural element to the contents of the window and decides if the output pixel should be active or inactive.

Figure 12 shows the block diagram for the morphological filter circuit. This implementation can be used for both erosion and dilation exploiting the fact that they are duals of each other. Additionally the structural element is programmable. This allows for reuse when multiple passes are desired as in the case of combining erosion and dilation to perform an open or close operation. This flexibility proved to be very helpful when testing to identify the combination of operations that best filtered the image. A structural element of a 3 × 3 disk and a single open operation were shown to best clean up the image while not harming the traffic signs. The first step, erosion, would serve to eliminate many of the stray pixels completely and the second dilation step would serve to repair any undesired degradation to the larger pixel clusters that may be traffic signs. When the resolution of the images

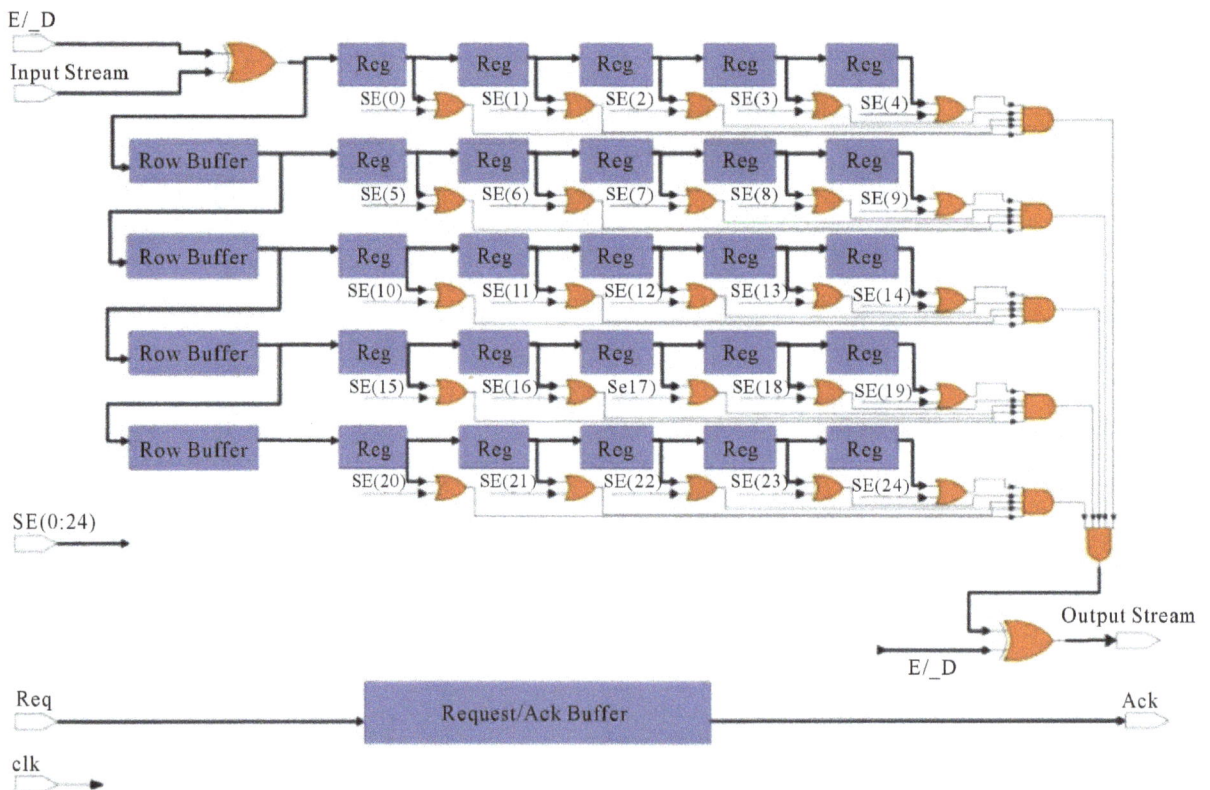

Figure 12. Morphological filter implementation with FPGA.

being processed is increased, larger structural elements may be needed to do adequate filtering.

Figure 13 shows the effect of performing morphological filtering on the test image after hue detection. The number of extraneous pixels has been greatly reduced from those in **Figure 11(b)**. This prepares the image for the next step in the algorithm; Labeling.

4.5. Labeling

Similar to the morphological filtering, labeling requires a window where a pixel's neighbors are observed. This window includes 5 pixels, the one current pixel and its 4 previously visited neighbors. A row buffer is needed to hold the labels of previously visited pixels until they come back into the window. **Figure 14** shows the block diagram for the labeling circuit. A Label Table and Record Table are used to store the bounding box coordinates for each label. As the pixels and their labels fill the window, the update logic evaluates what the current pixels label should be and how to update the Label Table and Record Table. These two tables are continually updated

to ultimately arrive at a list of labeled pixel groupings. The Update Logic block represents a behavioral process that determines the labels for each pixel and updates the Label and Record table. Given that the current pixel is active, this logic determines if this pixel will receive a new label, if it is the meeting of two labels, or simply if it is a pixel that should receive a previous label. After this, the logic will determine how to update the Label and

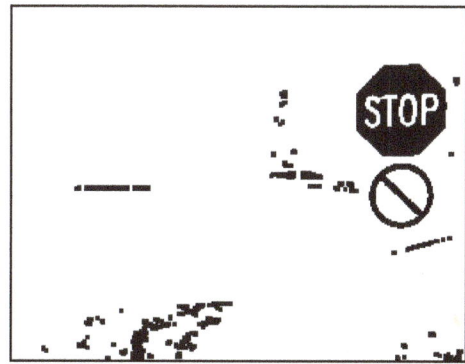

Figure 13. Test image after morphological filtering.

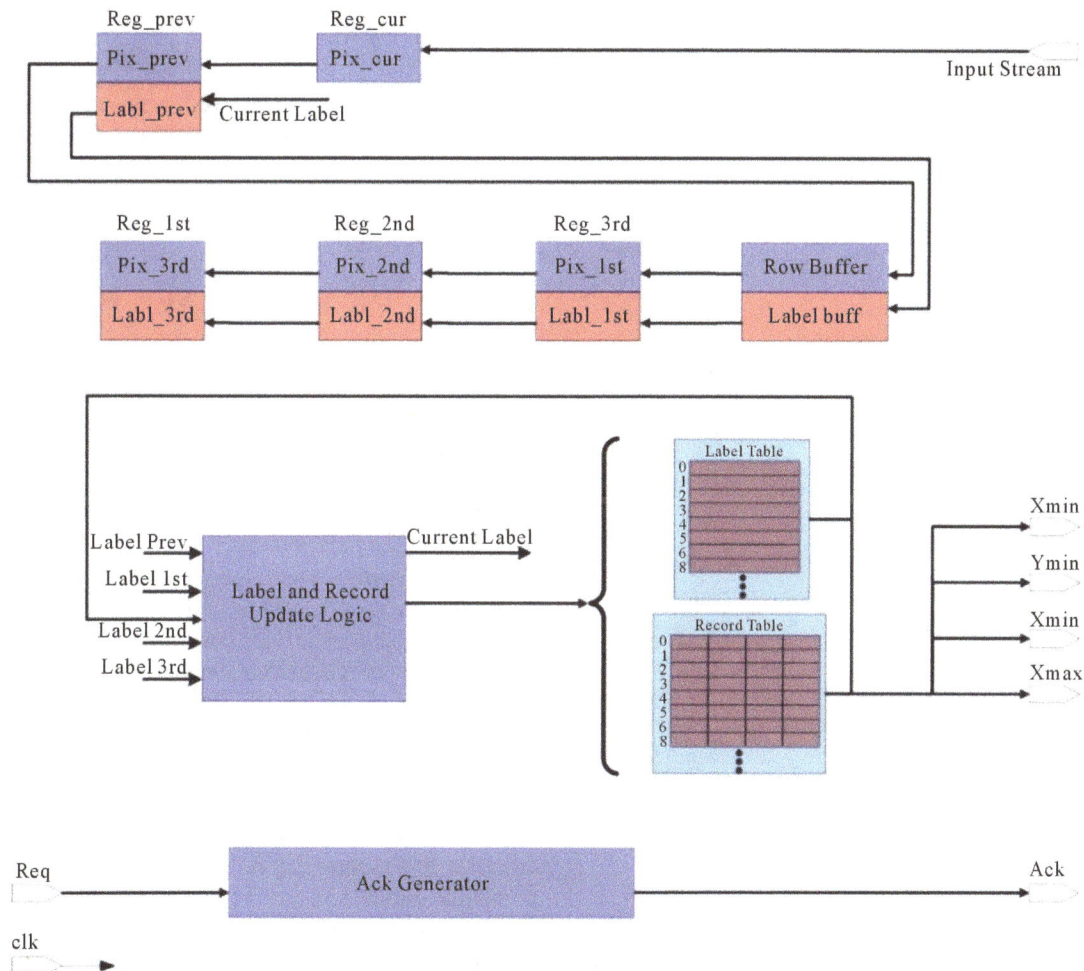

Figure 14. Labeling algorithm hardware implementation.

Record tables. New labels simple result in new entries in both tables. Pixel meetings result in Label Table entries pointing to different records as it is found that two entries should merge. Previously used labels result in the Record Table entry being grown to encompass the new pixel.

The Acknowledgement Generator (Ack) monitors the labeling process to determine when labeling is complete. This Ack signal goes high only when the Xmin, Ymin, Xmax, Ymax values become valid. For every clock cycle that this signal is high, these bounding box values indicate another labeled blob. This output is then fed back to the MicroBlaze via the FIFO for further processing. **Figure 15(a)** shows the test image fully labeled. At this point these are the candidates to be considered for traffic sign matching.

However if a simple filter is implemented this number of candidates is greatly reduced. By ignoring bounding boxes that have at least one dimension smaller than a specified threshold most of the candidates are rejected. It was experimentally found that a blob smaller than 16 × 16 pixels would not contain enough information to provide a good chance for any match. **Figure 15(b)** shows the test image after this simple filter is applied. Only 3 candidates remain. These 3 candidates will be the only ones considered during the scaling and matching steps.

4.6. Candidate Resizing

The last two algorithm steps are implemented in software running on the MicroBlaze CPU. At this point in the image processing algorithm, each traffic sign candidate must be extracted from the binary image, scaled to a resolution that will match that of the template images, and then compared to existing templates. To extract the candidate, the bounding box is passed to a routine that uses these boundaries as the edges of an image. The resulting extracted image is then scaled according to the ratio of its length and width to the length and width of the template images. The initial resolution chosen for the templates was 160 × 160 pixels.

The scaling algorithm calculates the ratio between the original size of the candidate and the template size by calculating an integer portion and a fractional part. This avoids the use of floating point arithmetic. A new memory location is used as the storage location of the new resized candidate. The algorithm scans the original size candidate and sets the pixels of the scaled candidate to active or inactive according to calculated ratios. **Figure 16** depicts the extraction of the candidate from the binary image and its scaling to match the template size.

4.7. Template Matching

Template Matching involves comparing each of these scaled candidates to the library of template images. As

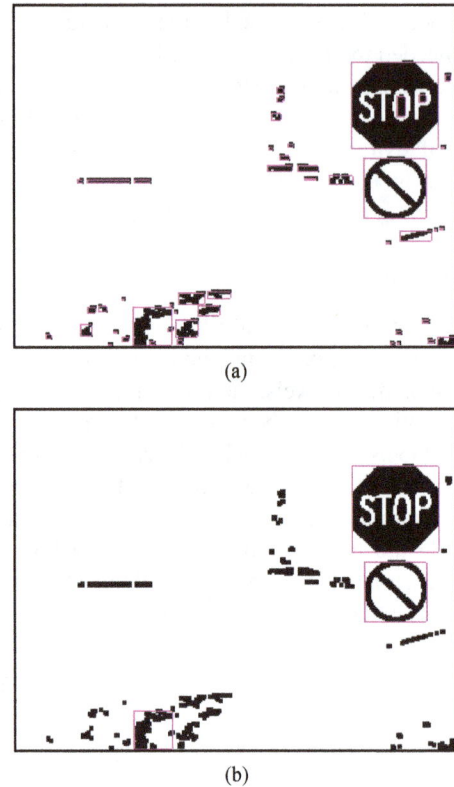

(a)

(b)

Figure 15. Labeling results. (a) Test image after labeling; (b) Test image after rejection of small labels.

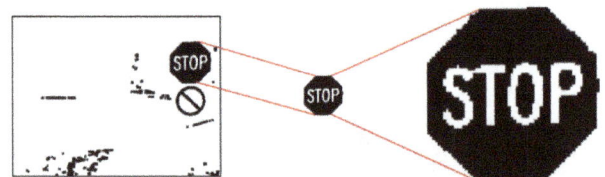

Figure 16. Extraction and scaling of candidate sign.

explained in Section 3, this comparison is done by calculating the Hausdorff distance between the candidate and the template images. **Figure 17** shows some of these template images for red color.

The implementation of the calculation of the Hausdorff distance is done according to the description of the algorithm in the previous section. For each active pixel at location (x,y) of the scaled candidate image, the same pixel location of the template image is examined. If the same (x,y) location of the template image contains an active pixel that the distance between that pixel and the template is 0. If it is not an active pixel, then a search begins for the nearest active pixel. This search is conducted with an increasing radius. When an active pixel is found, the radius is the distance between that pixel of the candidate image and the template. No more distances need to be calculated for that same pixel since all other distances will be equal or greater.

Once a Hausdorff distance is calculated for each template, they can be compared to determine if there is a good match. It was observed that at times there were high outliers and at times there was very little margin between the lowest and the next lowest value. It was determined experimentally that if a the minimum distance is found to be at least 10% lower than the average of the lowest three distances, this can be considered a good match. Because shifting to the right by 3 bits is a simpler calculation than dividing by 10, 12.5% was used instead of 10%. This method avoided the impacts of high valued outliers and clusters of low values. This provides a threshold to show that the lowest distance is actually significantly lower than the other values. If this threshold is not met, then no match is found and the image is declared to not contain any traffic signs. **Figure 18** shows the test image with all 3 good candidates highlighted. The two highlighted in magenta are the candidates that found matches according to the description above. The blue highlighted candidate did not find an adequate match.

5. Results

5.1. Device Utilization

Table 1 details the usage statistics of this design implemented on the Virtex 5 device xc5vlx110t with package ff1136 and a speed grade of −1.85% of slices have been used, but very little of the fabric RAM and DSPs have been used.

5.2. Performance: Timing

The key requirement for a traffic sign recognition system is that it should operate in real-time. Depending on the application, real-time may be defined very differently. The term real time refers to a system that is able to take

Figure 17. Template images.

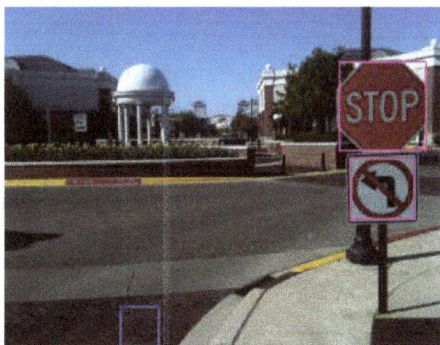
Figure 18. Test image after template matching.

data and process it sufficiently rapidly to be able to take the actions required of the system. For a system, whose purpose is to detect traffic signs while traveling on the road, a threshold for real time could be calculated as the time required for detecting a sign in 50 feet of travel at 65 mph. This time is 525 milliseconds. Therefore, a system of this nature can be considered real time if it can reliably detect a traffic sign in approximately half a second.

Initially the template images were set to a resolution of 160 × 160 pixels. The candidates would then also be scaled to this size. This resolution was chosen because it was large enough to contain a good amount of detail. **Table 2** shows the timing results of this system. The IP peripheral includes the all the algorithm steps that were implemented in the reconfigurable hardware namely, Hue Calculation and Detection, Morphological Filtering, and Candidate Labeling.

The timing shown here would likely not be adequate for the use case of the system designed here. It is clear, however, that the bulk of the time is contributed by the template matching part of the algorithm. The resolution of the templates and scaled candidates have a direct impact on this figure. **Tables 3** and **4** show the results after reducing the resolution of these templates to 80 × 80 pixels and then 40 × 40 pixels. The performance improvement is dramatic.

With these substantial improvements, results are on the same order of magnitude of the goal. With an average detection time of 777 milliseconds a car could travel at 44 miles per hour and still detect a traffic sign in 50 feet.

Table 1. FPGA device utilization summary.

Resource	Number Used	Percent Used
DSP48Es	3 out of 64	4%
RAMB36 EXPs	18 out of 148	12%
Slices	14,764 out of 17,280	85%
Slice Registers	12,023 out of 69,120	17%
Slice LUTS	42,330 out of 69,120	61%
Slice LUT-Flip Flop Pairs	44,712 out of 69,120	64%

Table 2. Algorithm performance details with 160 × 160 pixel templates.

Algorithm Step	Time (msec)
IP Peripheral	114
Candidate Scaling (avg)	114
Template Matching (avg)	93,900
Total Time (avg)	94,150

Table 3. Algorithm performance details with 80 × 80 pixel templates.

Algorithm Step	Time (msec)
IP Peripheral	114
Candidate Scaling (avg)	30.8
Template Matching (avg)	6500
Total Time (avg)	6641

Table 4. Algorithm performance details with 40 × 40 pixel templates.

Algorithm Step	Time (msec)
IP Peripheral	114
Candidate Scaling (avg)	8.3
Template Matching (avg)	655
Total Time (avg)	777

Given that the algorithm steps of Hue Calculation, Hue Detection, Morphological Filtering, and Labeling were done in the IP peripheral in only 114 milliseconds, whereas the Template Matching step, on the other hand, implemented in software running on a 100Mhz CPU, could only be performed in 655 milliseconds at best, it is clear that the parallel implementation significantly outperforms the serial software implementation.

5.3. Performance: Accuracy

An important factor to the accuracy of the system is the orientation of the traffic signs with reference to the camera. When signs are perpendicular to the viewer, the system detection performance is very accurate as shown in **Table 5**, even with the low resolution template. If the sign is skewed or tilted, the performance could be affected. To evaluate how sensitive this system was to these perturbations, images were processed that contained them. **Figure 19** shows an image similar to the one used in previous examples, but with the traffic signs at an angle to the viewer. **Figure 20** is an example where the traffic signs are tilted with respect to the viewer. **Table 6** details the results of processing these two example images. The stop sign is detected successfully in both of the higher resolution template versions but not with the lowest resolution.

6. Conclusion

Automotive technologies continue to explore new ways to keep drivers safe. Recently, technologies have emerged that monitor more complex parameters. This paper describes one such system. The implementation of the traffic sign recognition system uses Xilinx's EDK toolkit to

Figure 19. Test image with skewed signs.

Figure 20. Test image with tilted signs.

Table 5. Algorithm accuracy.

Sample Image	Res. 160^2	Res. 80^2	Res. 40^2
Do-Not-Enter	Det.	Det.	Not Det.
Prohibition	Det.	Det.	Det.
Wrong Way 1	Det.	Det.	Det.
Wrong Way 2	Det.	Det.	Det.
Yield	Det.	Det.	Det.

Table 6. Algorithm robustness.

Sample Image	Res. 160^2	Res. 80^2	Res. 40^2
Stop Sign (skewed)	Det.	Det.	Not Det.
Stop Sign (tilted)	Det.	Det.	Not Det.

create an image processing flow that is partitioned across hardware and software. An embedded processor is used to receive RGB pixel data from a PC and to forward it to a hardware IP peripheral. This peripheral in responsible for extracting red pixels, cleaning up the resulting binary image, and labeling possible candidates for matching to traffic sign templates. These candidates are passed back to the embedded processor where they are scaled and evaluated against an array of template images. The implementation described here is able to detect the sings within 50 feet of distance at a travel velocity of 44 miles per hour or less.

REFERENCES

[1] J. Urry, "The 'System' of Automobility," *Theory, Culture & Society*, Vol. 21, No. 4-5, 2004, pp. 25-39.

[2] E. Eckermann, "World History of the Automobile," Society of Automotive Engineers, Warrendale, 2001.

[3] "Transportation: Motor Vehicle Accidents and Fatalities," The 2012 Statistical Abstract, US Census Bureau, Suitland, 2011.

[4] "Population," The 2012 Statistical Abstract, US Census Bureau, Suitland, 2011.

[5] "EuroFOT Study Demonstrates How Driver Assistance Systems Can Increase Safety and Fuel Efficiency," EuroFOT, 2012.
http://www.eurofot-ip.eu/en/news_and_events/eurofot_stu dy_demonstrates_how_driver_assistance_systems_can_in crease_safety_and_fuel_efficiency_acr.htm

[6] M. Meuter, C. Nunn, S. M. Gormer, S. Muller-Schneiders and A. A. Kummert, "A Decision Fusion and Reasoning Module for a Traffic Sign Recognition System," *IEEE Transactions on Intelligent Transportation Systems*, Vol. 12, No. 4, 2011, pp. 1126-1134.

[7] C. Lai, "An Efficient Real-Time Traffic Sign Recognition System for Intelligent Vehicles with Smart Phones," *Proceedings of 2010 International Conference on Technologies and Applications of Artificial Intelligence*, Hsinchu, 18-20 November 2010, pp. 195-202.

[8] V. Andrey, "Automatic Detection and Recognition of Traffic Signs Using Geometric Structure Analysis," *Proceedings of SICE-ICASE International Joint Conference*, Busan, 18-21 October 2006, pp. 1451-1456.

[9] D. Soendoro and I. Supriana, "Traffic Sign Recognition with Color-Based Method Shape-Arc Estimation and SVM," *Proceedings of 2011 International Conference on Electrical Engineering and Informatics*, Bandung, 17-19 July 2011, pp. 1-6.

[10] Y. Liu, H. Yu, H. Yuan and H. Zhao, "Real-Time Speed Limit Sign Detection and Recognition from Image Sequences," *Proceedings of 2010 International Conference on Artificial Intelligence and Computational Intelligence*, Shenyang, 23-24 October 2010, pp. 262-267.

[11] R. Kastner, T. Michalke, T. Burbach, J. Fritsch and C. Goerick, "Attention-Based Traffic Sign Recognition with an Array of Weak Classifiers," *Proceedings of 2010 IEEE Intelligent Vehicles Symposium*, San Diego, 21-24 June 2010, pp. 333-339.

[12] M. A. Souki, L. Boussaid and M. Abid, "An Embedded System for Real-Time Traffic Sign Recognizing," *3rd International Design and Test Workshop*, IDT 2008,

Monastir, 20-22 December 2008, pp. 273-276.

[13] M. Muller, A. Braun, J. Gerlach, W. Rosenstiel, D. Nienhuser, J. M. Zollner and O. Bringmann, "Design of an Automotive Traffic Sign Recognition System Targeting a Multi-Core SoC Implementation," *Design, Automation & Test in Europe Conference and Exhibition* (*DATE*), Dresden, 8-12 March 2010, pp. 532-537.

[14] H. Irmak, "Real Time Traffic Sign Recognition System on FPGAs," M.S. Thesis, Middle East Technical University, Ankara, 2010.

[15] Frost & Sullivan, "Development of Low-cost DAS Technologies to Help Reach European Union's Target to Increase Road and Driver Safety," 2011.
http://www.frost.com/prod/servlet/press-release.pag?docid =251082001

[16] Continental AG, "MFC 2 Multi-Function Camera," Datasheet, 2009.
http://www.conti-online.com/generator/www/de/en/contin ental/industrial_sensors/themes/mfc_2/mfc_2_en.html

[17] Continental AG, "Traffic Sign Recognition," 2012.
http://www.conti-online.com/generator/www/de/en/contin ental/automotive/general/chassis/safety/hidden/verkehrszei chenerkennung_en.html

[18] J. Markoff, "Smarter than You Think: Google Car Drives Itself," The New York Times, October 2010.
http://www.nytimes.com/2010/10/10/science/10google.ht ml?_r=1

[19] A. C. Clark and E. N. Wiebe, "Color Principles—Hue, Saturation and Value," North Carolina State University, Raleigh, 2002.
http://www.ncsu.edu/scivis/lessons/colormodels/color_mo dels2.html

[20] D. M. Rouse and S. S. Hemami, "Quantifying the Use of Structure in Cognitive Tasks," *Proceedings of SPIE, Vol. 6492, Human Vision and Electronic Imaging XII*, 649210, San Jose, 12 February 2007.

[21] J. P. Serra, "Image Analysis and Mathematical Morphology," Academic Press, Inc., Orlando, 1983.

[22] D. G. Bailey, "Design for Embedded Image Processing on FPGAs," Wiley-IEEE Press, Singapore, 2011.

[23] S. Mignot, "A Hardware-Oriented Connected-Component Labeling Algorithm," Technical Report, GEPI—Observatoire de Paris, Paris, 2006.

[24] J. Chen, M. K. Leung and Y. Gao, "Noisy Logo Recognition Using Line Segment Hausdorff Distance," *Pattern Recognition*, Vol. 36, No. 4, 2003, pp. 943-955.

[25] E. Baudrier, F. Nicolier, G. Millon and S. Ruan, "Binary-Image Comparison with Local-Dissimilarity Quantification," *Pattern Recognition*, Vol. 41, No. 5, 2008, pp. 1461-1478.

[26] C. Y. Fang, S. W. Chen and C. S. Fuh, "Road-Sign Detection and Tracking," *IEEE Transactions on Vehicular Technology*, Vol. 52, No. 5, 2003, pp. 1329-1341.

[27] M. F. Hashim, P. Saad, M. R. M. Juhari and S. N. Yaakob,

"A Face Recognition System Using Template Matching And Neural Network Classifier," *Proceedings of* 1*st International Workshop on Artificial Life and Robotics*, Kangar, May 2005, pp. 1-6.

[28] H. Bay, A. Ess, T. Tuytelaars and L. Van Gool, "Speeded-Up Robust Features (SURF)," *Computer Vision and Image Understanding*, Vol. 110, No. 3, 2008, pp. 346-359.

[29] "Embedded System Tools Reference Manual EDK (v 13.2)," Xilinx, 2011. http://www.xilinx.com/support/documentation/sw_manuals/xilinx13_2/est_rm.pdf

2

Cost Deployment Tool for Technological Innovation of World Class Manufacturing

Luan Carlos Santos Silva, João Luiz Kovaleski, Silvia Gaia, Manon Garcia, Pedro Paulo de Andrade Júnior
Department of Production Engineering and Technology Transfer Research Group,
Federal University of Technology—Paraná (UTFPR), Ponta Grossa, Brazil

ABSTRACT

This article brings a discussion about using the Cost Deployment methodology for technological innovation in the World Class Manufacturing (WCM) systems at Fiat Group Automobiles Production System (FAPS). It aims to show how this tool acts in the technical pillars of the WCM, and its proper use as an alternative to innovate in production processes, achieving a drastic reduction in wastes and cost optimization during specific activities in production systems. The Cost Deployment builds a distinctive transversal method of WCM which helps to promote and provide extremely effectiveness in the activation of more specific methods that have been tried successfully in the Japanese manufacturing improvements. It also allows to link the operational performances, usually measured with indicators such as efficiency, providing number of defects, hours of desaturation. The used methodology was based on a literature review about the proposed topic. It ends up finding that the Cost Deployment tool is one of the most sophisticated technological innovations existing for the production systems of the World Class Manufacturing.

Keywords: Cost Deployment; Technological Innovation; World Class Manufacturing; Fiat Group Automobile Production Systems

1. Introduction

In the current scenario, the process of industrial system modernization is linked to the technological innovation and new technology of products, processes and services, demanding from companies a continuous seeking for innovation in their activities and production systems. Companies aim to increase optimization of costs, as well. In the role of these new parameters in the industrial sphere, many companies from various sectors have been involved in rethinking their old forms of production.

This paper was developed in the nucleus of Postgraduate in Management for Sustainability and Innovation at the State University of Santa Cruz (UESC), by the Research Group on Management Technology Transference of the Post-Graduation Program in Engineering of Production at the Federal University of Technology—Paraná (UTFPR). And it is under discussion here the use of the Cost Deployment Methodology as the new technological innovation of the World Class Manufacturing—WCM, at Fiat Group Automobiles Production Systems (FAPS). Therefore, it aims to describe the tool Cost Deployment inside the technical pillars of the WCM and the

appropriate use of this tool as an alternative to innovate in production processes, achieving a drastic reduction of wastes and costs. The optimization of specific activities in production systems enables an increasing on the identification of losses in the entire logistics chain. According to Yamashina [1], "The identification of losses depends on your eyes [...] people improve their eyes as they learn."

The Cost Deployment is a distinctive and transversal method of WCM which helps to promote and provide extremely effectiveness in the activation of more specific methods that have been successfully tried in the Japanese manufacturing improvements [2].

The used methodology is based on a literature review about some concepts on Technological Innovation, World Class Manufacturing, production systems at Fiat Group Automobiles Production System, and Cost Deployment. Considerations were discussed about the main relationships among the themes in order to explain its importance in the production chain.

The World Class Manufacturing is a set of concepts, principles and techniques for managing the operational processes of a company. The term World Class Manufacturing effectively captures the essence of the fundamental changes taking place in world industries in the 1970s in a very wide set of elements that characterize the production: quality management, industrial relations, the training, support staff, sourcing, relationships with suppliers and customers, product design, organization of establishments, the scheduling, maintenance, production line, the accounting system, automation and others [3].

The WCM is based on models created by the activities of Japanese manufacturing after World War II and the results obtained by the Japanese approach in order to organize production. It adapts the ideas used by the Japanese in automobile and electronics sectors to achieve significant competitive advantages.

The WCM was first presented as an organic approach by Schonberger [4]. It shows a series of American companies which have adopted and implemented the Japanese approach to production, adapting it to the Western context. From the text of Schonberger [4] it is possible to identify that this adaptation has not occurred by pure imitation, but the adoption of the Japanese ways of producing in the West helped to publicize a very different approach. However, these changes generate a need for creating continuous technological innovations stimulated and generated by people in companies, aiming to manage and keep working in the current competitive environment.

2. Technological Innovation

In a highly competitive market, with production processes filled with organizational bottlenecks, the generation of innovations becomes critical in this process; organizations can continuously improve their processes through products and services. Innovation is an evolving set of new evolutionary functions that change the production methods, creating new ways of organizing work, and in producing new products, stimulating the opening of new markets by creating new uses and consumptions [5].

Following the same analysis Canon [5], relates that the concept of technological innovation can be understood as the expansion of production (more machines) and the increase of new products. In other words, it means that the company should think about producing products which have been produced by other manufacturers or competitors. In this same perspective, innovate is create and is also improving in products or processes that are working or can make a good profit. Innovate is to produce what the company did not deliver before. Innovation is expanding industrial sheds, is to install more machines, and is to install more modern machines let that to produce more quantity of products. Many times is to increase the productivity and optimize costs, mostly is to

increase the production.

Therefore, technological innovation must be undermnstood as an activity that involves not only the industry in re- search and development in an organization, not even as an activity performed only by large companies. It must always be presented in all companies wishing to act in an innovative and competitive way in the market [6].

2.1. WCM Concepts through Its Evolution

There is no consistent definition of WCM, as many other new concepts related to management and supply chain. The term "World Class Manufacturing" was created by Hayes & Wheelwright [7] and Schonberger [4], to describe the technological capabilities that had been developed by Japanese and German companies, as well as U.S. companies that had competed on equal terms with Japanese and German companies. The term WCM was used because these companies have achieved an outstanding performance in their global competition, resulting in the concept described as "World Class". However, the term became popular only after Schonberger [4] discussed the issue as: "[...] The term captures well the breadth and the essence of the fundamental changes taking place in industrial companies." WCM is a major philosophies focusing primarily on production, with a level of excellence throughout the logistics and productive cycles, in reference to the methodologies applied and the performances achieved by the best companies worldwide, mostly based on the concepts of Total Quality (TQC), Total Productive Maintenance (TPM), Total Industrial Engineering (TIE) and Just Time (JIT).

It is noteworthy that the critical factors of WCM success during the recent years have received widespread attention. It also became one of the driving forces for business success. Huczynski and Buchanan [8], Escrig-Tena [9], Flynn et al. [10], McAdam and Henderson [11], Oakland [12], Salaheldin and Eid [13], Sharma and Kodali [14], Sinclair and Zairi [15], Sohal and Terziovski [16] and Svensson and Klefsjo [17] conducted extensive studies to understand the factors that strengthen and enforce WCM application. These cited authors concluded that companies need to understand how to identify the critical factors which affect the implementation process by analyzing their tools in order to solve them effectively. This procedure ensures that benefits can be realized and established so faults can be drastically avoided. Therefore the need for a more systematic and deliberate study on the critical factors of success on implementing WCM is still considered fundamental.

Critical factors of success can be defined as areas where things should go well for the business to flourish, Butler and Fitzgerald [18], Digman [19], Eid et al. [20]. Oakland [21] highlights the importance of observing such critical areas in which the organization must have a greater attention to achieve the organization's mission, through examination and classification of impacts. In terms of WCM, they can be viewed as the activities and practices that should be addressed to ensure its successful implementation.

According to Sinclair and Zairi [15] the quality department plays a key role in implementing WCM, since proper training is provided to this department, Escrig-Tena [9], McAdam and Henderson [11]. However, the support of the senior management team is essential to integrate systems with the implementation plan for WCM, Avlonitis and Karayanni, [22], Mora-Monge et al. [23]. Bose [24] reinforces that programs of people awareness management are important for WCM implementation.

In the current literature, there is a considerable discussion about the importance of the human dimension in WCM implementation. It is taken as a facilitator of the process, not just an adjuvant, Flynn et al. [10]. Sureshchandar et al. [25], Oakland [12] and Sinclair and Zairi [15] also added the factor of continuous improvement for effective implementation. Finally, Dubrovski [26] and Kasul and Motwani [27] indicated that for a successful implementation of WCM, an integration of the entire company is required.

2.2. Fiat Group Automobiles Production System, in the WCM Production System (FAPS)

As defined by Massone [3], responsible for Production System Development of Fiat Group Automobiles Production System (FAPS), the introduction of FAPS concept is a great program of technological innovation that has the intention to change profoundly the way of producing, in order to achieve the standards of excellence set by the World Class Manufacturing (WCM).

For Professor Yamashina, cited by Massone [3], from the University of Kyoto, the basic and fundamental principle of WCM with FAPS is to bring the man to the entire production process to think and act effectively, and each time, act like men's of thought and think as men of action.

The model of Fiat Group Automobiles Production System (FAPS) is a structured set of methodologies and tools whose application spread across the enterprise through the involvement of all employees. It allows a radical improvement for the performance of the production system, optimizing all production processes and logistics working in continuous improvement of key factors: Quality, Productivity, Security, and Delivery. The implementation support is done by a system of Audits and it is structured by goals whose achievement is measured by performance indicators [3].

This allows the product to be delivered within the required time and quality, simultaneously eliminating activities that generates losses and do not add value (manpower, machinery, materials and energy).

The FAPS is constituted by 10 technical pillars of the WCM as described below [3]:

1) Safety—aims to eliminate accidents;

2) Cost Deployment—aims to identify problems that increase costs;

3) Focused Improvement—aims to develop the know how to reduce costs by using appropriate methods;

4) Autonomous Activities—consists in autonomous activity such as TPM (Autonomous Maintenance e Workplace Organization);

5) Professional Maintenance—it

6) Objectives zero breaking of machines;

7) Quality Control—SPC utilization (Statistical Process Control), Project Six Sigma, aiming at zero defect;

8) Logistics—utilization of logistics based on Picking foundations, just in time, Kan Ban, with the commitment to fully satisfy customers;

9) Early Equipament Management—Launch products providing adequate manufacturing. It utilizes fundamentals from QFD, FMEA of product and FMEA of process;

10) People Development—aims to create a culture of results by discipline and improving training of people in the organization;

11) Environment—It is the development of activities in a sustainable working environment for all in the organization, concerned with the prevention of environmental pollution.

The technical implementation of the 10 pillars of WCM in the production system of FAPS had as main objective to pursue a mental attitude or relevant philosophy to their scope and its improvement, following a well-defined path based on the removal of all barriers concerning production to achieve maximum simplification. There are many indications showing which obstacles should be eliminated in production and which routes should be followed for the simplification [3].

However, the production system of FAPS, which itself is based on WCM, becomes a formidable element of competitiveness for enterprises, and an important and lasting contributor to improving customer satisfaction.

2.3. Describing the Cost Deployment

Cost Deployment is a method that innovate systems management and control of establishments, introducing a strong link between individualization of the areas to be improved and the results of the performance improvements obtained through application of technical pillars of the WCM, measured through the appropriate KPI [3]. Consequently, it constitutes a reliable means to program budgeting.

Cost Deployment allows defining improvement programs that have an impact in reducing losses, everything that can be classified as wastes or non-value added in a systematic way. It also ensures collaboration between units of production and function of Administration and Control [2]. This is accomplished through:

1) The study of reactions between the cost factors, the processes that generate wastes and losses in its various kinds of ways;

2) The relationship between demand for waste reduction and losses, and reduction of related costs;

3) The verification of know-how to reduce waste and losses: if it is already available or it should be acquired;

4) Establishing a priority of projects to reduce waste and losses in accordance with the priorities derived from an analysis of costs/benefits;

5) The continuous monitoring of progress and results of improvement projects.

Cost Deployment is the ability to transform losses costs, quantifying in physical measurements: hours, kW/h, unit numbers of material, etc… [2].

The foundation of the methodology is the systematic identification of waste and losses of the area under examination, its evaluation and transformation into values. This is possible because it relates waste and losses to their causes and origins, allowing a complete definition of the loss.

In addition, Cost Deployment guides the individuali-

zation of the best technical method to remove the cause and assess in detail the activity costs of removal and improving performance [2].

Figure 1 shows the detail of the logic route of Cost Deployment.

The completion of the logic route of Cost Deployment as shown in the figure is made up as follows:

1) From the establishment total costs of processing and from the analysis of its structure and composition, reduction targets and costs are established (step 1);

2) Losses and wastes are identified in a qualitative way, placing them in the processes in which they happen (The Matrix—Loss / Processes) (step 2);

3) It identifies the relationship between casual losses and all losses (Matrix B—Causal/Arising) (step 3);

4) The dimensions of losses and individualized wastes are transformed into costs, actual values (Matrix C—Costs/Losses) (step 4);

5) Methodologies (WCM Pillars) are selected in order to remove the original causes of losses and wastes and as well as priorities are established (Matrix D—Losses/Methods) (step 5);

6) The costs of projects implementation are estimated expecting the removal of the causes and the benefits in terms of cost reductions (Matrix E—Costs/Benefits) (step 6);

7) Finally, improvement plans are implemented by collecting the results (step7) following them up.

As steps 1 to 4 consist of preparatory activities which

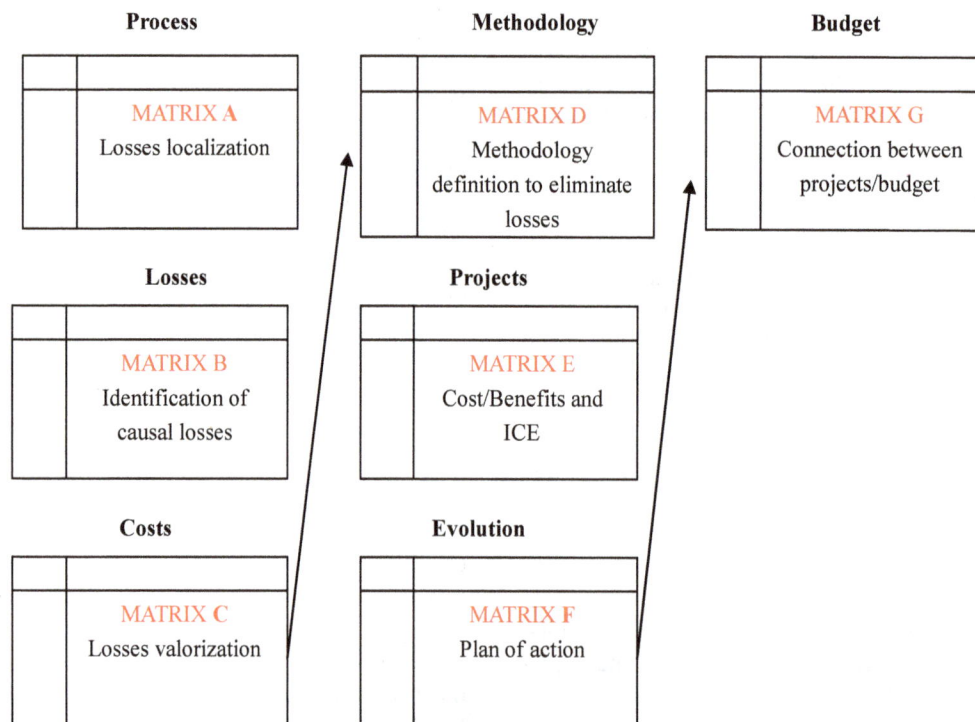

Figure 1. Logic route of cost deployment.

serve to set priorities to make value-added activities of steps 5 to 7 really effective.

The first three steps are specifically designed to calculate and quantify the losses from the establishment budget data and establishment costs and from the operating data, as well.

The fourth and fifth steps aim to define the economy program, through the layering of economies in terms of costs and impacts for the improvement of relative KPIs. This stands as the definition of projects plan.

The sixth and seventh steps are intended to ensure the reporting and monitoring of results analyzing the quarterly progress of operating performance and the calculation of savings in costs and improvements.

After the completion of step 7, Cost Deployment activities should start at step 5, again taking into consideration the matrix A and the costs and losses, with the purpose of selecting other losses which had not been evidenced before because of resources lack, bad judgment, etc. This procedure permits these losses to be attacked with other projects then. In the case of lack of resources, they can be provided from previous projects.

Projects usually last three months. If projects are complex and require a longer time, it is important to divide them into subprojects with intermediate and shorter targets [2]. The following **Figure 2** illustrates the seven steps of Cost Deployment methodology [2]:

The losses and wastes that occur during the execution of a production process are allocated to machines, people and materials.

Cost Deployment aims to determine the individualization of what is a loss and what is a waste, as well as its measurement, and the distinction between resulting cause and root cause.

In a production process which is characterized by generating an output from an input, the efficiency is the ability to produce an output (constant) and at least one input, so waste is defined as an excess input.

As the efficacy is given by the ability to produce a maximum output with a constant input, the loss is defined as unused input. In the imposing of Cost Deployment, first it's important to consider that in a production process 18 significant losses can be identified. They are grouped in terms of personnel and materials/energy [2].

Huge losses tied to machines operation are identified as losses that have impact on the overall efficiency of the equipment and as time losses, the time the equipment is off.

Regarding machines losses, Deployment Cost cannot always be seen immediately especially when a particular piece of equipment is critical in terms of effectiveness. By the way, it may be useful having Overall Equipment Effectiveness, OEE, as a reference. It lets to visualize equipment losses structure, taking into consideration technical efficiency, management and quality. OEE is an indicator that measures the overall rate of quality, the efficiency of delivery and the technical availability of the machine [2].

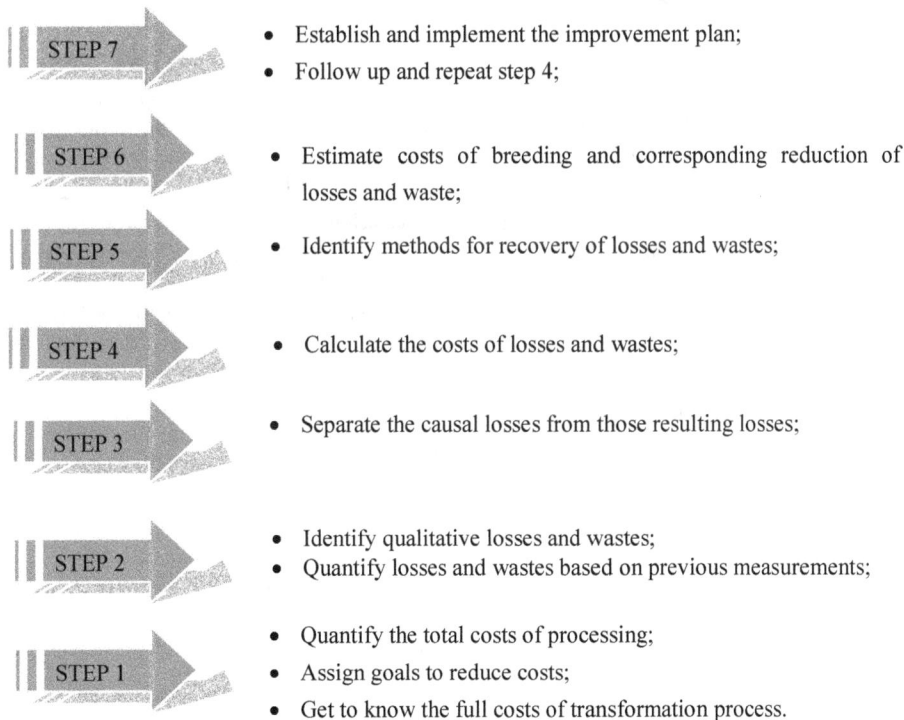

STEP 7
- Establish and implement the improvement plan;
- Follow up and repeat step 4;

STEP 6
- Estimate costs of breeding and corresponding reduction of losses and waste;

STEP 5
- Identify methods for recovery of losses and wastes;

STEP 4
- Calculate the costs of losses and wastes;

STEP 3
- Separate the causal losses from those resulting losses;

STEP 2
- Identify qualitative losses and wastes;
- Quantify losses and wastes based on previous measurements;

STEP 1
- Quantify the total costs of processing;
- Assign goals to reduce costs;
- Get to know the full costs of transformation process.

Figure 2. The seven steps of cost deployment.

Yamashina [2] describes losses of operations related to equipment, people and material/energy. Losses tied to machines that hinder the overall efficiency of the equipment are:

1) Losses that interfere in the technical availability or timing of actual production;

2) Losses that hinder the efficiency of delivery: they are losses which interfere in the effective production time;

3) Losses that impair the quality level: they are losses that affect the effective time of production value;

4) Losses of equipment which have no impact on OEE: losses are attributable to loss of time and theoretical availability of equipment.

Losses caused by people can be grouped into five major losses:

1) Losses of management: waiting time for instructions or for materials, absences, strike, training and education programs;

2) Losses of operating movements: observation, walking, crouching, controlling;

3) Losses of line organization: desaturation, losses due to automation lack;

4) Losses committed by employees affecting quality: rework, lack of automatic control, measurement and implementation, human errors.

Material losses are grouped into three major losses:

1) Losses in the use of direct and consumable materials;

2) Losses in energy use;

3) Losses during maintenance.

Deployment Cost goes deeper. It considers not only resulting in losses as in the traditional way of managing the manufacturing, but also it tries to search for the cause of such losses. For example, loss of manpower can come from downtime that may have been originated from problems with components. These events may be originated in processes or sub-processes even though they were indirectly affected [2].

Therefore, the application of Cost Deployment in the technical pillars of Fiat Group Automobiles Production System allowed a strong acceleration of the results and the achievement of important advantages in reducing losses. This method is the compass that directs and guides continuous improvement projects [3].

3. Conclusions

Industries are not unrelated to the changes that are occurring in the current competitive scenario. However, some management leaders who are part of the same scenario have not given proper treatment for issues related to the specificities of each organization [28].

The philosophy of World Class Manufacturing (WCM)

has proven to be sophisticated and efficient in order to operate in highly globalized and competitive markets, and that their current innovation process, Cost Deployment, radicalized the optimization of production costs and logistic processes.

By analyzing Cost Deployment methodology, it was possible to understand that it also allows linking operational performances which are usually measured with indicators such as efficiency, providing numbers of defect, hours of desaturation, etc... Normally, these indicators are non-comparable among them or with economic performance. They are valued at cost providing a common language to institutions allowing effective definition of the priorities for improvement.

Cost Deployment methodology also enables focus on areas where the greatest casual losses are placed providing opportunities for greater efficiency and effectiveness in reducing and eliminating them. It also facilitates the selection of methodologies and technical pillars to be activated in order to remove or correct the causes of such losses, allowing an easy evaluation of costs and benefits [2].

To face this new challenge of implementing Cost Deployment in operating activities of WCM, Fiat Group Automobiles Production System (FAPS) must be always innovating and successively acquiring new organizational knowledge in order to always present a competitive posture. Therefore, it is necessary to create a proper environment to create and implement innovative, flexible and non-rigid structures operational processes.

REFERENCES

[1] H. Yamashina, "Just-in-Time Production—A New Formulation and Algorithm of the Flow Shop Problem," *Computer—Aided Production Management*, Vol. 34 No. 3, 1998, pp. 120-140.

[2] H. Yamashina, "Manufacturing Cost Deployment," *Journal of the Japan Society for Precision Engineering*, Vol. 65, No. 2, 1999, pp. 260-266.

[3] L. Massone, "Fiat Group Automobiles Production System: Manual do WCM, Wold Class Manufacturing: Towards Excellence Class Safety, Quality, Productivity and Delivery," Ed. Fiat, Brazil, 2007.

[4] R. J. Schonberger, "The Vital Elements of World-Class Manufacturing," *International Management*, Vol. 41, No. 5, 1986, pp. 76-78.

[5] A. Caron, "Technological Innovations in Small and Medium Industrial Enterprises in Times of Globalization: The Case of Paraná," Thesis (Doctoral in Production Engineering), Curitiba, 2003.

[6] H. Chesbrougr, "Technological Innovation Requires Enterprise Networking," 2011. http://www.inovacaotecnologica.com.br/noticias/noticia.php?artigo=inovacao-tecnologica-pressupoe-networking-empresarial

[7] R. H. Hayes and S. C. Wheelwright, "Restoring Our Competitive Edge: Competing Through Manufacturing," Wiley, New York, 1984.

[8] D. Buchanan and A. Huczynski, "Organizational Behavior," 5th Edition, Prentice-Hall, Harlow, 2004.

[9] A. Escrig-Tena, "TQM as a Competitive Factor: A Theoretical and Empirical Analysis," *International Journal of Quality & Reliability Management*, Vol. 21 No. 6, 2004, pp. 612-637.

[10] B. B. Flynn, R. G. Schoroeder and E. G. Flynn, "World Class Manufacturing: An Investigation of Hayes and Wheelwright's Foundation," *Journal of Operation Management*, Vol. 17, No. 3, 1999, pp. 249-269.

[11] R. Mcadam and J. Henderson, "Influencing the Future of TQM: Internal and External Driving Factors," *International Journal of Quality & Reliability Management*, Vol. 21, No. 1, 2004, pp. 51-71.

[12] J. S. Oakland, "Total Organizational Excellence: Achieving World Class Performance," Elsevier, Oxford, 2001.

[13] I. S. Salaheldin and R. Eid, "The Implementation of World Class Manufacturing Techniques in Egyptian Manufacturing Firms: An Empirical Study," *Journal of Industrial Management & Data Systems*, Vol. 107, No. 4, 2007, pp. 551-566.

[14] M. Sharma and R. Kodali, "Development of a Framework for Manufacturing Excellence," *Measuring Business Excellence*, Vol. 12, No. 4, 2008, pp. 55-60.

[15] D. Sinclair and M. Zairi, "An Empirical Study of Key Elements of Total Quality Based Measurement Systems: A Case Study Approach in the Service Industry Sector," *Total Quality Management*, Vol. 12, No. 4, 2001, pp. 535-550.

[16] A. S. Sohal and M. Terziovski, "TQM in Australian Manufacturing: Factors Critical to Success," *International Journal of Quality & Reliability Management*, Vol. 17, No. 2, 2000, pp. 158-168.

[17] M. Svensson and B. Klefsjo, "Experiences from Creating a Quality Culture for Continuous Improvements in Swedish School Sector by Using Self-Assessments," *Total Quality Management*, Vol. 11, No. 4-6, 2000, pp. 800-807.

[18] T. Butler and B. Fitzgerald, "Unpacking the Systems Development Process: An Empirical Application of the CSF Concept in a Research Context," *The Journal of Strategic Information Systems*, Vol. 8, No. 4, 1999, pp. 351-371.

[19] A. L. Digman, "Strategic Management: Concepts, Decisions, Cases," 2nd Edition, Irwin, Homewood, 1990.

[20] R. Eid, I. Elbeltagi and M. Zairi, "Making B2B International Internet Marketing Effective: A Study of Critical Factors Using a Case-Study Approach," *Journal of International Marketing*, Vol. 14, No. 4, 2006, pp. 87-109.

[21] J. S. Oakland, "Total Quality Management-Text with Cases," Butterworth-Heinemann, Oxford, 1995.

[22] G. J. Avlonitis and A. D. Karayanni, "The Impact of Internet Use on Business-to-Business Marketing: Examples from American and European Companies," *Industrial Marketing Management*, Vol. 29, No. 5, 2000, pp. 441-459.

[23] C. A. Mora-Monge, M. E. Gonz´lez, G. Quesada and S. S. Rao, "A Study of AMT in North America: A Comparison between Developed and Developing Countries," *Journal of Manufacturing Technology Management*, Vol. 19, No. 7, 2008, pp. 812-829.

[24] R. Bose, "Customer Relationship Management: Key Components for IT Success," *Industrial Management & Data Systems*, Vol. 102, No. 2, 2002, pp. 89-97.

[25] G. S. Sureshchandar, C. Rajendran and R. N. Anantharaman, "A Conceptual Model for Total Quality Management in Service Organizations," *Total Quality Management*, Vol. 12, No. 3, 2001, pp. 343-363.

[26] D. Dubrovski, "The Role of Customer Satisfaction in Achieving Business Excellence," *Total Quality Management*, Vol. 12, No. 7, 2001, pp. 920-925.

[27] R. Kasul and J. Motwani, "Performance Measurement in World-Class Operations," *Benchmarking for Quality Management & Technology*, Vol. 2, No. 2, 1995, pp. 20-36.

[28] L. C. S. Silva, J. L. Kovaleski, S. Gaia, E. A. S. A. de Matos and A. C. de Francisco, "The Challenges Faced by Brazil's Public Universities as a Result of Knowledge Transfer Barriers in Building the Technological Innovation Center," *African Journal of Business Management*, Vol. 6, No. 41, 2012, pp. 10547-10557.

Effect of Snow, Temperature and Their Interaction on Highway Truck Traffic

Hyuk-Jae Roh[1], Sandeep Datla[2], Satish Sharma[3*]
[1]Saskatchewan Ministry of Highways and Infrastructure, Regina, Canada
[2]City of Edmonton, Edmonton, Canada
[3]Faculty of Engineering, University of Regina, Regina, Canada

ABSTRACT

Based on statistical amount of traffic and weather data sets from three weigh-in-motion sites for the study period of from 2005 to 2009, permanent traffic counters and weather stations in Alberta, Canada, an investigation is carried out to study impacts of winter weather on volume of passenger car and truck traffic. Multiple regression models are developed to relate truck and passenger car traffic variations to winter weather conditions. Statistical validity of study results are confirmed by using statistical tests of significance. Considerable reductions in passenger car and truck volumes can be expected with decrease in cold temperatures. Such reductions are higher for passenger cars as compared to trucks. Due to cold and snow interactions, the reduction in car and truck traffic volume due to cold temperature could intensify with a rise in the amount of snowfall. For passenger cars, weekends experience higher traffic reductions as compared to weekdays. However, the impact of weather on truck traffic is generally similar for weekdays and weekends. Interestingly, an increase in truck traffic during severe weather conditions is noticed at one of the study sites. Such phenomenon is found statistically significant. None of the past studies in the literature have presented the possibility of traffic volume increases on highways during adverse weather conditions; which could happen due to shift of traffic from parallel roads with inadequate winter maintenance programs. It is believed that the findings of this study can benefit highway agencies in developing such programs and policies as efficient monitoring of passenger car and truck traffic, and plan for efficient winter roadway maintenance programs.

Keywords: Highway Truck Traffic; Weigh-in-Motion; Traffic-Weather Models; Highway Design; Vehicle Classification; Highway Operations; Traffic Volume Studies; Transportation Statistics

1. Introduction

Traffic volumes on highways vary with both time and space. Temporal variation occurs with respect to hour, day and month of the year. The highway type, location and route choice behavior of road users cause spatial variation. Even if traffic streams are investigated for the same time and location, the variations of traffic volumes could differ substantially when each vehicle class travelling in the traffic stream is analyzed separately. Adverse weather conditions (in any season or climate) add another dimension to variations of traffic stream. Understanding of such temporal and spatial variations of truck and passenger car traffic under different weather conditions is very useful for both macroscopic and microscopic modelling of highway traffic.

Literature shows that extreme weather conditions cause travel disruptions resulting in slower speeds, travel delays, trip adjustments (for example, cancellation or de-

laying of trips) and increased collisions [1-11]. Owing to the geographical location, the winter weather conditions in Canada and many other northern regions in the World are very severe with heavy snowfall, and extremely cold temperatures. Such winter weather conditions cause increased travel disruptions and trip adjustments which could result in significant changes to truck and passenger car traffic patterns on highways. Datla and Sharma's [1,12] study showed that winter weather causes variations in highway traffic volumes, and the magnitude of such variations depends on time of the day, day of the week, location, highway type and severity of the weather. However, their study and other similar studies in the literature were conducted solely on the basis of total traffic volume data which includes a mix of passenger cars and trucks. None of the past studies in the literature provided detailed information regarding impacts of winter weather on temporal and spatial variations of truck traffic. Impact of weather on route choice behavior of truck and passenger car drivers have also not been studied in the literature.

---

*Corresponding author.

More importantly, none of the studies in the literature has presented the possibility of traffic volume increases on high standard highways during adverse weather conditions; which could happen due to shift of traffic from low standard highways. This study is the first to recognize and model such occurrences.

The main purpose of this study is to investigate and model the impacts of winter weather on temporal and spatial variations of truck and passenger car traffic. Multiple regression models are developed to relate truck and passenger car traffic variations to winter weather conditions. This study uses Weigh-in-Motion (WIM) data from six different sites located on the highways of the province of Alberta, Canada. Hourly traffic volume data from several permanent traffic counters located in the vicinity of WIM stations have also been used in this study. Climate data is obtained from the 598 weather stations (operated by Environment Canada) in the province of Alberta. Modeling work is carried out using CARPACKAGE from the statistical software R [13,14].

2. Literature Review

Past studies on the impact of weather on traffic flow can be broadly categorized into two groups: 1) studies focusing on the impact of weather on traffic parameters such as volume, speed, and headway and 2) studies focusing on the impact of weather events on the quality of traffic flow (e.g. operating level of service, crash rates, traffic delays, start-up delays at intersections, and traffic congestion). As the main focus of this study is on highway traffic volumes, most relevant studies in literature that deal with the impact of weather on highway traffic volumes are only presented in this section.

A majority of studies in literature reported quantitatively the association of traffic volumes with weather conditions *i.e.*, magnitude of reduction in traffic volumes due to different weather conditions. Datla and Sharma [1, 12] studied the impact of cold and snow on daily and hourly traffic volumes on provincial highways of Alberta, Canada. Their study showed that the impact of cold and snow on traffic volumes vary with the day of the week, the hour of the day, the month of the year, the highway type, and the severity of cold temperature. They reported that average winter daily traffic volume reduces by about 30% during extremely cold weather (below −25°C). Their study also showed daily traffic volume reductions ranging from 7% to 17% for each centimeter of snowfall and up to 51% reduction during severe snowstorms (a snowfall of 30 cm or above). Effect of temperature and snowfall interactions on traffic volume was also investigated in their study. Hanbali and Kuemmel [3] studied the average traffic volume reductions due to snow storms on rural highways in the United States. They reported

that traffic volume reductions up to 56% might occur depending on the adversity of snow storm. For Lothian region, Scotland, Hassan and Barker [6] studied the association traffic with meteorological parameters such as minimum and maximum temperatures, snow and rain fall, snow on ground and sunshine hours were considered in their study. Their study concluded that the average traffic reductions were less than 5% under extreme weather conditions but there was a reduction of 10% - 15% in traffic activity when snow was lying on the ground. Knapp and Smithson [2] analyzed the average traffic reductions on interstate highways in Iowa State during winter storms. Their study reported average reductions ranging from 16% to 47% for different storm events. They considered only the storm events having air temperature below freezing, wet pavement surface, pavement temperature below freezing and a snowfall of at least 4-h duration with intensity higher than 0.51 cm/h. Keay and Simmonds [8] reported the association of rainfall and other weather variables with traffic volume on urban arterials in Melbourne, Australia. The traffic reductions during wet days were 1.35% in winter and 2.11% in spring. A maximum reduction of 3.43% was reported for a rainfall ranging from 2 mm to 5 mm in spring. On Inter-state Highway 35 in northern rural Iowa, Maze *et al.* [9] reported a strong correlation between the percentage reduction in traffic volume and wind speed and visibility during snowy days. They have reported 20% reduction in traffic during snowy days with good visibility and low wind speed. The reductions are about 80% when the visibility is less than one-quarter mile and high wind speed (as high as 40 miles per hour). Changnon [15] studied the impact of variations in summer precipitation patterns on travel patterns in Chicago. Using the matched-pair technique, he compared the traffic volume patterns during the days with and without rainfall. Several other studies [3-5,7,10,16-19] have also reported reductions in traffic volume levels and changes in traffic patterns during adverse weather conditions.

Previous paragraph mainly discussed the impact of adverse weather conditions on traffic volumes without providing the details regarding travelers' decision-making behaviour before and during trips. In this section, some studies that consider driver behaviour that directly influence the temporal and spatial variation of highway traffic are reviewed. Hanbali and Kuemmel [3] indicated that the traveling characteristics of trip makers during adverse weather conditions depend on four factors: 1) the trip maker's willingness to travel; 2) the importance of the particular destination; 3) the difficulty of moving from the origin to the destination; and 4) other related factors. They indicated that a reduction in traffic movement occurs due to a traveler's desire to avoid travel during wet or snowy weather. Maki [20] reported that the traveling characteristics of trip makers are highly dependent on the severity of weather conditions and their driving comfort in adverse weather conditions. He mentioned that the driver behaviour during adverse weather conditions varies from place to place. He pointed out that traffic volume reductions occur during adverse weather conditions due to trip adjustments such as leaving for work early, staying late before coming back, and an avoidance of unnecessary and discretionary trips. Khattack [21] and ITT Industries [22] also reported that extreme weather conditions can cause drivers to change their mode of travel, route, departure time, or cancellation of the trip entirely. A recent survey conducted by Markku and Heikki [23] on driver behaviour during adverse weather conditions indicated that trip adjustments during these conditions could be associated with limited driving experience, increasing age, female gender, and the length of the trip. According to their survey, discretionary trips would be underrepresented on highways during poor driving conditions due to the cancellation or postponement of trips. Datla and Sharma [1] showed that the reduction in highway traffic volume due to snow and severe cold is related to the proportion of discretionary trips. Their study also indicted that social-recreational trip makers have more choices as to whether or not they will make a trip with respect to weather conditions. Datla and Sharma's study also showed more trip adjustments during the start of the winter season as compared to middle or end of winter season due to psychological adaptation of drivers to winter weather conditions.

As indicated previously, there is a serious lack of detailed information in the literature regarding impacts of winter weather on truck traffic. Differential impact of winter weather on passenger cars and trucks are not available in the existing literature. Also, there is little or no past research that investigated the interaction of the amount of snowfall and winter temperatures on passenger cars and truck traffic during severe winter conditions. Moreover, none of the studies in the literature have presented the possibility of traffic volume increases on high standard highways during adverse weather conditions; which could happen due to shift of traffic from parallel low standard highways with inadequate winter maintenance programs. Based on large traffic and weather data from weigh-in motion sites, permanent traffic counters and weather stations in the province of Alberta, a detailed investigation is carried in this study to address the research topics identified in this paragraph.

3. Study Data

Traffic data for this study were obtained from Alberta Transportation (AT), the agency that is responsible for provincial transportation in Alberta, Canada. They cur-

rently operate nearly 350 permanent traffic counters (PTCs) to monitor traffic volumes on a 30,875-kilometer highway network. Each PTC represents a particular highway segment. The most recent data available at the start of this study were hourly traffic volume data for 16 years from year 1995 to 2010. AT also collects vehicle classification and weight data at six key highway sections using Weigh-in-Motion (WIM) sites. These WIM sites were installed in July 2004 and have continuously been collecting vehicle classification and load data for programs such as Alberta's Strategic Highway Research and Long Term Pavement Performance Programs. Calibration verification is conducted by a contractor on a monthly basis and it is at 79.4% [24]. The weather data for this study were obtained from Environment Canada's National Climate Data and Information Archive [25]. Environment Canada collects climatologically data from nearly 8000 weather stations (including many intermittent ones) across the country. Each of these weather stations provide detailed weather parameters such as maximum, minimum, and mean temperature (measured in degrees centigrade (°C)), total rain (millimeters), total snow (centimeters), total precipitation (millimeters) and snow on ground (centimeters) on a daily basis. Details of raw data format and measuring methods for each of these weather parameters are available in Environment Canada website [25]. There were about 600 weather stations operated by Environment Canada in the province of Alberta.

Historically, the locations of WIM and PTC sites have been selected without considering the locations of weather stations in the province of Alberta. In some cases, weather stations are located close to the WIM or PTC site, and in other cases, they are far away. Therefore, one of the criteria used to select the study sites was based on the availability of weather stations within acceptable distance from the study sites. Literature review was carried out to define approximate distance within which weather conditions could be similar. From the research done by Andrey and Olley [19] and Datla and Sharma [1], it was found that weather conditions could be considered homogeneous within the area of 16 - 25 km radius around the weather station. A Geographical Information Systems (GIS) base map with 600 weather stations, 6 WIM sites and 350 PTC sites was developed. WIM and PTC sites with weather stations within the 16 - 25 km radius were identified using Proximity analysis module provided by GIS software Arc Map 9.9 [26].

Based on the availability of complete data from different data sources (WIM sites and weather stations), traffic and weather data from three WIM sites over a span of 5 years from 2005 to 2009 were selected for the present study.

After thorough research on different vehicle classification methods, Federal Highway Administration (FHWA) 13-category classification scheme was selected to classify the vehicles from WIM data. This scheme was developed in 1985 by Maine Department of Transportation (Maine DOT) [27] and is called "scheme F". This classification scheme is a standard scheme adopted by several highway agencies in North America.

Figure 1 shows location of study WIM sites and nearby Environment Canada weather stations. Two of the three study sites are located on Alberta Highway 2, which connects the province's two largest cities Edmonton and Calgary. One study site is located on Highway 2 just south of the City of Red Deer while the other is south of the City of Leduc. The third site is located on Highway 2A south of the City of Leduc. Highway 2 is the longest highway in Alberta. This corridor is among the top five in Canada in terms of traffic volume. The highway serves as the major economic corridor in western Canada, and it is part of the CANAMEX TRADE

CORRIDOR which connects Mexico to Alaska through Nevada, Utah, Montana, Alberta, British Columbia, and the Yukon. A significant portion of the heavy vehicle traffic on this facility is oil field related, such as preassembled oil processing equipment, pipes and construction machinery. Agricultural and consumer goods are the next significant goods movements on Highway 2.

The Calgary to Edmonton section of Highway 2 is a four lane divided freeway facility with a 110 km/hr posted speed limit. Highway 2A parallels the east side of Highway 2 north of the City of Red Deer. Highway 2A is a two-lane undivided highway with varying speed limits. The posted speed limit on Highway 2A at the study site is 100 km/hr.

4. Methodology

To investigate the impact of weather on highway traffic by vehicle type, the raw data obtained from the WIM

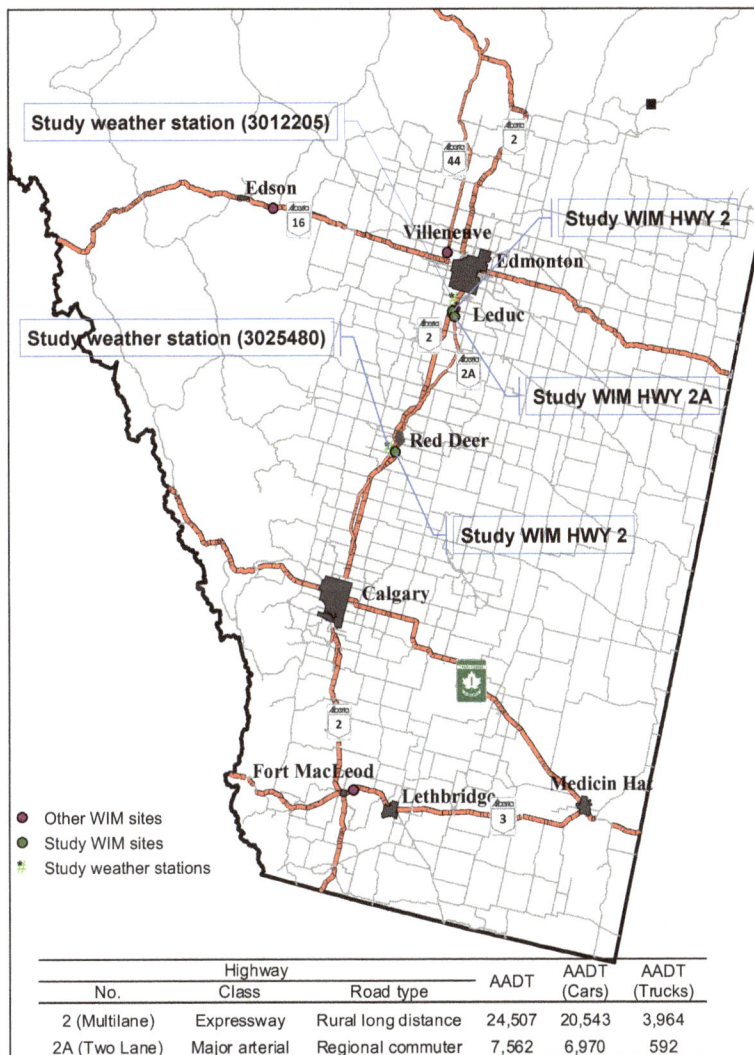

	Highway		AADT	AADT (Cars)	AADT (Trucks)
No.	Class	Road type			
2 (Multilane)	Expressway	Rural long distance	24,507	20,543	3,964
2A (Two Lane)	Major arterial	Regional commuter	7,562	6,970	592

Figure 1. Thematic map showing cities in Alberta, study WIM site, weather station, and highway network.

sites were classified using the FHWA classification method described in the previous section. This process resulted in 13 vehicle classes. However, due to lower number of total trucks in general and too many truck classes, sample data couldn't generate sufficient samples to carry out detailed statistical analysis by each vehicle class. Therefore, the 13 vehicle classes were aggregated into two classes *i.e.*, passenger cars and trucks.

Historical weather records from Environment Canada climate database indicate that the province of Alberta experiences severe snowfall and cold conditions from November to March. Based on these observations study period for this research was limited to November to March from 2005 to 2009. Liu and Sharma [28] indicates that traffic patterns during long weekend holidays are very unique and special attention is needed to conduct research using data from holidays and their neighboring days. For this reason, the three holidays observed in the study period (New Year Day, Alberta Family Day (3rd Monday of February), and Christmas Day) were excluded from the study analysis.

A thorough analysis was carried out to understand the hourly, daily traffic patterns of the classified vehicles (trucks and passenger cars). **Figure 2** shows the typical daily and hourly variations of total traffic, passenger cars and trucks at the three study sites during winter months. The typical weekday hourly volume to AADT ratios (hourly factors) are shown in **Figures 2(a), (c) and (d)**. The typical daily traffic volume to AADT ratios (daily factors) versus day of week plots are shown in **Figures 2(b), (d) and (f)**. Because major portion of total traffic is passenger cars (80% to 85%), the traffic patterns of total traffic and passenger cars are very similar. However truck patterns are unique and different to passenger car patterns. The hourly passenger car factor variations show a single peak from 5 PM to 7 PM for the highway 2 sites (**Figure 2(a)** and **Figure 2(c)**). For the highway 2A site two distinct peaks are observed each day (**Figure 2(e)**): the morning peak from 7 AM to 9 AM and the evening peak from 4 PM to 7 PM. There is a general upward trend in weekday (Monday to Friday) passenger car traffic for all the study sites (see **Figures 2(b), (d) and (f)**). There is only a slight increase from Monday to Thursday; however, traffic on Friday traffic is substantially higher in all cases. The weekend passenger car traffic volumes are higher than the regular weekday (Monday to Thursday) passenger car traffic volumes and slightly less than Friday volumes for highway 2 sites (**Figures 2(b)** and **(d)**). For the highway 2A (**Figure 2(f)**), the Friday volumes are generally the highest, followed by Sunday, Saturday, and other weekdays. The differences in traffic patterns at the study sites show the need to conduct separate analysis. For trucks, the weekday hourly traffic patterns are different from passenger cars. Trucks show a

single peak from 2 PM to 6 PM. The daily truck volume patterns are also different as compared to passenger cars with weekday volumes significantly higher than weekends and unlike passenger cars, Friday volumes are similar to other weekdays. Also truck traffic decreases significantly (more than 50%) during weekends as compared to weekdays. The traffic patterns in **Figures 2(a)** and **(b)** are very similar to the patterns in **Figures 2(c)** and **(d)**. This indicates that the traffic conditions are very consistent on highway 2 with a similar traffic patterns for the two study sites located on highway 2. This is very helpful for testing the study findings for consistency. As per the road type classification system proposed by Sharma *et al.* [29] and traffic patterns observed in **Figure 2**, highway 2 sites can be classified as Regional Long Distance Roads and highway 2A site as a Commuter road. Because the traffic patterns for passenger cars and trucks are very distinct (as observed in **Figure 2**), modeling work has been carried out to develop separate models for the cars and trucks for Highway 2 and Highway 2A sites.

Literature indicates that a standard regression analysis is appropriate to quantify the association between highway traffic volumes and weather. For example, Keay and Simmonds [8] successfully used regression analysis to quantify the impact of rainfall on road traffic volume. Furthermore, Knapp and Smithson [2] showed the appropriateness of the relationship between traffic volume and total snowfall using regression analysis. Based on these previous successful experiences, regression models are developed in this study to relate passenger car and truck traffic volumes with temperature and snowfall.

It is a well known fact that highway traffic volumes generally follow a similar trend year after year. This is the reason why highway agencies across the World rely so much on historical data for applications such as imputation of missing data, traffic monitoring program plans, and development of daily, hourly and monthly adjustment factors. Therefore the expected traffic volume which is historical average traffic volumes at the same time and location is used as an independent variable in the model. Such average volumes represent the range of traffic volumes under different cold and snow conditions. Before proceeding to modeling, scatter plots were developed to get a better understanding of the relationship between dependent and independent variables identified for the modeling. The traffic volume factors (volume to AADT ratios) rather than traffic volumes (number of vehicles) were used in the analysis to account for yearly variations in traffic volumes. Although other weather factors (e.g., wind, pavement condition, etc.,) cause variations in daily traffic volume, we limited in this research to two weather factors (*i.e.*, snow, and temperature) that commonly result severe driving conditions during winter season in North America.

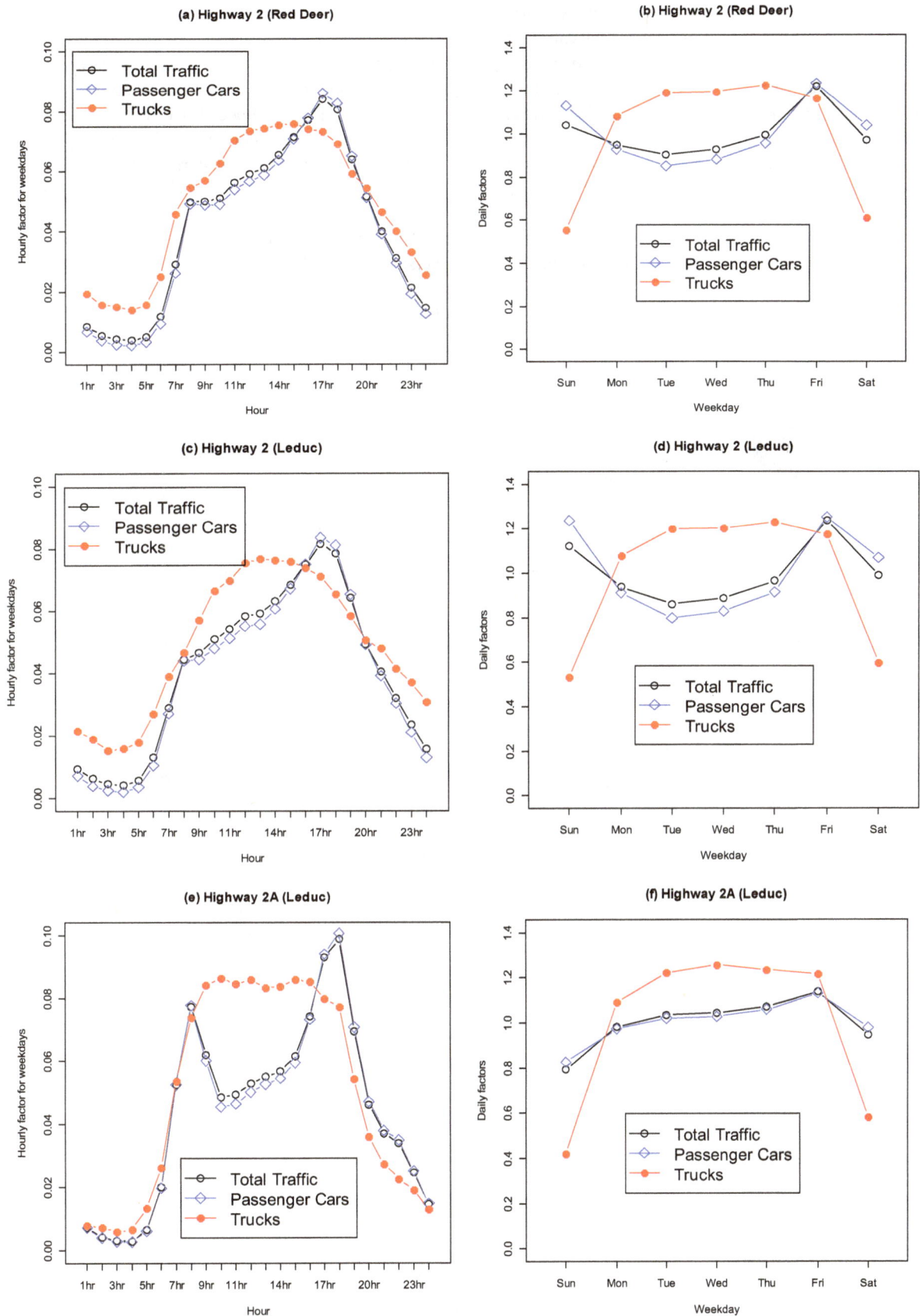

Figure 2. Typical daily and hourly variations of total traffic, passenger cars and trucks.

Figures 3 and **4** show the scatter plots between car and truck traffic volume factors versus expected volume factors, snowfall and temperature. An estimated regression line is also added in the same plot to show the level of closeness between the two values. The blue circles in these plots represent weekdays and red squares represent weekends. The scatter plots between daily volume factors and expected daily volume factors show a very good cluster of sample data along the fitted regression line. This shows a strong positive linear relationship between daily volumes and historical average daily volumes (*i.e.*

expected daily volume factor). The blue circles and red squares overlap considerably in case of total traffic and car traffic plots. For trucks, the red squares are always on the lower side as compared to blue circles. This is because of significantly lower truck traffic volumes on the weekends. It is worth noting that the weekday truck traffic factors are generally clustered between the values of 0.90 to 1.30. However, the weekend daily truck factors were found to spread between 0.30 and 0.70.

The regression lines of total vehicles and car volumes versus snowfall show moderate negative linear relationship,

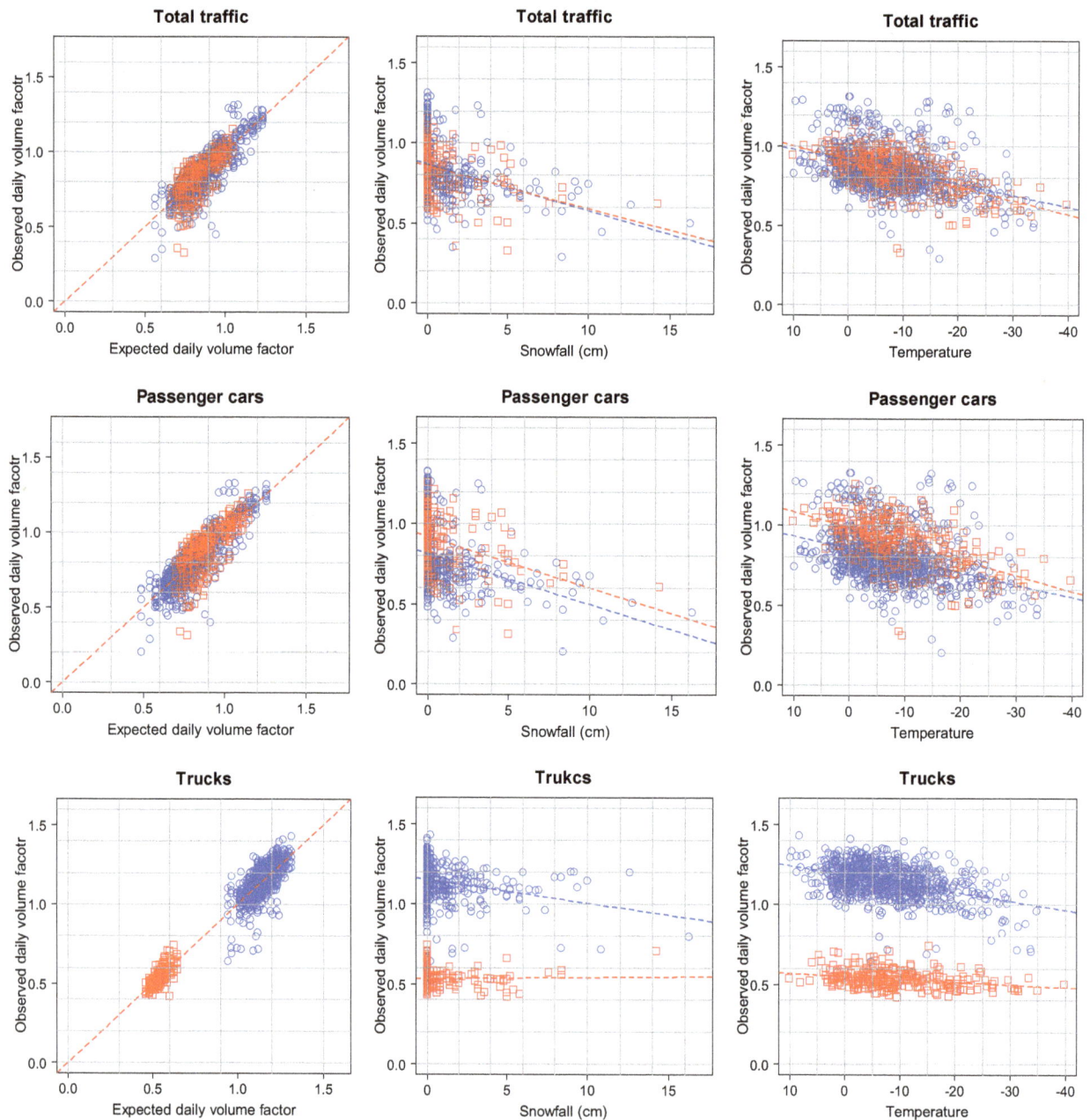

Figure 3. Scatter plots between car and truck traffic volume factors versus snowfall and temperature in highway 2 WIMs site (weekday traffic indicated by blue circle, weekend traffic by red square).

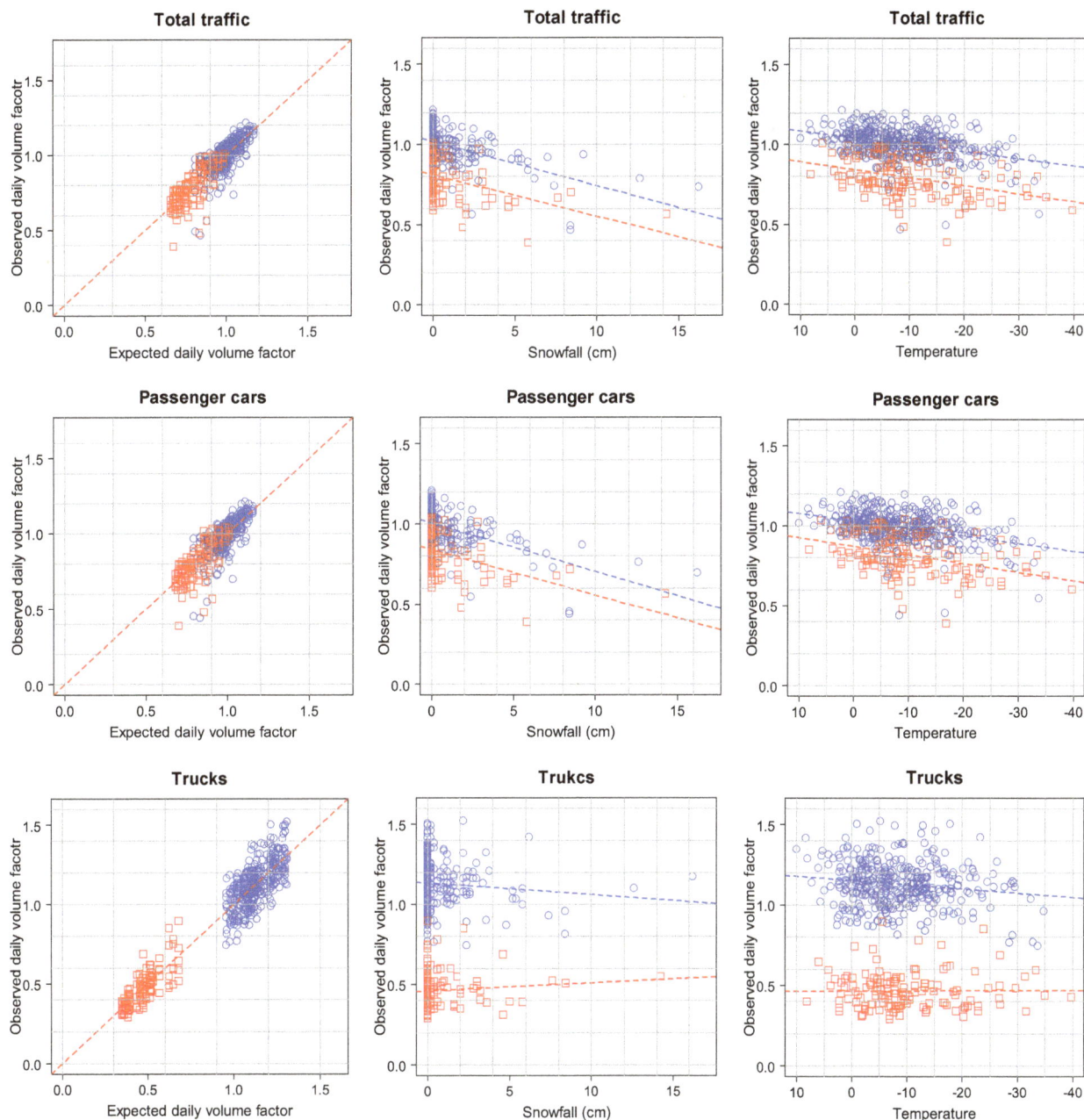

Figure 4. Scatter plots between car and truck traffic volume factors versus snowfall and temperature in highway 2A WIM site (weekday traffic indicated by blue circle, weekend traffic by red square).

i.e., traffic volumes decrease with increase in amount of snowfall. A positive linear relationship is observed between daily volumes and temperature, *i.e.*, daily passenger car volumes increase with increase in daily average temperature. The slope of regression lines of both snowfall and temperature plots are very similar between weekdays and Weekends. In the case of trucks, the clusters of observed data and the fitted regression lines follow similar very different as compared to cars. In case of trucks, the regression lines are very different for weekdays and weekends. Similar patterns are observed for the

Highway 2A site in **Figure 4**. These observations support the need for the development of separate models for cars and trucks. The differences in weekday and weekend scatter plots, especially for trucks, stresses development of separate models for weekdays and weekends.

The histograms of temperature for the days with snow and for the days without snow were constructed separately using weather data. The results indicated that average temperature is colder during the days with snowfall (–10.94˚C, 160 days over the five year period) than the no snowfall days (–8.08˚C, 350 days during the study period). A correlation analysis was also conducted between snowfall and temperature. The correlation coefficients ranged from 0.03 to −0.18, which means that little to no correlation exists between snowfall and temperature. This observation justifies the inclusion of snow and cold as independent variables in model specification.

5. Regression Model

5.1. Without Interaction of Snow and Temperature

Regression analysis has long been recognized as a most flexible and widely used technique to explain variation of quantitative dependent variable by establishing the relationships between dependent variable and a specified set of independent variables in a form of additive and linear mathematical functions [30]. In this research, an attempt has been made to model the impact of weather factors on daily traffic volumes. For the purpose of initial mapping the relationships between daily traffic volume and weather factors, a regression model has been designed with three independent variables i.e., EDVF (expected daily volume factor), snow, and temperature. The additive regression model formulated for this research is:

$$y_i = f(\text{expected daliy volume factor,}$$
$$\text{snow fall, temperature}) \quad (1)$$
$$= \beta_1 \text{EDVF}_i + \beta_2 \text{SNOW}_i + \beta_2 \text{TEMP}_i$$

where,

i : refers to the ith observation

$\beta_1, \beta_2, \beta_3$: regression coefficients estimated for the respective independent variable

y_i : estimated value of daily traffic volumes factor for different vehicle classes (cars, trucks etc.)

EDVF : expected daily volume factor

SNOW : amount of snowfall per day (cm)

TEMP : average daily temperature (°C).

Normalized daily total volumes (traffic volume to AADT ratio), passenger car volumes (passenger car volume to passenger car AADT ratio), and truck traffic volume (truck volume to truck AADT ratio) are used instead of actual volumes to take into consideration the yearly variations in traffic volumes. The normal traffic volume and weekly traffic trends are reflected by adding an expected daily volume factor (EDVF) as an independent variable in model. EDVF is calculated using historical data of the same month, week, and day. The variables SNOW and TEMP represent the weather conditions.

5.2. With Interaction of Snow and Temperature

The additive regression model (Equation (1)) with EDVF, SNOW, and TEMP as independent variables would take into account the individual impact of cold and snowfall

on highway traffic (i.e., the impact of temperature on traffic volumes after impact of snowfall is taken into account and the impact of each centimeter of snowfall after the effect of temperature is taken into account). A more complex model design is required to investigate whether the impact of snowfall on traffic volumes is the same at all temperature ranges or varies with the severity of temperature. Such analyses would verify the existence or non-existence of cold and snow interaction impacts on traffic volumes Therefore the model structure shown in Equation (1) is modified to accommodate interactions between cold (temperature) and snow. The differential impacts of snowfall by cold or, equivalently, the differential impacts of cold by snowfall were captured in the model by including $\text{SNOW}_i * \text{TEMP}_i$ interaction terms to the earlier model design in Equation (1). The final form of the model used in this paper is shown in Equation (2).

$$y_i = f(\text{expected daliy volume factor,}$$
$$\text{snow fall, temperature}) \quad (2)$$
$$= \beta_1 \text{EDVF}_i + \beta_2 \text{SNOW}_i + \beta_3 \text{TEMP}_i$$
$$+ \beta_4 \text{SNOW}_i * \text{TEMP}_i$$

where,

β_4 : regression coefficients estimated for the snow-temperature interaction variable.

Except interaction terms, all other terms serve the same purposes as defined in Equation (1).

6. Results and Analysis

The modeling process was carried out using the classified WIM data and corresponding weather records from the three study WIM sites (two WIMs site for highway 2 and one WIM site for highway 2A). Regression models were calibrated using Equation (2). In total, 18 models (2 study highway sites × 3 day groups × 3 vehicle classes) were developed. These include separate models for all days, weekdays and weekends and for total traffic, passenger cars and trucks for both highways. **Table 1** shows the 18 calibrated models along with the statistical test results.

The overall goodness of fit of the regression model to sample data is evaluated by the squared multiple correlation coefficient (R^2). The R^2 values for all models are over 0.98. The value of the F test is also reported to assess the overall adequacy of the model. The results show that all the 18 models are statistically valid by rejecting the null hypothesis at better than the 0.001 level. The statistical significance for individual coefficients is evaluated by t-statistic. The significance levels of t-tests are shown in **Table 1** using "*" symbol. More number of stars ("*") represent better model. Incremental F-statistic and R^2_{Naive} are also calculated to test the null hypothesis of no interaction between snow and cold and the test results are shown in **Table 1**.

Based on the results of the four statistical tests (*i.e.*, R^2, *F*-test, incremental *F*-test, and *t*-test) conducted to evaluate the study models, it is apparent that the models fit well to the sample data and the structure of the proposed model is appropriate.

A partial effect analysis between the dependent variable and an independent variable could be determined by varying the selected independent variable between its lower and upper boundaries (based on the study samples) and holding the other independent variables at a certain value. The partial effects of cold temperatures on daily traffic volumes are investigated in this section *i.e.*, for example, partial effect of cold temperatures on daily traffic for a specific pre-defined snowfall amount. **Figures 5** and **6** show the graphs developed to study the partial impact of cold temperatures on car and truck volumes. Results are presented separately for weekdays and weekends.

The solid line gives estimated daily traffic from the models shown in **Table 1**. The dotted lines give the 95% envelop for the upper and lower thresholds of dependent variable estimates. Each plot in **Figures 5** and **6** shows the partial impact of cold temperatures on daily traffic

Table 1. Calibrated traffic weather models by vehicle class, day of the week and highway.

Variables	All Days Models (Highway 2A)			Weekdays Models (Highway 2A)			Weekend Models (Highway 2A)		
	Total Traffic	Passenger Cars	Trucks	Total Traffic	Passenger Cars	Trucks	Total Traffic	Passenger Cars	Trucks
EDVF	1.0329102***	1.0374549***	1.0010000***	1.0266381***	1.0309509***	1.0028914***	1.0605530***	1.0637035***	0.9907***
SNOW	−0.0170572***	−0.0182616***	−0.0108900*	−0.0155569***	−0.0169710***	−0.0085150	−0.0222481***	−0.0226961***	−0.11626**
TEMP	0.0020028***	0.0022982***	0.0000494	0.0015939***	0.0018881***	0.0001073	0.0032571***	0.0035110***	−0.00001412
SNOWTEMP	0.0004917*	0.0006044*	−0.0009029*	0.0006064*	0.0006976**	−0.0003994	0.0001235	0.0003061	−0.001941***
R^2	0.9971	0.9968	0.9916	0.9977	0.9974	0.9923	0.9952	0.9948	0.9823
Statistic	43840***	39220***	14280***	39500***	35200***	11290***	6750***	6287***	1814***
Change from R^2_{Naive}	-	0.0001	0.0001	0.0001	0.0001	-	-	-	0.0017
Incremental *F*-Statistic	-	15.8125***	5.7857*	16.1304***	14.2692***	-	-	-	28.9220***
Number of Sample Days	510	510	490	375	375	355	135	135	135

Variables	All Days Models (Highway 2)			Weekdays Models (Highway 2)			Weekend Models (Highway 2)		
	Total Traffic	Passenger Cars	Trucks	Total Traffic	Passenger Cars	Trucks	Total Traffic	Passenger Cars	Trucks
EDVF	1.0419390***	1.0463226***	1.0160252***	1.0373627***	1.0405659***	1.0211869***	1.0558368***	1.0606082***	1.0085723***
SNOW	−0.0125431***	−0.0135735***	−0.0052943*	−0.0144013***	−0.0153455***	−0.0080829**	−0.0078173	−0.0102099	0.0051253
TEMP	0.0028608***	0.0030388***	0.0015290***	0.0026138***	0.0026134***	0.0022545***	0.0035069***	0.0040640***	0.0004533
SNOWTEMP	0.0009854***	0.0011379***	0.0002958	0.0007948**	0.0008879***	0.0003974	0.0014021***	0.0015639***	0.0004921*
R^2	0.9949	0.9936	0.9962	0.9953	0.994	0.9964	0.994	0.993	0.9957
Statistic	49690***	39670***	66770***	39520***	31070***	52080***	11030***	9497***	15520***
Change from R^2_{Naive}	0.0001	0.0001	-	0.0001	0.0001	-	0.0003	0.0003	0.0001
Incremental *F*-Statistic	19.92***	15.875***	-	15.8723***	12.4333***	-	13.3***	11.4***	6.186*
Number of Sample Days	1020	1020	1020	750	750	750	270	270	270

***Coefficient is statistically significant at the 0.001 level; **0.01 level; *0.05 level.

volumes for a specific combination of vehicle type, day of the week, location and pre-defined amount of snowfall. First row of plots in **Figure 5** show the change in daily volume factor of passenger cars due to winter temperatures during weekdays for the Highway 2 sites for different pre-defined amounts of snowfall. These plots are generated by fixing EDVF at its average value calculated from sample data set. It should be noted that the estimated daily volume factor at the point where the two dotted lines cross each other shows the average EDVF for that particular subset of data. Also note that the

crossing point generally happened at about –10°C temperature which represents average and also most frequently occurring winter weather conditions in the vicinity of study sites. Comparison of slopes of plotted lines from left to right show change in partial impact of temperature on daily traffic with increase in amount of snowfall. Second row of plots in **Figure 5** shows similar plots for trucks. Third and fourth rows of plots show the partial impact of cold on daily volume factor of passenger cars and trucks during weekends. The plots in **Figure 6** are similar to **Figure 5** except for different road site

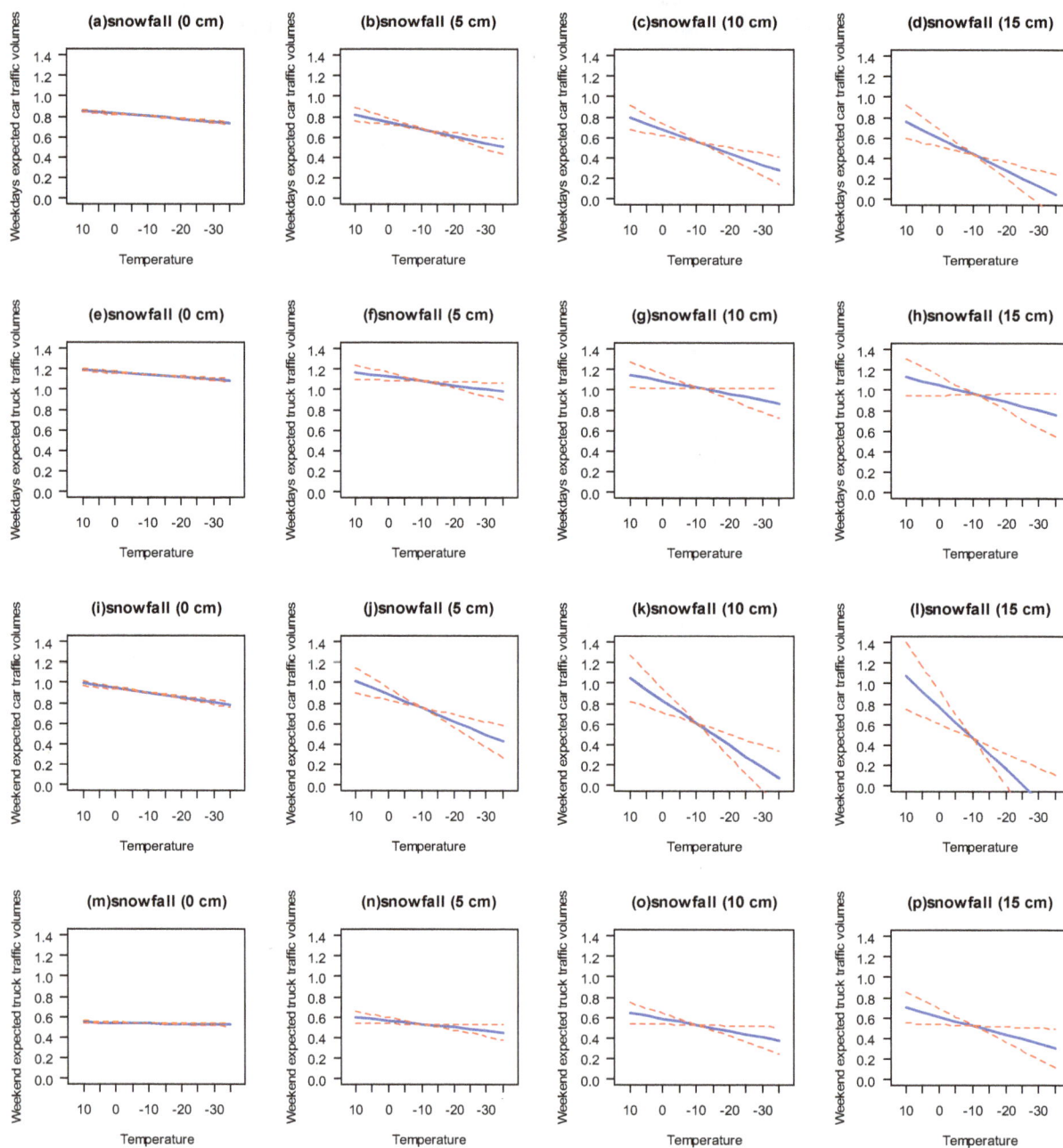

Figure 5. Partial impact of cold temperatures and snowfall on car and truck volumes for highway 2 sites (95% envelopes of volume estimates indicated by red dotted line, expected volume by blue solid line).

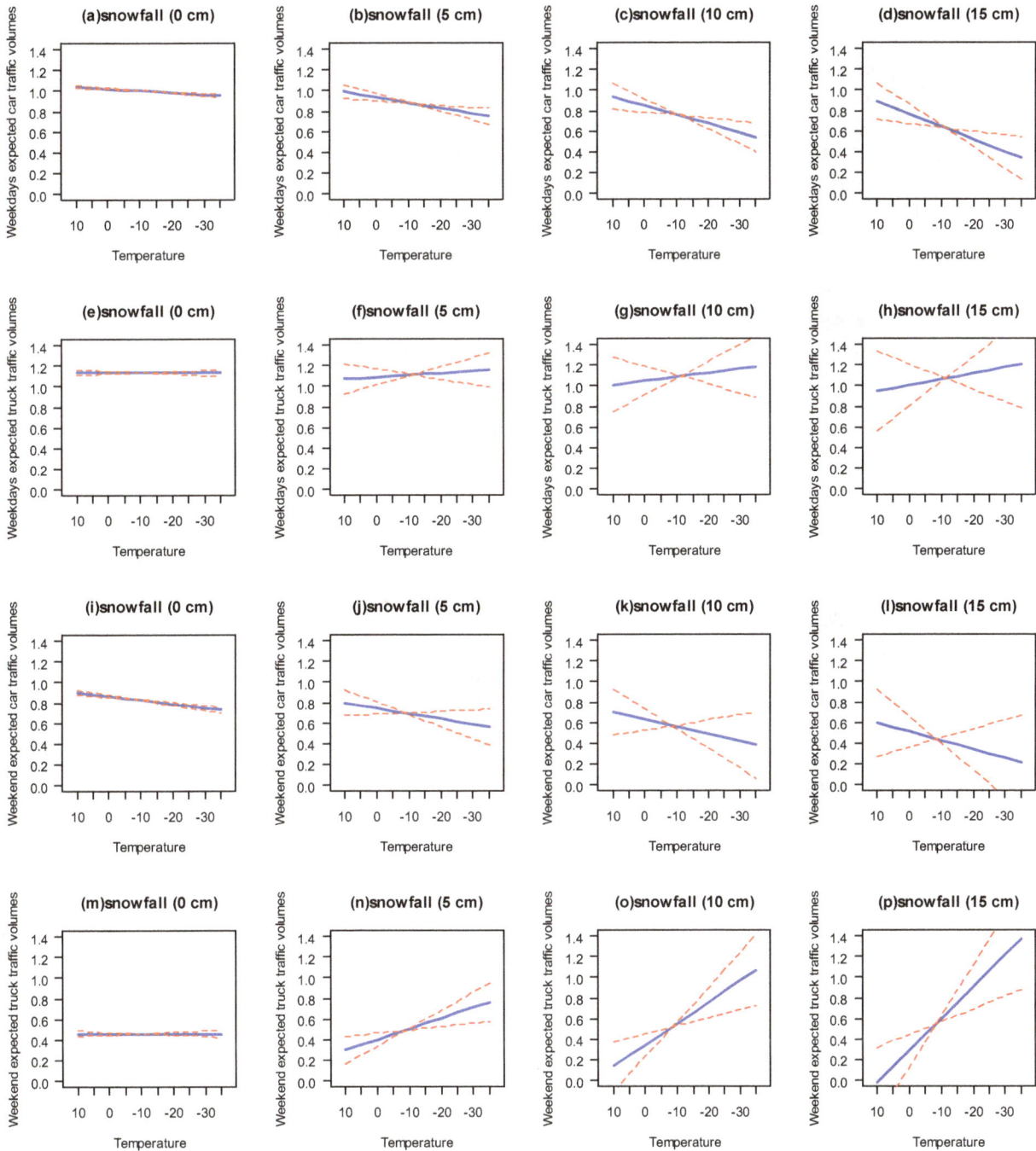

Figure 6. Partial impact of cold temperatures and snowfall on car and truck volumes for highway 2A Site (95% envelopes of volume estimates indicated by red dotted line, expected volume by blue solid line).

(Highway 2A site).

Based on the traffic weather patterns of **Figure 5**, the following interpretations can be made for Highway 2 sites, which serve regional long distance trips:

- There is a clear indication that traffic volume on a given day depends on severity of cold and amount of snowfall.
- A reduction in passenger car and truck volumes can be expected with increase in severity of cold tem-

peratures. The amount of decrease in traffic volume depends on severity of cold.

- When there is no snowfall the impact of cold temperature on both car and truck volumes is very marginal (see plots in **Figures 5(a)**, **(e)**, **(i)** and **(m)**).
- As the amount of snowfall increases the steepness of the regression lines increases, which means that the reduction in traffic volume due to cold temperature would intensify with a rise in amount of snowfall.

This clearly shows the existence of cold and snowfall interactions.

- A snowfall of 15 cm or higher during severe cold conditions (–20°C or lower) would result in a dramatic decrease in passenger car traffic on the road (see plots in **Figures 5(d)** and **(l)**).
- With higher amounts of snowfall, passenger cars experience higher reductions due to cold and snowfall as compared to trucks (as shown in plots d and l for cars; and plots h and p for trucks)
- For passenger cars, the weekend plots (**Figures 5(i)**, **(j)**, **(k)** and **(l)**) are steeper than weekdays (a, b, c, and d) indicating higher car traffic reductions in weekends as compared to weekdays. In the case of trucks, the regression lines seem parallel, which means the impact of cold and snow on truck traffic is similar for weekdays and weekends.
- It should be noted that the width of the 95% envelop (shown by dashed lines) also increases from left to right indicting that the reliability of model estimate deplete with increase in severity of weather conditions. This could be because of low sample size due to lesser number of days of heavy snowfall during the winter season.

The amount of reduction in traffic volume may be attributed to the proportion of discretionary trips in the traffic stream. The existence of more discretionary trips results in higher trip adjustments and, hence, higher traffic reductions. The reason of higher weekend traffic reductions may be due to the large proportion of discretionary trips during weekends. Trucks (or commercial vehicles) are usually required to follow rigid schedules to complete their mandatory travel irrespective of severe weather conditions. These kinds of business-oriented mandatory movements are most likely to generate unique truck travel patterns even when weather is unfavorable for making a trip. This could be the reason for (i) lower truck reductions in general and (ii) similar impact of weather on weekdays and weekends truck traffic

Figure 6 shows the partial impact of cold on car and truck traffic on Highway 2A site, which serves as a regional commuter road in Leduc region. Interpretation of plots is the same as in **Figure 5**. Similar to Highway 2, passenger car traffic experiences reduction in volume with increase in severity of cold and amount of snowfall. However, for both weekdays and weekends, the regression lines seem to be less steep than those observed for Highway 2. This indicates lesser impact of cold on car traffic at Highway 2A site as compared to Highway 2. It is interesting to note that the width of the 95% envelops of volume estimates for Highway 2A site are larger as compared to the plots in **Figure 5**, especially for weekend car traffic. This could be due to a much smaller sample size available for used in the analysis for Highway

2A site, as shown in **Table 1**. Except for these differences, all the remaining observations made earlier based on passenger car traffic-weather relation plots in **Figure 5** are valid for Highway 2A site also.

The behavior of truck traffic at Highway 2A site during adverse weather conditions is quite different from the Highway 2 site. It is very interesting to see the regression lines of truck traffic showing a reverse trend *i.e.*, increase in truck traffic volume with increase in severity of weather conditions. Such an increase in truck traffic is marginal for weekdays and high for weekends. This is contradictory to what has been observed for truck traffic at Highway 2 site. Moreover, none of the studies in the literature reported increase in traffic volumes during severe weather conditions. A further investigation was conducted to understand the possible reasons for such a controversial truck traffic patterns.

After reviewing the highway network near the study site, it was found that Highway 814 runs parallel to the Highway 2A. Highway 814 is a secondary highway with lower priority for winter road maintenance (snow removal etc.). Therefore it is possible that truck drivers may change their travel route by shifting to Highway 2A. A careful analysis of hourly traffic data from permanent traffic counter site located on Highway 814 showed significantly lower traffic volumes during the days with severe weather conditions (*i.e.*, higher than normal traffic reductions that were seen at other similar highway segments). Brief communications with local municipalities in the vicinity of the study sites also indicated the possibility of traffic shifting. Therefore it could be concluded that there is a possibility of traffic volume increases on high standard highways during adverse weather conditions, which could happen due to shift of traffic from parallel low standard highways.

7. Summary and Conclusions

The literature study presented in this paper indicates clearly that severe weather conditions trigger variations in highway traffic. However, none of the past studies in the literature provided detailed information regarding impacts of winter weather on temporal and spatial variations of truck traffic. It is believed that understanding of truck traffic variations (or behaviour) under severe weather conditions can benefit highway agencies in developing such programs and policies as efficient monitoring of passenger car and truck traffic, and plan for efficient winter roadway maintenance programs.

An attempt has been made in this study to quantify the traffic variations under different weather conditions. Vehicle classification data from the WIM sites operated in the province of Alberta were used in this study. Climate data (2 weather stations) were obtained from the 598 weather stations (operated by Environment Canada) in the province of Alberta. Multiple regression models were formulated and calibrated to relate truck and passenger car traffic variations to winter weather conditions. All the calibrated models were tested for statistical significance of the independent variables using the standard test statistics such as F-test, incremental F-test, and t-test.

A number of conclusions were drawn from the study results. Firstly, both car and truck traffic volumes on highways vary with severity of cold and amount of snowfall. The impact of cold temperature on both car and truck volumes is marginal during no snowfall days. The reduction in traffic volume due to cold temperature would intensify with a rise in amount of snowfall indicating the existence of cold and snowfall interactions. With higher amounts of snowfall, passenger cars experience higher reductions due to cold and snowfall as compared to trucks. A snowfall of 15 cm or higher during severe cold conditions (–20°C or lower) would result in very few Cars travelling on the road. Moreover passenger cars experience higher traffic reductions in weekends as compared to weekdays. In the case of trucks the impact of cold and snow on truck traffic is similar for weekdays and weekends.

It is evident from this study that passenger cars are more vulnerable to adverse weather conditions than trucks. This vulnerability to severe weather conditions could be attributed to such behavior of drivers as choosing flexible departure times, changing routes, or canceling travel entirely and being able to make trip adjustments by avoiding discretionary trips. Trucks are not as greatly affected as passenger cars by adverse weather conditions. Trucks (or commercial vehicles) are usually required to follow rigid schedules to complete their mandatory travel irrespective of severe weather conditions. These kinds of business-oriented mandatory movements are most likely to generate unique truck travel patterns even when weather is unfavorable for making a trip.

Interestingly, the modelling results for one of the study sites reveal that higher truck traffic volumes can result during heavy snowfall (or other adverse weather conditions) in winter months. This is contradictory to observations from other similar studies in the literature. None of the studies in literature have reported such an increase in traffic volumes during severe weather conditions. A further investigation carried out for this study to understand the reasons for such behavior indicated shifting of trucks from secondary highways to primary highways due to poor winter maintenance programs. Therefore it can be concluded that there is a possibility of traffic volume increases on high standard highways during adverse weather conditions; which could happen due to shift of traffic from parallel low standard highways.

8. Acknowledgements

The authors are grateful towards the Natural Science and Engineering Research Council of Canada (NSERC), the Faculty of Graduate Studies at the University of Regina, and Saskatchewan Government Insurance (SGI) for their financial support. The authors also thank Alberta Infrastructure and Transportation for providing the WIM data used in this study.

REFERENCES

[1]   S. Datla and S. Sharma, "Variation of Impact of Cold Temperature and Snowfall and Their Interaction on Traffic Volume," In: *Transportation Research Record: Journal of the Transportation Research Board*, No. 2169, Transportation Research Board of the National Academies, Washington DC, 2010, pp. 107-115.

[2]   K. K. Knapp and L. D. Smithson, "Winter Storm Event Volume Impact Analysis Using Multiple-Source Archived Monitoring Data," In: *Transportation Research Record: Journal of the Transportation Research Board*, No. 1700, Transportation Research Board of the National Academies, Washington DC, 2000, pp. 10-16.

[3]   R. M. Hanbali and D. A. Kuemmel, "Traffic Volume Reduction Due to Winter Storm Conditions," In: *Transportation Research Record: Journal of the Transportation Research Board*, No. 1387, Transportation Research Board of the National Academies, Washington DC, 1993, pp. 159-164.

[4]   L. C. Goodwin, "Weather Impacts on Aterial Traffic Flow," The Road Weather Management Program, FHWA, US Department of Transportation, Washington DC, 2002.

[5]   F. L. Hall and D. Barrow, "Effects of Weather and the Relationship between Flow and Occupancy on Freeways," In: *Transportation Research Record: Journal of the Transportation Research Board*, No. 1194, Transportation Research Board of the National Academies, Washington DC, 1988, pp. 55-63.

[6]   Y. A. Hassan and J. J. Barker, "The Impact of Unseasonable or Extreme Weather on Traffic Activity within Lothian Region, Scotland," *Journal of Transport Geography*, Vol. 7, No. 3, 1999, pp. 209-213.

[7]   A. T. Ibrahim and F. L. Hall, "Effect of Adverse Weather Conditions on Speed-Flow-Occupancy Relationships," In: *Transportation Research Record: Journal of the Transportation Research Board*, No. 1457, Transportation Research Board of the National Academies, Washington DC, 1994, pp. 184-191.

[8]   K. Keay and I. Simmonds, "The Association of Rainfall and Other Weather Variables with Road Traffic Volume in Melbourne, Australia," *Accident Analysis and Prevention*, Vol. 37, No. 1, 2005, pp. 109-124.

[9]   T. H. Maze, M. Agarwal and G. D. Burchett, "Whether Weather Matters to Traffic Demand, Traffic Safety, and Traffic Operations and Flow," In: *Transportation Research Record: Journal of the Transportation Research*

Board, No. 1948, Transportation Research Board of the National Academies, Washington DC, 2006, pp. 170-176.

[10] B. L. Smith, K. G. Byrne, R. B. Copperman, S. M. Hennessy and N. J. Goodall, "An Investigation into the Impact of Rainfall on Freeway Traffic Flow," *Proceedings of the Annual Meeting of the Transportation Research Board*, Washington DC, 2004.

[11] J. Andrey, B. Mills, M. Leahy and J. Suggett, "Weather as a Chronic Hazard for Road Transportation in Canadian Cities," *Natural Hazards*, Vol. 28, No. 3, 2003, pp. 319-343.

[12] S. Datla and S. Sharma, "Impact of Cold and Snow on Temporal and Spatial Variations of Highway Traffic Volumes," *Journal of Transport Geography*, Vol. 16, No. 5, 2008, pp. 358-372.

[13] J. Fox, "An R Companion to Applied Regression," SAGE Publications, Inc., Thousand Oaks, 2011.

[14] R Foundation for Statistical Computing, "R Development Core Team: A Language and Environment for Statistical Computing," R Foundation for Statistical Computing, Vienna, 2010.

[15] S. A. Changnon, "Effects of Summer Precipitation on Urban Transportation," *Climatic Change*, Vol. 32, No. 4, 1996, pp. 481-495.

[16] J. C. McBride, M. C. Benlangie, W. J. Kennedy, F. R. McCornkie, R. M. Steward, C. C. Sy and J. H. Thuet, "Economic Impacts of Highway Snow and Ice Control," FHWA, US Department of Transportation, Washington DC, 1997.

[17] V. P. Shah, A. D. Stern, L. C. Goodwin and P. Pisano, "Analysis of Weather Impacts on Traffic Flow in Metropolitan Washington DC," *Proceedings of the Annual Meeting of Institute of Transportation Engineers*, Washington DC, 2003.

[18] L. I. Zang, P. Holm and J. Colyar, "Identifying and Assessing Key Weather-Related Parameters and Their Impact on Traffic Operations Using Simulation," FHWA, US Department of Transportation, Washington DC, 2004.

[19] J. Andrey and R. Olley, "Relationships between Weather and Road Safety, Past and Future Directions," *Climatological Bulletin*, Vol. 24, No. 3, 1990, pp. 123-137.

[20] P. J. Maki, "Adverse Weather Traffic Signal Timing," *Proceedings of the Annual Meeting of Institute of Transportation Engineers*, Las Vegas, 1999.

[21] A. J. Khattak, P. Kantor and F. M. Council, "Role of Adverse Weather in Key Crash Types on Limited Access Roadway—Implications for Advanced Weather System," In: *Transportation Research Record: Journal of the Transportation Research Board, No.* 1621, Transportation Research Board of the National Academies, Washington DC, 1998, pp. 10-19.

[22] ITT Industries, Inc., System Division, "Identifying and Assessing Key Weather-Related Parameters and Their Impacts on Traffic Operations Using Simulation," FHWA, US Department of Transportation, Washington DC, 2003.

[23] M. Kilpeläinen and H. Summala, "Effects of Weather and Weather Forecasts on Driver Behavior," *Transportation Research Part F: Traffic Psychology and Behaviour*, Vol.10, No. 4, 2007, pp. 288-299.

[24] P. Kilburn, "Alberta Infrastructure & Transportation Weigh in Motion Report," Government of Alberta Ministry of Transportation, Calgary, 2008.

[25] Weather Office, Environment Canada, Gatineau, Quebec, Canada, 2010. www.climate.weatheroffice.gc.ca/climateData/canada_e.html.

[26] "ArcGIS 10 Help Library: Geographic Information System (GIS)," ArcGIS 10, Environmental System Research Institute, Inc., Redlands, 2010.

[27] J. H. Wyman, G. A. Braley and R. I. Stephens, "Field Evaluation of FHWA Vehicle Classification Categories," FHWA, US Department of Transportation, Washington DC, 1985.

[28] Z. B., Liu and S. Sharma, "Statistical Investigations of Statutory Holiday Effects on Traffic Volumes," In: *Transportation Research Record: Journal of the Transportation Research Board, No.* 1945, Transportation Research Board of the National Academies, Washington DC, 2006, pp. 40-48.

[29] S. Sharma and A. Werner, "Improved Method of Grouping Province Wide Permanent Traffic Counters," In: *Transportation Research Record: Journal of the Transportation Research Board, No.* 815, Transportation Research Board of the National Academies, Washington DC, 1981, pp. 12-18.

[30] J. Fox, "Applied Regression Analysis and Generalized Linear Models," SAGE, Inc., Thousand Oaks, 2008.

4

Using OD Estimation Techniques to Determine Freight Factors in a Medium Sized Community

Michael D. Anderson[1], Jeffrey P. Wilson[1], Gregory A. Harris[2]
[1]Department of Civil and Environmental Engineering, University of Alabama in Huntsville, Huntsville, USA
[2]Center for Management and Economic Research, University of Alabama in Huntsville, Huntsville, USA

ABSTRACT

Developing an understanding of the socio-economic factors that can be used to generate truck trip productions and attractions in small and medium sized communities can be used to improve travel models and provide better information upon which infrastructure decisions are made. Unfortunately, it is difficult to collect this data in a timely, cost-effective manner. This paper presents a methodology that uses matrix estimation techniques from existing traffic counts to develop origin/destination pairs that can be used to statistically develop truck trip generation models. A case study is presented and a model is presented for one smaller urban community.

Keywords: Freight Modeling; Origin/Destination Estimating

1. Introduction

Freight transportation has become an increasingly important issue for small and medium sized communities [1-3]. Understanding which factors influence truck productions and attractions, and the magnitude of the influence within the modeling environment, could improve the accuracy of the models and potentially lead to better decisions regarding transportation infrastructure investment. These factors, if understood, could be developed into a truck trip generation model similar to other models that have been proposed [3], but developed for a local area. Unfortunately, it is difficult to directly measure and identify those factors that specifically predict truck volumes. Surveying individual companies provides a detailed view but lacks the transferability to the entire region and large scale cordon-line studies are expensive and impractical to perform. Thus, a new methodology to develop a truck trip generation equation would potentially be beneficial to transportation planners in smaller urban communities.

An alternative method to determine the number of trucks produced or attracted to locations within a smaller community is proposed in this paper. The method proposed uses existing truck traffic counts on a network and then working backwards to determine the origin/destination matrix that could have produced the specific counts observed. The method is presented in a paper by Van Zuylen and Willumsen where the origin/destination matrix estimation procedure was tested on a sample network

and determined to produce accurate results [4]. The method is outlined in the text by Ortuzar and Willumsen, and demonstrates how there are a limited number of trip matrices that can be used to replicate the traffic volumes [5]. Once the origin/destination pattern is identified, the total quantity of trucks exiting and entering the zones can be used as actually production and attraction values. A regression analysis can be performed to understand the contributions of the socio-economic data within the zone and a prediction model for truck trip generation developed.

The methodology is used to determine the origin/destination pairs that would likely use the roadways upon which the count was collected. If a roadway indicates truck volume, it is assumed that the volume is developed from the origin/destination pairs that have that roadway along the shortest path, as it is unlikely that a truck would intentionally take a longer path through the network. The use of multiple roadways within the study area allow for distribution of truck traffic to alternate origin/destination pairs until all the roadways are used and a final potential solution is reached. The final potential solution is used as the input instead of basic initialization beginning principle and another complete run of the roadways is performed. Iterative runs are performed until the difference between iterations reaches a desired level of closure and the most likely origin/destination matrix is obtained.

In this paper, an approach to determine the factors that contribute to truck generation models for a medium-sized community through the use of an origin/destination matrix estimation technique requiring existing truck traffic counts as the input is used. The paper discusses the application of the methodology to convert the truck counts into an origin/destination matrix for a case study community and then applies statistical techniques to determine the appropriate factors for truck trip generation using the number of truck produced and attracted to areas of the network and the corresponding socio-economic data for those areas. The contribution of the paper is the application of the methodology and a model that potentially can be utilized by other communities.

2. Methodology and Case Study

The focus of this paper is the development of a methodology capable of extracting an origin/destination matrix from a set of existing traffic counts; then, develop a statistical relationship to determine the trip production and attraction values for trucks in a small or medium sized community. The location for the case study was Mobile, AL, a medium sized community of approximately 300,000 people. This location was ideal as the Alabama Department of Transportation had recently conducted a set of manual truck counts at 42 locations within the Mobile area to be used for validating a freight model.

A methodology was developed to develop the estimated origin/destination matrix using traffic counts was developed to support the work and is shown in **Figure 1**.

The application of the methodology utilized CUBE/TRANPLAN (Citilabs Corp.) to develop the origin/destination pairs that would use the roadway in question and

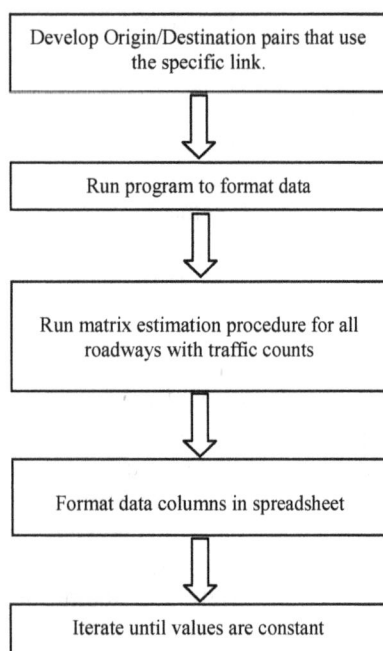

Figure 1. Flowchart of methodology.

EXCEL (MicroSoft Corp.) to run the matrix estimation technique. The approach used to create the EXCEL file to perform the analysis is as follows:

1a. Locate all the roadways in the Mobile CUBE/TRANPLAN network that had a truck count conducted. This information was recorded to ensure all the roadways were properly used.

2a. Perform a Select Link Analysis within CUBE/TRANPLAN for a roadway location corresponding to the truck count.

3a. Perform a Build Selected Link Trip Table within CUBE/TRANPLAN using the selected link to identify the specific origin/destination pairs that had that links along the shortest path.

4a. Export the trip table for the selected link for use outside of CUBE/TRANPLAN.

5a. Run a FORTRAN program that was developed by the authors to format the data for later use.

6a. Copy the data output from the FORTRAN program into EXCEL. The data entered into EXCEL included all of the origin/destination pairs available within the model, however, the formatting was such that pairs that used the roadway on the shortest path were coded with a 1 and those not on the shortest path were coded with a 0.

7a. Repeat Steps 2 - 6 until all roadways where a truck count was available were used.

To execute the matrix estimation in EXCEL the following steps were taken:

1b. Develop a column of all possible origin/destination pairs in the community.

2b. Add a column with a value of 1 for all possible origin/destination pairs to serve as an initializing value.

3b. Add a column from the previous set of steps that has the truck count for a link, and the 1 or 0 representing the origin/destination pairs likely to use that roadway.

4b. Divide the total truck count by the total number of pairs using the roadway and distribute all the traffic evenly to these origin/destination pairs.

5b. Add a column from the previous set of steps that has the truck count for the next link and the 1 or 0 representing the origin/destination pairs likely to use the next roadway.

6b. Sum the total expected traffic, either a value from step 4 or a 1 and divide the total truck count by the total usage on the roadway and distribute all the traffic evenly to these origin/destination pairs.

7b. Repeat Steps 5b - 6b until all of the roadways are used. This will give an initial solution.

8b. Repeat Steps 2b - 7b until reaching closure. Observing the difference that the initial solution developed in Step7b will serve as the initial value used in Step 2b. This will allow for the first potential solution to serve as an input to next iteration as opposed to starting over. Additionally, Steps 3b - 4b are unnecessary as they will also cause the knowledge gained from the previous iteration to be lost.

The steps presented were used in the case study with Mobile, AL using 42 truck count locations (**Figure 2**) and the Mobile CUBE/TRANPLAN network (**Figure 3**). The Selected Link Analysis and Build Selected Link Trip Table modules were executed and the shortest paths between origin/destination pairs were identified. As the truck counts used to build the origin/destination matrix contained internal, internal/external and pass-through trips, the origin/destination matrix contained all internal zones and zones representing external stations. These values were entered in EXCEL as directed and a total of fifty complete iterations, each one using all the truck count locations, were performed to ensure that a final solution was reached. To verify that the iterations reached a converging solution; the percent change in the total trips between iterations was tracked. Between iteration 49 and iteration 50, less than 0.01 percent change occurred in the total number of trips assigned in the network. A graph showing the change in total trips for each iteration is shown in **Figure 4**.

The final results contained the total number of trips expected between each origin/destination pair in the model. Thus, the matrix estimation technique was capable of taking the 42 truck count locations and developing a single origin/destination matrix that, when assigned to the network, would result in the counts collected. The final step in the process was to aggregate all the volume for each origin and use that value as the production for the zone and all of the volume for each destination and use that as the attraction for the zone. This process therefore was capable of identifying the number of trucks produced and attracted to zones within the community without direct sampling of businesses in the zone and using a cordon line to capture trucks crossing into and out of a zone.

Figure 3. Mobile CUBE/TRANPLAN network.

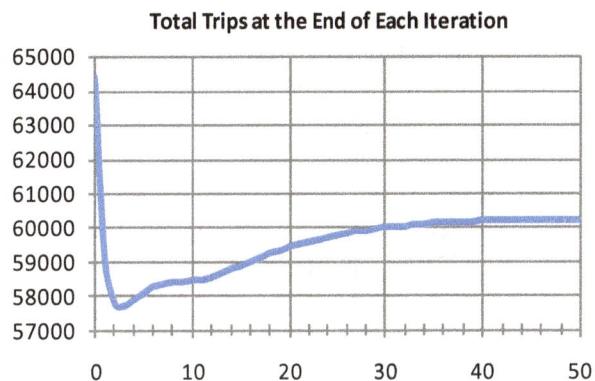

Figure 4. Total Trip Ends showing convergence.

3. Statistical Analysis

After completing the EXCEL runs, the data was combined into trips entering and exiting traffic analysis zone in the Mobile network. Truck trips were assigned into production values and attraction values. A statistical analysis was performed to develop a predictive model to create the number of trucks trips produced and attracted from the zones.

The variables used in the statistical analysis were provided by the Mobile MPO and represent a list of variables likely collected in the planning agencies of small and medium sized communities to perform trip generation for passenger trips. The variables were:

- Total Households
- Low Income Households
- Medium Income Households
- High Income Households
- Number of Students
- Number of Residents in Dormitories
- Total Employment
- Retail Employment

Figure 2. Map of mobile county showing truck count locations.

- Service Employment
- Other Employment
- Area in Acres
- Employment Density

These variables, along with the production and attraction values were entered into MINITAB software and a regression analysis performed.

A stepwise regression analysis was performed and three variables were selected for inclusion in the model at the alpha = 0.05 level: Other Employment, Retail Employment and High Income Households. The coefficients for the model were determined by the software to minimize the error. The model developed was:

$$\text{Truck Trips} = -518 + 0.6 * \text{Other Employment}$$
$$-0.724 * \text{Retail Employment} \qquad (1)$$
$$+1.41 * \text{High Income Households}$$

In an attempt to improve the model, Other Employment was subdivided into Manufacturing Employment and Non-Manufacturing Employment. The Manufacturing Employment replaced the Other Employment in the stepwise regression analysis and the model developed was:

$$\text{Truck trips} = -512$$
$$+0.986 * \text{Manufacturing Employment}$$
$$-0.634 * \text{Retail Employment} \qquad (2)$$
$$+1.49 * \text{High Income Households}$$

Examining the equations developed in the process, the value of the parameters (positive and negative) show the interactions between the variables. The equations are not intended to be additive, such that an X number of additional employees or households will directly relate to a Y increase in trucks. Therefore, the application of the equation is based on the collection and distribution of the socio-economic data across the area.

As there was not a specific truck volume study performed for selected areas of Mobile due to time and budget limitations, a secondary validation technique was undertaken. The process involved applying the developed equations to the community and modeling the truck traffic that would result from the equations. These values were then compared to the actual truck counts on the roadway as observed during the traffic count process. **Figure 5** presents a scatter plot of the model in predicting the number of truck trips in each zone versus the number of truck trips expected in each zone from the origin/destination estimation methodology. From the figure, it can be seen that the data trends to follow the 1:1 slope line indicating that the estimated truck trips for each zone are closely matched to the number of truck trips expected from the model in Equation (2). The correlation coefficient for the data is 0.6.

For comparison purposes, an analysis was performed using only the total estimated truck trips and employees by type of employment and households to give a one to one relationship (an X increase in employment or households results in a Y increase in trucks). For the different possible socio-economic data points, **Table 1** presents the total rate. This table is presented as an alternative means to utilize the data collected in this study. For each socio-economic variable presented, the total number of trucks expected for the zone would simply be the aggregate value of socio-economic variable times the rate calculated. In this fashion, a direct value for truck trips could be calculated using a single variable, such as manufacturing employment.

4. Conclusions

This paper examined the use of a matrix estimation technique to develop a trip generation model for truck trips in a medium sized community. The methodology used had the benefit of operating using existing software and truck count data and socio-economic data that would be available at most smaller communities. The model developed for the case study community contained manufacturing employment and high income households as positive factors for truck trip generation, indicating that if a greater number of manufacturing employees or high income households were in an area, a greater number of

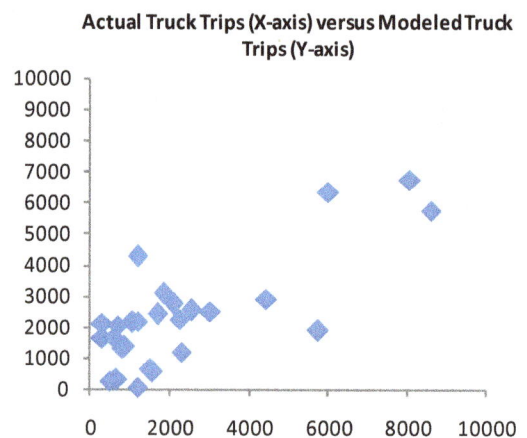

Figure 5. Comparison plot for model accuracy.

Table 1. Rate for socio-economic factor as single variable.

Socio-Economic Variable	Rate (Truck Trips/Units)
Manufacturing Employment	0.3588
Non-Manufacturing Employment	0.3267
Service Employment	0.7675
Retail Employment	0.3721
Households	1.5838

truck trips would be expected. This result make sense if you consider that manufacturing employees would be directly related to good movement and high income households might lead to increases in home deliveries.

The model could be improved by increasing the number of factors used in the analysis. The limitation of using household and employment data values at the sub-county level may have influenced the model and other potential factors that should be included in a truck trip generation model were missed. Overall, this paper presents a methodology and result for one smaller community that could be used add truck trips as an explicit trip purpose in their modeling efforts.

5. Acknowledgements

This research was sponsored by the U.S. Department of Transportation, Federal Transit Administration, Project No. AL-26-7262-03.

REFERENCES

[1] National Cooperative Highway Research Program (NC HRP), "Report 570: Guidebook for Freight Policy, Planning, and Programming in Small- and Medium-Sized Metropolitan Areas," 2007. http://www.nap.edu/catalog.php?record_id=14036

[2] Cambridge Systematics, Inc., "Quick Response Freight Manual," Federal Highway Administration, 1996. http://media.tmiponline.org/clearinghouse/quick/quick.pdf

[3] Cambridge Systematics, Inc., "Quick Response Freight Manual II," Federal Highway Administration, 2007. http://ops.fhwa.dot.gov/freight/publications/qrfm2/qrfm.pdf

[4] H. J. Van Zuylen and L. G. Willumsen, "The Most likely Trip Matrix Estimated from Traffic Counts." *Transportation Research Part B: Methodological*, Vol. 14, No. 3, 1980, pp. 281-293.

[5] J. de D. Ortuzar and L. G. Willumsen, "Modeling Transport," 2nd Edition, John Wiley and Sons, New York, 1994.

An Induced Demand Model for High Speed 1 in UK

Francesca Pagliara[1], John Preston[2]

[1]Department of Civil, Architectural and Environmental Engineering, University of Naples Federico II, Naples, Italy
[2]School of Civil Engineering and the Environment, University of Southampton, Southampton, UK

ABSTRACT

Induced travel is an important component of travel demand and increasing attention has been paid to building analytical model to get more precise travel demand forecasting. In general, induced demand can be defined in terms of additional trips that would be made if travel conditions improved (less congested, lower vehicle costs or tolls). In this paper the induced demand resulting from higher design speeds and, therefore by less travel time, for the High Speed 1 in UK will be modelled on the basis of the relationship between existing High Speed Rail demand (dependent variable) to existing High Speed Rail travel times and costs. The covariates include socioeconomic variables related to population and employment in the zones connected by the High Speed Rail services. This model has been calibrated by mean of a before and after study carried on the corridor, when the new High Speed Rail services was introduced. Elasticities of induced travel (trips and VMT) have been computed with respect to fares, travel time and service frequency.

Keywords: High Speed Rail; Induced Demand; Regression Models; Elasticities

1. Introduction

Investments in High Speed Rail (HSR) systems are currently being undertaken in many countries around the world. These systems represent a closer to optimal solution to meet challenges of increasing mobility demand while simultaneously addressing the greater attention of citizens to sustainability issues. Europe, together with Asia, is already the leader in HSR systems; in fact the development of HSR has been one of the central features of recent European Union transport infrastructure policy. In fact, the programme for the trans-European transport network (TEN-T), introduced under the Treaty of Maastricht, was designed to guarantee optimum mobility and coherence between the various modes of transport in the Union, establishing the key links needed to facilitate transport, optimize the capacity of existing infrastructure, produce specifications for network interoperability, and integrate the environmental dimension [1]. The TEN-T focuses very closely on the development of HSR transport: of the 30 priority projects put forward under this programme, among them 14 concern HSR lines.

In the literature several contributions have been proposed on the empirical analysis of the change of users'travel behaviour after the introduction of HSR services. Most of them have focussed on modelling the mode choice. However very few, as far as the authors know, concentrate on induced demand by HSR [2].

It is well reported in the literature that induced demand is an important component of travel demand, and increasing attention has been paid to building analytical models to get more precise travel demand forecasting. In general, induced demand can be defined in terms of additional trips that would be made if travel conditions improved (less congested, lower vehicle costs or tolls). In this paper the induced demand resulting from higher design speeds and, therefore by less travel time, for High Speed 1 (HS1) in UK will be modelled.

This contribution is organised as follows. In Section 2 an overview on induced demand is presented reporting some case studies present in the literature. Section 3 deals with the case study of HS1 in UK, while Section 4 reports some implications for further research.

2. Induced Demand: An Overview

Modelling induced travel demand is not an easy task due to the high number of variables playing which make the analysis complicated and difficult to generalize. A transportation system is a set of elements interconnected by complex relationships, such as supply sub-system, demand sub-system, residences and activities sub-system; whenever an action is planned on a part of a transportation system, there are unavoidable impacts on other parts, positive or negative. Improvement within the supply sub-system, such as the introduction of a new road infrastructure, of faster/cheaper services, more comfortable vehicles, or in any case actions that increase the utility and/or the satisfaction of the customer about the possibility of moving, create a new share in travel demand [3,4]. Any intervention on a given transport link, producing a mode shift, determines a number of trips which can be split in two parts:

- A diverted demand, that is the number of trips previously carried out by other transport modes, by other route with the same transport mode, or by other services in the same route;
- An induced demand, that is a number of shifts previously not existed and generated directly by the intervention performed [5].

It is difficult to estimate induced traffic with the conventional four-step models, because generation-distribution sub-models are not able to cleave diverted and induced shares; they are also generally not sensitive to changes in the level of service, therefore not able to capture the related effects.

Litman's contribution [5] provides a comprehensive literature review on the importance of evaluating induced demand brought by road transport. Cervero [6] used data on freeway capacity expansion, traffic volumes, demographic and geographic factors from California between 1980 and 1994. He estimated the long-term elasticity of vehicles-mile-travelled (VMT) with respect to traffic speed to be 0.64, meaning that a 10% increase in speed results in a 6.4% increase in VMT, and that about a quarter of these results from changes in land use (e.g., additional urban fringe development). He estimated that about 80% of additional roadway capacity is filled with additional peak-period travel, about half of which (39%) can be considered the direct result of the added capacity. Duranton and Turner [7] investigated the relationship between interstate highway lane kilometres and highway vehicle-kilometres travelled (VKT) in US cities. They found that VKT increases proportionately to highways and identify three important sources for this extra vehicle travel: increased driving by current residents, an inflow of new residents, and more transport intensive production activity. Time-series travel data for various roadway types indicated an elasticity of vehicle travel with respect to lane miles of 0.5 in the short run, and 0.8 in the long run [8]. This means that half of increased roadway capacity was filled with added travel within about 5 years, and that 80% of the increased roadway capacity will be filled eventually. Noland and Quddus [9] found that increases in road space or traffic signal control systems that smooth traffic flow induced additional vehicle traffic which quickly diminished any initial emission reduction benefits. Small [10] concluded that 50% - 80% of increased highway capacity was soon filled with generated traffic, based on a detailed review of previous studies.

A comprehensive study of the impacts of urban design factors on US vehicle travel found that a 10% increase in urban road density (lane-miles per square mile) increased per capita annual VMT by 0.7% [11]. In a study of eight new urban highways in Texas over several years. Schiffer et al. [12] performed a meta-analysis of induced travel studies to identify short- and long-term elasticities of VMT with respect to changes in traffic lane-miles and other variables. They predicted the amount of VMT induced by regional highway expansion in the Wasatch Front (Salt Lake City region). They concluded that induced travel effects generally decreased with the size of the unit of study.

It is evident how in all the studies stated above, induced traffic has had a strong impact on the expectations of the projects. Most of the benefits estimated, specially about interventions of capacity increase, have been strongly restricted due to filling by induced traffic. It is important within the decision-making process to take into account an induced demand estimation model, for which due to the high number of factors playing (such as kind of roadway, land use, regional or urban area, short or long run effect), the proper adaptation is necessary. Yao and Morikawa [13] developed a model of induced demand resulting from the introduction of a HSR service linking Tokyo, Nagoya and Osaka metropolitan areas in Japan. They proposed an integrated intercity travel demand model with nested structure, including trip generation, destination choice, mode choice and route choice models. Induced demand was estimated introducing an accessibility measure, as an expected maximum utility able to capture the short run behavioural effects such as changes in travel departure times, routes switches, modes switches, longer trips, changes of destination, and new trip generation. They calculated elasticities of induced travel (trips and VMT) with respect to fares, travel time, access time and service frequency for business and non-business travel.

Ben-Akiva et al. [2] modelled the induced demand by HSR in Italy on the basis of the relationship between existing HSR demand (dependent variable) to existing HSR travel times and costs. The covariates include socioeconomic variables related to population and employment in the zones connected by the HSR services. The model was calibrated by mean of a before and after study carried out in the Naples-Rome corridor, where the new HSR services was introduced.

3. The Case Study

High Speed 1 (HS1) is a 108 km (67 mile) HS railway line running from London through Kent to the British end of the Channel Tunnel. Section 1, opened on 28 September 2003, is a 74 km section of HS track from the Channel Tunnel to Fawkham Junction in north Kent. The section's completion cut the London-Paris journey time by around 21 minutes, to 2 h 35minutes. Section 2 opened on 14 November 2007 and is a 39.4 km stretch of track from the newly built Ebbsfleet station in Kent to London St Pancras. Completion of the section cut journey times by a further 20 minutes (London-Paris in 2 h 15 minutes; London-Brussels in 1 h 51 m) (see **Figure 1**).

HS 1 railways hosts international services to continental Europe and domestic services to Kent area. Eurostar is the service connecting London with Paris and Brussels across the Channel Tunnel and is provided by Eurostar International Limited, a company owned jointly by London and Continental Railways (40%), the French national railway company Société Nationale des Chemins de fer français (55%), and the Belgian railway operator Nationale Maatschappij der Belgische Spoorwegen (5%). Eurostar services, started in November 2007, have St. Pancras International station as terminal in London, and other calling point in UK are Ebbesfleet International and Ashford International.

Domestic HS services (see **Figure 2**) are provided by Southeastern Railway Limited, the train operating company holder of the "Integrated Kent Franchise" (1st April 2006 – 31st March 2014). It operates with three kinds of services, which are High Speed, Mainline, and Metro. HS services, begun on December 2009 and they connect London with the Kent using the High Speed 1 to Ebbsfleet and Ashford from which branch off the routes throughout south east of England. Mainline represents

the conventional services connecting London with Dover, Canterbury, Maidstone, Thanet, Tunbridge and East Sussex. There are also metropolitan services serving south east and south of London, split in four lines:
- Dartfort and Gravesend metro line
- Hayes metro line
- Orpington and Sevenoaks via Grove Park Metro lines
- Orpington and Sevenoaks via Bromley South Metro lines

The number of stations is 179 of which 173 operating and the passenger journeys in 2010/11 were 162.3 million [14].

The introduction of a HSR line naturally leads to travel time savings and may also provide congestion relief for

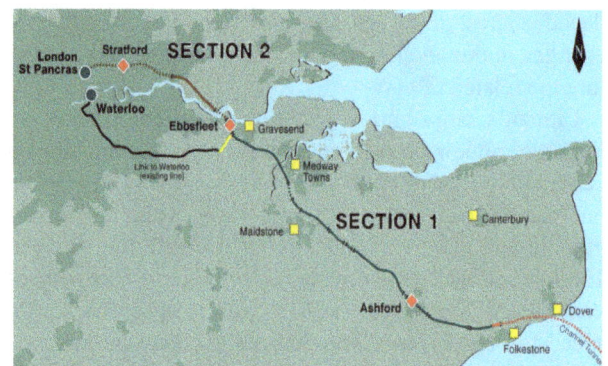

Figure 1. High Speed 1 route map; source: www.lcrhq.co.uk.

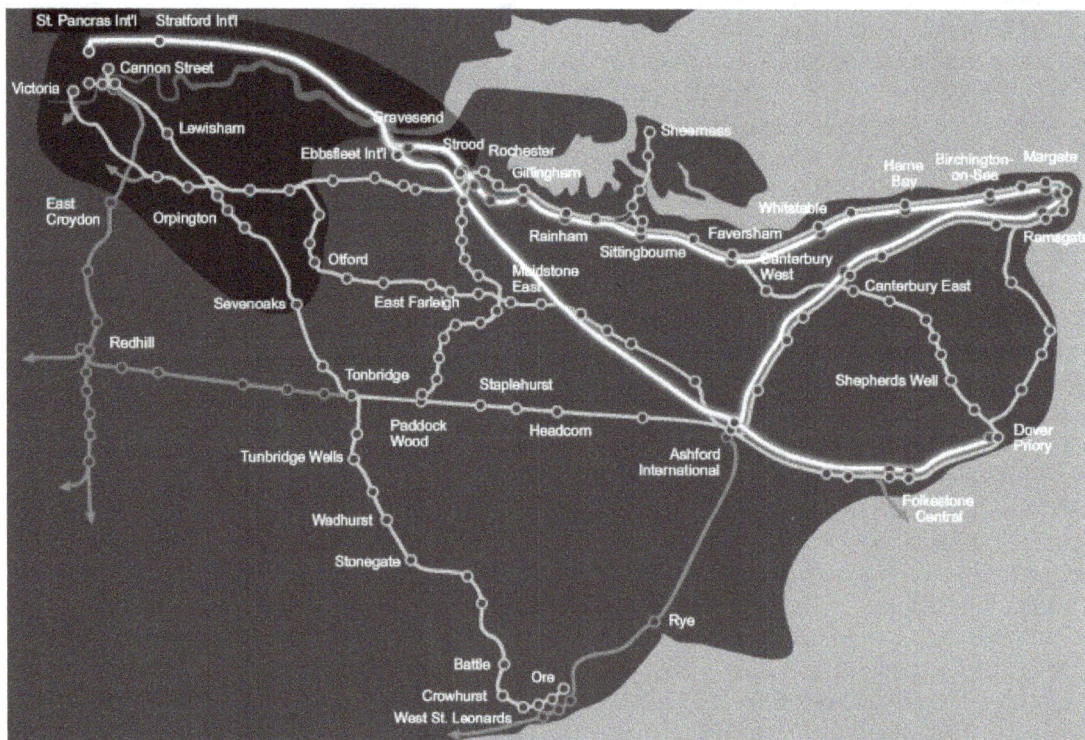

Figure 2. Southeastern route map (grey line is the HS route and the white line represents mainline routes); source: www.southeasternrailway.co.uk.

other transport users on the same corridor. These are, together with the capital and operating costs deriving by the new service, the main impacts on the transport system that several studies have monetised within cost-benefit analysis [15,16]. Travel time saving for international trip between London and Paris-Brussels due to the introducing of the HS track is 35 minutes for domestic services between London and Kent using the HS track.

The total time saving benefits, estimated in 2009 by London & Continental Railways (LCR) using the value of time in accordance with WebTAG (the Department for Transport's website for guidance on the conduct of transport studies) and expressed as a Present Value over 60 years, is 2500 and 1200 for international and domestic respectively.

Concerning congestion relief, LCR assumed a valuation of 40 pence per trip for passengers who switched to the new HS domestic services, and 20 pence for remaining passengers, with an annual growth rate of 1% applied to those values. This approach indicates that, over 60 years as a Present Value, the congestion relief benefit would be £113.6 million. The total construction cost of the line was £5.2 billion (Section 1: £1.9 billion; Section 2: £3.3 billion) and the annual operating cost of around £75 million [16]. The HS fares are on average 30% higher than the conventional, and it was estimated an additional revenue as a Present Value over 60 years of £3.53 billion [10]. The actual use of it is very low, being the service actually running up to 16 trains per day. It is evident that HS1 led to only modest improvements in services between London and Paris/Brussels. The National Audit Office [17] indicated that on transport grounds alone, the case for the HS1 is quite weak, with a Benefit Cost Ratio (BCR) of only 1.1. If regeneration benefits are included (estimated at the time as around £500 million), the BCR is estimated to increase to 1.35, still below the UK Government's current Value for Money threshold of 1.5. However, recent estimates of regeneration benefits have been as high as £8 billion [18]. Preston and Wall [16] concluded that wider socio-economic impacts are crucial in the justification of HS1 and HSR services more generally.

Data and Methodology

Gravity models are used in social sciences to predict and describe behaviours that mimic gravitational interaction as described in Isaac Newton's law of gravity [19]. A general application of these models to transport systems analysis concerns demand forecasts in a corridor, expressed as a function of a mass of the origin, a mass of the destination, and a cost function of the shift. Once calibrated applying to all origin/destination (O/D) pairs, these models are able to reproduce generated traffic on a

certain corridor in which there has been a change in one or more variable values. Obviously they are not able to cleave diverted and induced demand, because they only take into account what happens on the relations considered. They forecast travel demand changes only on the O/D pairs where a change of the conditions occurred. Therefore the right application of this kind of model depends on the expected transport intervention effects. If diverted traffic, is expected, a generation-distribution model is preferred; when generation effects are expected, a gravity model applied to all O/D pairs is more appropriate. Introducing a new HSR system produces both diverted (from conventional services) and induced traffic; the objective here is to investigate to which results a gravity model leads applied to forecast changes in the number of trips before and after Britain's HS domestic services introduction. A gravity model to be calibrated needs real travel demand data for conventional and HS services referred to a proper catchment area. Being quite difficult to collect historical travel demand data, the station usage information have been collected, provided by the Office of Rail Regulation (ORR), the independent safety and economic regulator for Britain's railways. Station usage data represent the total number of trips occurred yearly for each Britain's railway station. Being the vast majority of the travels concerning HS services towards London metropolitan area, it has been considered a catchment area made up by a number of origins equal to the number of stations served by HS services, and London as the only destination. Socioeconomic variables have been at first considered, but the low sample size and the unavailability of a precise catchment area led to very low statistical significant results.

The first step has been the analysis of the HSR services running from London to Kent, provided by the Southeastern Train Operating Company. These services involve 24 stations throughout Kent area, each of which is served by both HS and conventional trains.

Referring to the current timetable available on the Southeastern website, the following information have been computed for each station:

- The number of HS trains per day to London;
- The mean HS journey time to London;
- The mean HS fare to London;
- The number of conventional trains per day to London;
- The mean conventional journey time to London;
- The mean conventional fare to London.

The mean fare is the average of the full fares ranging in the day and the same values are deducted from the travelcard discount (34%).

The second step has been that of finding a correlation between the station usage provided by ORR, and a set of variables linked by the following function:

$$T_i = f\left(JT_{ij}, F_{ij}, NT_{ij}, Pop_i, CP_i\right) \qquad (1)$$

where:

i, j is the considered station and London respectively;

T_i is the number of trips in station i;

JT_{ij} is the journey travel time between i, j;

F_{ij} is the fare between i, j;

NT_{ij} is the number of trains between i, j;

Pop_i is the population within 4 minutes uncongested driving time from station i;

CP_i is the number of car parking spaces available in station i.

$$\ln T_i = \delta + \beta_1 \ln CP_i + \beta_2 \ln\left(\sum_{HS;C} \exp\left(U_{ij}\right)\right)_i$$
$$+ b_3 \ln\left(\text{DummyHSonly}\right)_i \qquad (2)$$
$$+ b_4 \ln\left(\text{DummydistrictHub}\right)_i$$

where:

δ is the intercept;

T_i is the number of trips in station i;

$$\ln\left(\sum_{HS;C} \exp\left(U_{ij}\right)\right)_i \quad \text{is the Logsum} \qquad (3)$$

The Logsum is a measure of the user's satisfaction with respect to the available alternatives. It has been computed (for each station toward London) as a generalized journey time (see Equation 4), then used together with the fare for obtaining a utility value (see Equation 6) both for HS and conventional services.

$$\left(GJT_{ij} = JT_{ij} + a\right)_{HS;C} \qquad (4)$$

where:

GJT_{ij} is the generalized journey time;

JT_{ij} is the in vehicle time (mean journey time)

$$a = 17 * 60/2NT_{ij} \qquad (5)$$

α is the average waiting time in minutes considering 17 hours of service per day;

$$\left(U_{ij}\right)_{HS;C} = (-1)\left(F_{ij} + \beta GJT_{ij}\right) \qquad (6)$$

β = 11.7247 pounds/hour (value of time derived by Webtag).

($DummyHSonly$)$_i$ is a dummy variable which assumes the value 1 for the stations served by HS services only, and the value 0 for the stations served by both HS and conventional services.

($DummydistrictHub$)$_i$ is a dummy variable which assumes the value 1 for the expected main District station, in terms of number of customer services, opportunity to make interchange, centrality in respect to the District area; the value 0 for the other stations. The regression outputs are shown **Table 1**.

The coefficients of the regression estimated are all statistically significant and to test the model accuracy a comparison has been made between the predictions of the model and the real values. Given the station usage data for the years 2008/09, 2009/10, and 2010/11, it has been calculated the real percentage change of the number of trips before the domestic HS (2008/09-2009/10) and after its introduction (2009/10-2010/11). It should be pointed out that 2009/10 data are considered before the HS introduction because they are referred to the financial year, and so they include only three months of HS services (started in December 2009). Forecast of the number of trips for 2008/09 is calculated considering in the model the Logsum related to 2008 timetable from where the mean journey time and the number of trains to London for each station have been calculated. The fares have been estimated starting by the current values considering the Retail Price Index and the annual fares increase. The value of time annual change (see **Table 2**) has been obtained from WebTAG.

Concerning the 2009/10 model prediction, the Logsum is calculated with the approximation of considering the current conventional timetable in place of 2009/10 values. As it can be seen in the **Table 3**, the real change in the number of trips from 2008/09 to 2009/10 is generally negative, except for Rochester and Canterbury West.

Table 1. Regression results.

Variable	Coefficient	t-statistic
Intercept	13.126	19.170
Ln Car Park	0.358	3.401
Logsum	0.031	3.453
Dummy HS only	3.953	−7.163
DDummy District Hub	1.093	4.540
R^2	0.8	
Adjusted R^2	0.75	
N. Obs.	24	

Table 2. Value of time adopted.

Year on Year	Annual RPI (Based on Previous July Figure)	Annual Fare Increase RPI+3%	VOT Mean Change	Value of Time
2008				12.1405
2009	−1.40%	1.600%	−5%	11.5646
2010	4.80%	7.800%	0.51%	11.6239
2011	5%	8.000%	0.86%	11.7247
total	16.530%		−3.5%	

Table 3. Comparison between model forecast and real data before High Speed 1.

STATION NAME	Model Change 2008/09-2009/10	Real Change 2008/09-2009/10	Model Change—Real Change
Stratford International		Didn't Exist	
Ebbsfleet International		No Data Available	
Gravesend	−2.79%	−7.75%	4.95%
Strood	−4.25%	−4.54%	0.29%
Maidstone West	−3.94%	−33.90%	29.95%
Rochester	−9.00%	1.29%	−10.29%
Chatham	−10.07%	−8.82%	−1.25%
Gillingham (Kent)	−10.67%	−2.76%	−7.92%
Rainham (Kent)	−7.38%	−7.89%	0.51%
Sittingbourne	−6.71%	−8.14%	1.43%
Faversham	−7.75%	−6.08%	−1.67%
Whitstable	−7.45%	−5.24%	−2.21%
Herne Bay	−7.72%	−4.95%	−2.77%
Birchington-on-Sea	−8.95%	−1.79%	−7.16%
Margate	−5.92%	−8.12%	2.19%
Broadstairs	−2.93%	−5.80%	2.87%
Ramsgate	−11.60%	−4.25%	−7.35%
Ashford International	−10.97%	−0.10%	−10.87%
Folkestone West	−10.75%	−3.91%	−6.84%
Folkestone Central	−14.19%	−4.69%	−9.50%
Dover Priory	−11.56%	−11.78%	0.21%
Deal	−11.14%	−9.99%	−1.15%
Sandwich	−11.03%	−12.73%	1.70%
Canterbury West	−4.04%	5.30%	−9.34%
Mean Standard Deviation	3.22%	7.38%	

Recession is certainly one of the reasons, but the prediction of the model reflects well the negative trend even if it does not take into account crisis effects. That is because the Logsum decreases from 2008/09 to 2009/10, first because of the fare increases, second because of the number of stops increase that has led to higher travel times.

As it can be seen the lower accuracy is for Maidstone West with a 30% error, and for Canterbury Westand Rochester, where the model predicts respectively a decrease of 4% and 9% against an increase of 5.3% and 1.3%, and for Ashford and Folkestone Central, with 10.87% and 9.5% error respectively. Besides the approximation concerning the 2009/10 Logsum, that likely would have led to more precise results, there are other factors which could have been taken into account. Firstly, the more or less influence of the crisis on the areas, difficult to take into account without reliable local GDP data; secondly, it could be a substantially different value of time from that adopted; thirdly, socioeconomic variable as the population, employees etc have not been considered. A further comparison has been made between the change in number of real trips and those predicted from 2009/10 to 2010/11 (see **Table 4**). This could also represent an approximate measure of the extent to which HSR has grown the market.

Even in this case there are values in which the forecast is substantially different from the real data, such as Strood, Rochester, Folkestone West and Canterbury West, and the reasons are similar to those stated above.

Table 4. Comparison between model forecast and real data before and after high speed 1.

STATION NAME	Model Change 2009/10-2010/11	Real Change 2009/10-2010/11	Model Change—Real change
Stratford International	No Data Available	487%	
Ebbsfleet International	No Data Available	152%	
Gravesend	5.54%	5.80%	–0.26%
Strood	4.74%	19.52%	–14.79%
Maidstone West	1.16%	5.27%	–4.11%
Rochester	1.25%	17.08%	–15.83%
Chatham	0.90%	–1.23%	2.13%
Gillingham (Kent)	2.75%	4.49%	–1.74%
Rainham (Kent)	0.14%	1.60%	–1.47%
Sittingbourne	0.56%	0.81%	–0.24%
Faversham	0.27%	0.21%	0.06%
Whitstable	0.02%	0.56%	–0.53%
Herne Bay	0.02%	–0.10%	0.12%
Birchington-on-Sea	0.005%	–2.50%	2.51%
Margate	0.04%	0.92%	–0.87%
Broadstairs	0.01%	5.43%	–5.42%
Ramsgate	7.40%	5.30%	2.10%
Ashford International	6.40%	13.32%	–6.91%
Folkestone West	1.41%	37.13%	–35.72%
Folkestone Central	3.74%	13.42%	–9.67%
Dover Priory	1.59%	12.03%	–10.44%
Deal	0.48%	2.03%	–1.55%
Sandwich	0.50%	–2.01%	2.51%
Canterbury West	6.82%	31.19%	–24.37%
Mean Standard Deviation	2.50%	10.62%	

Ebbsfleet and Stratford stations are charecterized by high real changes because 2009/10 data include only three months of HS service, which is the only service available for them. The model is therefore not able to predict the number of trips for 2009/10 because there is not the Logsum measure related to the conventional service. This is a consequence of the approximation of considering the station as origin, instead of the classic area's centroids.

As the model is linear with respect to the Logsum variable, the elasticity related to the latter has been computed as:

$$\varepsilon = (100\beta\text{Logsum})\% = 3.15\% \qquad (7)$$

It follows that increasing the Logsum of one unit, the number of trips will increase of 3.15%. A more immediate measure is the elasticity with respect to the fare and the generalized journey time, from where the Logsum is derived. Obviously there are two elasticity values, for both conventional HS services.

Concerning the generalized journey time, which is the sum of in-vehicle time and average waiting time, the elasticity is –0.06 for conventional trains and –0.14 for HS trains; this means that a 10% increase in journey time results in a 0.6% decrease in conventional number of trips, and a 1.4% decrease in a HS number of trips. Elasticity with respect to the fare is –0.05 for conventional and 0.34 for HS; meaning that a 10% increase in journey time results in a 0.5% decrease in the number of trips for conventional and a 3.4% decrease for the HS number of trips.

4. Conclusions and Further Perspectives

Modelling induced demand by HSR systems is a topic not well-established in the literature, this paper attempts to provide a contribution to the international research. The model specified has been used to predict the change in the number of trips between two years and with the purpose of comparing the results with the real available data. There are two possible implications for this model:

- It could be used to forecast the change in the number of trips deriving by the change in time and fare on a corridor where two services, HS and conventional are available;
- It could be used to predict the number of trips deriving by the introduction of a new HS service.

In any case, an improvement in the level of services variables is reflected in a customer satisfaction increase (modelled by the Logsum variable), and the related elasticity indicates that a 1 unit increase results in a 3.15% increase in the number of trips.

The main constraint of this study is related to the unavailability of a conventional catchment area, to which always transport analysis is referred, besides the unavailability of the real travel demand data. These limitations also can be found in the exclusion in the regression models of some socio-economic variables such as population and employment, as well as other transport variables such as car journey times and costs. Further perspectives will consider the inclusion of such variables in the models.

REFERENCES

[1] European Union, "High-Speed Europe, a Sustainable Link between Citizens," 2010. http://ec.europa.eu/research/horizon2020/index_en.cfm

[2] M. Ben-Akiva, E. Cascetta, P. Coppola, A. Papola and V. Velardi, "High Speed Rail Demand Forecasting: Italian Case Study," *Proceedings of the WCTR Conference*, Lisbon, 11-15 July 2010, 9 p.

[3] M. Ben-Akiva and S. Lerman, "Discrete Choice Analysis," The MIT Press, Cambridge, 1985.

[4] E. Cascetta, "Transportation Systems Analysis: Models and Applications," Springer, New York, 2009.

[5] T. Litman, "Generated Traffic and Induced Travel, Implications for Transport Planning—Report for the Victoria Transport Institute," 2011. http://www.vtpi.org/gentraf.pdf

[6] R. Cervero, "Road Expansion, Urban Growth, and Induced Travel: A Path Analysis," *Journal of the American Planning Association*, Vol. 69, No. 2, 2003, pp. 145-163.

[7] G. Duranton and M. A. Turner, "The Fundamental Law of Road Congestion: Evidence from US Cities," Working Paper 370, University of Toronto, Toronto, 2008.

[8] R. B. Noland, "Relationships between Highway Capacity and Induced Vehicle Travel," *Transportation Research Part A: Policy and Practice*, Vol. 35, No. 1, 2001, pp. 47-72.

[9] R. Noland and M. A. Quddus, "Flow Improvements and Vehicle Emissions: Effects of Trip Generation and Emission Control Technology," *Transportation Research Part D: Transport and Environment*, Vol. 11, No. 1, 2006, pp. 1-14.

[10] K. A. Small, "Urban Transportation Economics," Harwood (Chur), Anne Arundel, 1992.

[11] L. C. Barr, "Testing for the Significance of Induced Highway Travel Demand in Metropolitan Areas," *Transportation Research Record*, Vol. 1706, 2000, pp. 1-8.

[12] R. G. Schiffer, M. W. Steinvorth and R. T. Milam, "Comparative Evaluations on the Elasticity of Travel Demand, Committee on Transportation Demand Forecasting," 2005. www.trbforecasting.org/papers/2005/ADB40/05-0313_Schiffer.pdf

[13] E. Yao and T. Morikawa, "A Study on Integrated Intercity Travel Demand Model," *Transportation Research Part A: Policy and Practice*, Vol. 39, No. 4, 2005, pp. 367-381.

[14] Office of Rail Regulation, "National Rail Trends 2010/11 Yearbook," 2011. http://www.rail-reg.gov.uk/

[15] LCR (London & Continental Railways), "Economic Impact of High Speed 1, Report," 2009. http://www.lcrhq.co.uk/

[16] J. M. Preston and G. Wall, "The Ex-Ante and Ex-Post Economic and Social Impacts of the Introduction of High-Speed Trains in the South-East of England," *Planning Practice and Research*, Vol. 23, No. 3, 2008, pp. 403-422.

[17] National Audit Office, "The Channel Tunnel Rail Link. HC 302," The Stationery Office, London, 2008. http://www.nao.org.uk/

[18] Dft, Transport Analysis Guidance, "UK Department for Transport," 2007. http://www.dft.gov.uk/

[19] A. G. Wilson, "Entropy in Urban and Regional Modelling: Retrospect and Prospect," *Geographical Analysis*, Vol. 42, No. 4, 2010, pp. 364-394.

6

Evaluation of In-Use Fuel Economy for Hybrid and Regular Transit Buses

Shauna L. Hallmark, Bo Wang, Yu Qiu, Robert Sperry
Iowa State University, Ames, USA

ABSTRACT

Fuel costs are a significant portion of transit agency budgets. Hybrid technology offers an attractive option and has the potential to significantly reduce operating costs for agencies. The main impetus behind use of hybrid transit vehicles is fuel savings and reduced emissions. Laboratory tests have indicated that hybrid transit buses can have significantly higher fuel economy and lower emissions compared to conventional transit buses. However, the number of studies is limited and laboratory tests may not represent actual driving conditions since in-use vehicle operation differs from laboratory test cycles. Several initial studies have suggested that the fuel economy savings reported in laboratory tests may not be realized on-road. The objective of the project described in this paper was to evaluate the in-use fuel economy differences between hybrid-electric and conventional transit buses for the Ames, Iowa (USA) transit authority. On-road fuel economy was evaluated over a 12-month period for 12 hybrid and 7 control transit buses. Fuel economy comparisons were also provided for several older in-use bus types. Buses other than the control and hybrid buses were grouped by model year corresponding to US diesel emission standards. Average fuel economy in miles per gallon was calculated for each bus group overall and by season. Hybrid buses had the highest fuel economy for all time periods for all bus types. Hybrid buses had a fuel economy that was 11.8% higher than control buses overall and was 12.2% higher than buses with model years 2007 and higher, 23.4% higher than model years 2004 to 2006, 10.2% higher than model years 1998 to 2003, 38.1% higher than for model years 1994 to 1997, 36.8% higher for model years 1991 to 1993, and 36.8% higher for model years pre-1991. Differences between groups of buses also varied by season of the year.

Keywords: Hybrid Vehicle; Fuel Economy; Transit

1. Introduction

Fuel costs are a significant portion of transit agency budgets. Hybrid buses offer an attractive option and have the potential to significantly reduce operating costs for agencies. Hybrid technology has been available in the transit market for some time. There are over 1200 hybrid buses in regular service in North America in over 40 transit agencies as of 2009 [1]. The majority are regular 40-foot buses although some smaller (20-foot) shuttle buses and larger articulated (60-foot) buses are also in service.

Hybrid technology offers an attractive option and has the potential to significantly reduce operating costs for agencies. The main impetus behind use of hybrid transit vehicles is fuel savings and reduced emissions. Wayne *et al* [2] estimated that use of diesel-electric hybrid buses in 15% of the US transit fleet could reduce fuel consumption by 50.7 million gallons of diesel annually.

However, purchase of hybrid transit buses requires a significant investment for transit agencies since a hybrid bus costs approximately 50% to 70% more than a con-

ventional diesel transit bus [3]. Additionally, early estimates of fuel savings were based on laboratory studies which demonstrated significant fuel savings and actual in-use savings may not have may not have materialized to the extent transit agencies expected.

2. Background

Laboratory tests in general have indicated that the fuel economy of hybrid transit buses is significantly better than for regular buses. Chassis dynamometer tests were conducted for 10 low-floor hybrid buses and 14 conventional high-floor diesel transit buses run by New York City Transit [4]. Buses were evaluated over three driving cycles including the Central Business District (CBD), New York bus cycle, and the Manhattan cycle. The operating costs, efficiency, emissions, and overall performance were also compared while both types of buses were operating on similar routes. They found that fuel economy was 48% higher for the hybrid buses.

A study by Battelle [5] tested emissions using a dynamometer for one diesel hybrid-electric bus and two regular diesel buses (with and without catalyzed diesel particulate filters [DPF]). The researchers reported that fuel economy for the hybrid bus was 54% higher than the two regular diesel buses. In another study, two buses were tested using a dynamometer at the National Renewable Energy Laboratory's (NREL's) Refuel facility in Golden, Colorado [4]. One bus was a conventional diesel and the other was a hybrid bus and both were tested over several drive cycles including Manhattan, Orange County Transit A (OCTA), CBD, and King County Metro (KCM). Results indicate 30.3% lower fuel use for the KCM cycle, 48.3% lower for the CBD cycle, 50.6% for the OCTA cycle, and 74.6% for the Manhattan cycle. Fuel economy was reported in miles per gallon (mpg).

In another study, Clark et al. [6] evaluated six transit buses with traditional diesel engines, two powered by spark-ignited compressed natural gas (CNG), and one hybrid transit bus in Mexico City using a mobile heavy-duty emissions testing lab. Buses were tested over a driving cycle representative of Mexico City transit bus operation, which was developed using GPS data from in-use transit buses. Depending on how fuel economy was evaluated, the hybrid bus ranked 4th and 1st in fuel economy.

Transport Canada [1] summarized several studies which compared fuel economy for several transit agencies in Canada. In one laboratory study, using the Manhattan Test Cycle, a reduction in fuel consumption of 36% resulted for hybrid buses as compared to regular buses. Results of a test track study showed a 28% fuel reduction for hybrid transit buses compared to regular buses when the buses were operated at an average speed of 10 km/h with 10 stops per kilometer. As average speed increased the differences in fuel consumption were smaller.

Barnitt and Gonder [7] collected school bus drive cycle data for a first-generation PHEV school bus and a conventional school bus. Both were tested over several different driving cycles to represent a range of driving activity. When in charge-depleting mode, the PHEV had a fuel savings of more than 30% for the Ruban Dynamometer Driving Schedule for Heavy Duty Vehicles and Rowan University Composite School Bus drive cycles. Fuels savings of over 50% were noted for the hybrid school bus for the Orange County Bus cycle. When in charge sustaining mode, smaller fuel savings were noted.

As noted, many of the studies which have demonstrated significant improvements in fuel economy were based on laboratory studies. Fuel economy varies and is correlated to a number of factors, including number of stops per unit distance, road grade, surrounding traffic volume and conditions, environmental conditions, driving style, type of hybrid technology (parallel versus series) [8], roadway type, and passenger load [9]. As a result, actual in-use fuel economy may vary from what has been reported for laboratory studies.

A few studies are available which have evaluated in-use fuel economy for hybrid transit vehicles. Transport Canada [1] reported that the Toronto Transit Corporation which has around 500 hybrid buses was only achieving a reduction of 10% for in-use fuel consumption. The Federal Transit Administration [10] published a report on the status, current issues, and benefits of hybrid bus technology. They summarized the experiences from four transit agencies in the US and found the hybrid buses had better fuel efficiency, accelerating, and handling experience. The overall fuel economy increases for New York, Cedar Rapids, and Los Angeles transit agencies were 18%, 15%, 5% respectively. However, these values were significantly lower than fuel economy estimates which have been reported in laboratory studies.

In a related study, Hallmark et al. [11] evaluated two plug-in hybrid-electric school buses and 2 regular school buses. They found that fuel economy was 29.6% and 39.2% higher for the hybrid buses than the control buses.

3. Project Objectives

Laboratory few tests have indicated that hybrid transit buses can have significantly higher fuel economy compared to conventional transit buses. However, the number of studies is limited and laboratory tests may not represent actual driving conditions since in-use vehicle operation differs from laboratory test cycles. Several initial studies have suggested that the fuel economy savings reported in laboratory tests may not be realized on-road.

This report summarizes the results of a study which evaluated fuel economy for 12 hybrid transit and 7 control buses which are part of the Ames Transit Agency, CyRide. Ames is a community with a population of almost 60,000 and is located in the state of Iowa, a Midwestern state in the US.

The fuel economy of the hybrid buses was compared to a set of buses with similar characteristics. Fuel economy comparisons were also provided for several older in-use bus types. This information is important because CyRide needed to compare the impact of the hybrid buses compared to the buses they replaced.

4. Study Background

CyRide, the city bus system for Ames, Iowa, is operated through collaboration between the city and Iowa State University (ISU). CyRide reported 5,447,289 passengers for fiscal year 2011 and posted 1,185,089 revenue miles [12].

CyRide regularly operates around 79 buses and purchased 12 hybrid transit buses using a Transit Investments for Greenhouse Gas and Energy Reduction (TIGGER) grant. In addition to the 12 hybrid buses, 7 regular diesel buses were selected from among the regular diesel buses in the CyRide fleet. These control buses were selected since they had similar bus specifications in terms of manufacture, model year, and engine size as the hybrid buses as shown in **Table 1**. Fuel economy was also tracked for other buses in the CyRide fleet although they vary in characteristics. Comparison to other buses was done so that CyRide could compare the hybrids to the rest of their fleet as well as assess how much improvement was gained over buses that were replaced by the hybrid buses.

Other buses were grouped into model years corresponding to US diesel emission standards since engines from years with the same standards would have similar emission control technology which can affect fuel consumption. US diesel engine standards cover 1991-1993, 1994-1997, 1998-2003, and 2004-2006, and 2007 and higher [13]. Buses were included in the analysis if they were operated for the majority of a season. Several buses were only used for a very limited period during the

analysis period and were not included in the analysis. Other bus types are shown in **Table 2**.

CyRide operates with 12 fixed routes. The fixed routes operate every day of the year except Thanksgiving, Christmas, and New Year's Day. CyRide rotates buses into and off the system to meet peak travel demands. Buses are driven over several routes according to a prescribed schedule (route pattern), depending on when the bus comes into and leaves the system. Drivers are typically assigned to a route pattern. In general the buses are randomly cycled through the various route patterns. As a result, it was assumed that buses were randomly deployed across routes and any individual bus was equally likely to be assigned to a particular route and driver. Consequently it was assumed that over time, differences due to route, driver characteristics, and passenger loading were minimized.

The hybrid-electric buses were in use by Fall 2010. Initially CyRide was not finding that the hybrid buses were performing as well as expected. The hybrid buses were the first ones in production and as a result experi-

Table 1. Specifications of both hybrid diesel buses and conventional diesel buses.

	Hybrid Diesel Buses	**Conventional Diesel Buses**
Bus Number	118, 119, 120, 121, 122, 123, 124, 429, 430, & 431	819, 820, 821, 822, 126, 127, & 128
	2010	2008/2010
Capital Cost	$521,970	$367,115
Manufacture	Gillig Hybrid	Gillig
Bus Type	Low Floor	Low Floor
Engine	Cummins '10 ISL 280 HP, in line six cylinders	Cummins '10 ISL 280 HP, in line six cylinders
Transmission	Voith DIWA Parallel Hybrid	Voith D864.5 4-speed
After treatment	Particular Filter (CPF)	Particular Filter (CPF)
Governed Speed	65 mph	65 mph
Start Date	6/28/2010	6/28/2010
Frontal Area	113.5 × 102 ft	113.5 × 102 ft
Length × Height	40 ft × 138 in	40 ft × 138 in
Curb Weight	29,500 lbs	25,000 lbs

Table 2. Specifications for other buses tracked for fuel economy.

Standard Year	**Number of Buses**	**Bus Types**
Pre-1991	7	GMC—1973; Orion V—1989
1991 to 1993	6	Gillig Phantom—1993
1994 to 1997	7	Gillig Phantom—1996; Gillig Phantom—1997
1998 to 2003	18	Gillig Lowfloor—1999; Orion V—2000; Orion V—2002
2004 to 2006	12	Gillig Lowfloor—2008; Orion V—2005; Orion VII—2006

enced some early issues. Consequently, the manufacturer made several adjustments to the brake pedals and made programming adjustments. The first adjustments made in early summer 2011 to fix the brake pedals, which directly affects the capture of regenerative energy. Programming was conducted to increase fuel economy. The team collected data prior to July 2011 but since fairly significant changes had occurred, they did not include that data in the final analysis.

5. Data Collection

CyRide buses are usually fueled at the end of their service day when they return to the garage. Each time a bus is fueled is referred to as a fueling period. CyRide technicians note the amount of fuel used (in gallons) and the odometer reading for each fueling period. Differences in odometer readings between fueling periods represents the operating mileage during the period between fueling which in most cases was one day. CyRide uses an average of a 2% blend of bio-diesel. Buses are all fueled from the same tank.

Fuel and odometer data were provided to the team weekly or bi-weekly in a hard copy format. Data were entered into a spreadsheet with date, bus number, fuel used, and odometer reading. One row of data represents on fueling period. Data were manually entered by fueling period by bus. Consequently each row of data (observation) was one fueling period for a one particular bus. Data were collected and entered for 50 buses. This resulted in 5746 fueling periods (observations) for the one-year analysis period.

Data were reviewed to ensure data quality. Outliers were identified and if they appear to be erroneous data,

were removed. For instance, odometer readings or fuel economy that clearly did not make sense were removed.

Since environmental conditions can impact fuel economy, weather was approximated by using season of the year. Quarters were designated using the following convention, which aggregates months where weather conditions were the most likely to be similar in Iowa:
- Winter (December, January, February);
- Spring (March, April, May);
- Summer (June, July, August);
- Fall (September, October, November).

6. Analysis and Resutls

In addition to the hybrid and control buses, other buses were grouped by model year corresponding to US diesel standards as shown in **Table 2**. Data were further disaggregated by season. Fuel economy (in miles per gallon {mpg}) for each observation was calculated by using Equation (1):

$$FE_i = m_i / F_i \tag{1}$$

where:

FE_i = fuel economy in miles per gallon for fueling interval i,

m_i = miles driven during fueling period i,

F_i = fuel used for fueling period i.

Average fuel economy was calculated for each bus group overall and by season. Results are shown in **Figure 1**. Average fuel economy by bus group by season is shown along with the standard error. Results were also compared using a t-test. Standard error was calculated using Equation 2. Standard error is shown as error bars in **Figures 1** and **2**.

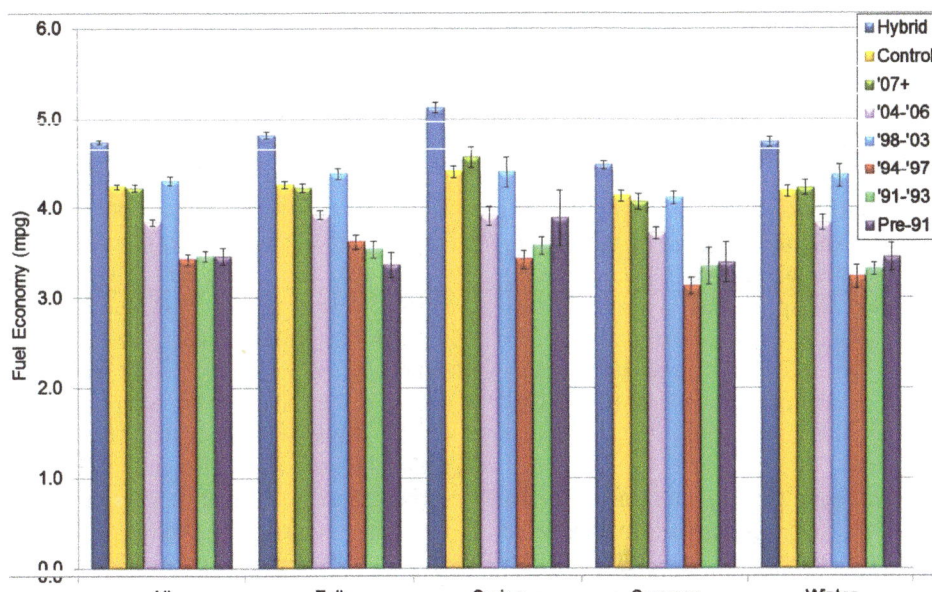

Figure 1. Average fuel economy (mph) by season by bus group.

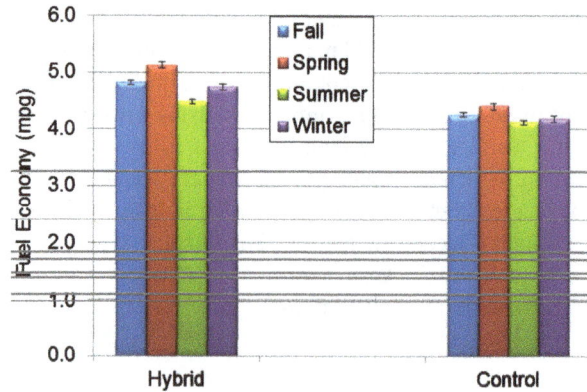

Figure 2. Fuel economy for hybrid and control buses by season.

$$SE = \frac{S}{\sqrt{N}} \qquad (2)$$

As shown, hybrid buses had the highest fuel economy for all time periods combined. Hybrid buses had a fuel economy that was 11.8% higher than control buses (p << 0) for all times periods and was 12.2% higher than buses with model years 2007 and higher (p << 0), 23.4% higher than model years 2004 to 2006 (p << 0), 10.2% higher than model years 1998 to 2003 (p << 0), 38.1% higher than for model years 1994 to 1997 (p << 0), 36.8% higher for model years 1991 to 1993 (p << 0), and 36.8% higher for model years pre-1991 (p << 0).

Differences between groups of buses also varied by season. The hybrid bus had average fuel economy that was 1.3% higher than control buses for fall (p << 0), 16.4% higher for spring (p << 0), 8.6% higher for summer (p << 0), and 13.3% higher for winter (p << 0). Fuel economy was highest for almost all bus types in the spring. Additionally, as shown average fuel economy was lower in the summer for most bus types than for the other seasons. In most cases, fuel economy was highest in the spring

Fuel economy was compared by season as shown in **Figure 2** for hybrid buses. Fuel economy was highest in the spring (4.8 mpg) and lowest in the summer (4.5 mpg). Fuel economy was 6.3% higher in the spring than fall (p << 0), 14.4% higher in the spring than in the summer (p << 0), and 8.1% higher for spring than winter (p << 0). Fuel economy was 7.6% higher in the fall than summer (p << 0) and 1.7% higher in the fall than winter although the difference was not statistically significant (p = 0.116). Finally fuel economy was 5.7% higher for the winter than summer (p = 0.0001).

Similar results were found for control buses as shown in **Figure 2**. However differences between seasons were less pronounced than for the hybrid buses. The highest fuel economy occurred in the spring and the lowest fuel economy occurred in the summer (6.7% higher, p =

0.0014). Fuel economy was also higher in the spring than in the fall (3.5% higher, p = 0.0315) and winter (5.3% higher, p = 0.009). Similarly fall has the next highest fuel economy by season for the control buses and was 3.1% higher than summer (p = 0.038) and 1.7% higher than winter (p = 0.157) although the difference is not statistically significant. Wintertime fuel economy is 1.3% higher than summer (p = 0.744) but the difference is not statistically significant.

7. Summary

Hybrid transit wbuses require a significant investment for transit agencies with purchase price currently being approximately 50% to 70% higher than a conventional diesel bus. Early estimates of cost savings may not have materialized to the extent transit agencies expected. In order to justify the expenditure, agencies require more quantitative information about the on-road fuel economy for hybrid buses.

In-use fuel economy differences were compared for hybrid-electric and conventional transit buses for the Ames, Iowa (USA) transit authority, CyRide. On-road fuel economy was evaluated over a 12-month period for the 12 hybrid transit and 7 control buses. The hybrid diesel buses are 2010 Gillig low floor buses with 280 HP and in line six cylinders. The control buses are 2008/2010 Gillig low floor diesel buses also with 280 HP and in line six cylinders. Fuel economy comparisons were also provided for several older bus types. Buses other than the control and hybrid buses were grouped by model year corresponding to US diesel standards.

Average fuel economy in miles per gallon was calculated for each bus group overall and by season. A total of 5,746 observations were available for the 12 month analysis period.

Hybrid buses had the highest fuel economy for all time periods for all bus types. Hybrid buses had a fuel economy that was 11.8% higher than control bus overall and was 12.2% higher than buses with model years 2007 and higher, 23.4% higher than model years 2004 to 2006, 10.2% higher than model years 1998 to 2003, 38.1% higher than for model years 1994 to 1997, 36.8% higher for model years 1991 to 1993, and 36.8% higher for model years pre-1991. Differences were statistically significant.

Fuel economy for the hybrid buses was highest in the spring at 4.8 mpg and lowest in the summer at 4.5 mpg (14.4% higher). Fuel economy for the hybrid buses was 6.3% higher in the spring than fall and 8.1% higher for spring than winter. Fuel economy was 7.6% higher in the fall than summer and 1.7% higher in the fall than winter although the difference was not statistically significant. Finally fuel economy was 5.7% higher for the winter than summer.

Fluctuations were slightly higher over season for the hybrid versus control buses. This may be due to power requirements for heating and air condition.

Similar results were found for other bus types with the highest fuel economy in the spring and lowest in the summer.

Overall, hybrid buses reported fuel economy that was around 12% higher than for similar control buses. This is similar to results found from other in-use studies which have reported fuel economy improvements from 5% to 18%. These estimates are somewhat lower than the estimates from laboratory studies which have shown improvements of 30% to 75%.

8. Study Limitations

Fuel economy is highly correlated to vehicle operation (amount of time spent in a particular speed/acceleration range). Fuel economy is also related to vehicle load (number of passengers). Ideally, the study would have collected instantaneous fuel economy and vehicle operation by instrumenting vehicles. This type of data could have been collected for a short analysis period; however, it was felt that tracking the buses over a longer period of time was more representative of in-use fuel economy. Additionally, it was not within project resources to instrument buses and collect and reduce these data daily over a 12-month period.

The team did attempt to collect passenger load. However, CyRide only collects total passengers who utilized a particular bus on a daily basis (no information about where passengers embark or disembark). As a result, collection of passenger load would require data collectors to be present on the buses, which was not feasible.

Since CyRide does rotate buses over route patterns, it was assumed that buses would randomly be assigned over all route patterns over time. Since each route has similar drivers, route characteristics and passenger loading, rotation across routes would minimize differences. The team acknowledges that this is not a perfect solution and assumes no bias on the part of CyRide in assigning buses. Additionally it cannot account for daily differences in traffic operations and ridership. These issues, however, are inherent with any large scale, uncontrolled data collection study. And the team feels that collection of the data in-use does provide valuable information to transit agencies that are likely to compare differences in a similar manner.

9. Acknowledgements

The team would like to thank the Iowa Energy Center for funding this project. We would also like to thank CyRide for all of their assistance in giving feedback, providing data, and allowing us to conduct the study. In particular we would like to thank Sheri Kyras, James Rendall, Rich Leners, and all of the technicians who provided data for their assistance.

REFERENCES

[1] Transport Canada, "Hybrid Buses," 2011. http://www.tc.gc.ca/eng/programs/environment-utsp-case study-cs71e-hybridbuses-272.htm

[2] S. W. Wayne, J. A. Sandoval and N. N. Clark, "Emission Benefits from Alternative Fuels and Advanced Technology in the US Transit Bus Fleet," *Energy and Environment*. Vol. 20, No. 4, 2009, pp. 497-515.

[3] Hybrid Center, "Hybrid Watchdog: Hybrid Transit Buses Are They Really Green?" 2010. http://www.hybridcenter.org/hybrid-transit-buses.html

[4] K. Chandler and K. Walkowicz, "King County Metro Transit Hybrid Articulated Buses: Final Evaluation Report," Technical Report NREL/TP-540-40585, National Renewable Energy Laboratory, Golden City, 2006.

[5] Battelle, "Technical Assessment of Advanced Transit Bus Propulsion Systems," Dallas Area Rapid Transit, Dallas, 2002.

[6] N. N. Clark, E. R. Borrell, D. L. McKain, V. H. Paramo, W. S. Wayne, W. Vergara, R. A. Barnett, M. Gautam, G. Thompson, D. W. Lyons and L. Schipper, "Evaluation of Emissions from New and In-Use Transit Buses in Mexico City, Mexico," *Journal of the Transportation Research Record*, Vol. 1987, 2006, pp. 42-53.

[7] R. A. Barnitt and J. Gonder, "Drive Cycle Analysis, Measurements of Emissions and Fuel Consumption of a PHEV School Bus," *The SAE* 2011 *World Congress*, Detroit, 12-14 April 2011, 13 p.

[8] X. Liang, C. Wang, C. Chapelsky, D. Koval and A. M. Knight, "Analysis of Series and Parallel Hybrid Bus Fuel Consumption of Different Edmonton Transit System Routes," *5th IEEE Vehicle Power and Propulsion Conference*, VPPC '09, 7-10 September 2009, Dearborn, pp. 1470-1475.

[9] H. C. Frey, N. M. Rouphail, H. Zhai, T. L. Farias and C. A. Goncalves, "Comparing Real-World Fuel Consumption for Diesel and Hydrogen-Fueled Transit Buses and Implications for Emissions," *Transportation Research Part D: Transport and Environment*, Vol. 12, No. 4, 2007, pp. 281-291.

[10] Federal Transit Administration, "Hybrid-Electric Transit Buses: Status, Issues, and Benefits," TCRP Report 59, Transportation Research Board, Washington DC, 2000.

[11] S. Hallmark and R. Sperry, "Comparison of In-Use Operating Costs of Hybrid-Electric and Conventional School Buses," *Journal of Transportation Technologies*, Vol. 2, No. 2, 2012, pp. 158-164.

[12] CyRide, "Statistics," 2012. http://www.cyride.com/index.aspx?page=1240

[13] US Emission Standards Reference Guide (USEPA), "Heavy-Duty Highway Compression-Ignition Engines and Urban Buses—Exhaust Emission Standards," 2012. http://www.epa.gov/otaq/standards/heavy-duty/hdci-exhaust.htm

7

Life Cycle Assessment of CCA-Treated Wood Highway Guard Rail Posts in the US with Comparisons to Galvanized Steel Guard Rail Posts

Christopher A. Bolin[1], Stephen T. Smith[2*]
[1]Division of Sustainability, AquAeTer, Inc., Centennial, USA
[2]Division of Sustainability, AquAeTer, Inc., Helena, USA

ABSTRACT

A cradle-to-grave life cycle assessment is done to identify the environmental impacts of chromated copper arsenate (CCA)-treated timber used for highway guard rail posts, to understand the processes that contribute to the total impacts, and to determine how the impacts compare to the primary alternative product, galvanized steel posts. Guard rail posts are the supporting structures for highway guard rails. Transportation engineers, as well as public and regulatory interests, have increasing need to understand the environmental implications of guard rail post selection, in addition to factors such as costs and service performance. This study uses a life cycle inventory (LCI) to catalogue the input and output data from guard rail post manufacture, service life, and disposition, and a life cycle impact assessment (LCIA) to assess anthropogenic and net greenhouse gas (GHG), acidification, smog, ecotoxicity, and eutrophication potentially resulting from life cycle air emissions. Other indicators of interest also are tracked, such as fossil fuel and water use. Comparisons of guard rail post products are made at a functional unit of one post per year of service. This life cycle assessment (LCA) finds that the manufacture, use, and disposition of CCA-treated wood guard rails offers lower fossil fuel use and lower anthropogenic and net GHG emissions, acidification, smog potential, and ecotoxicity environmental impacts than impact indicator values for galvanized steel posts. Water use and eutrophication impact indicator values for CCA-treated guard rail posts are greater than impact indicator values for galvanized steel guard rail posts.

Keywords: Life Cycle Assessment; LCA; LCI; Environmental Impact; Treated Wood; Chromated Copper Arsenate; CCA; Guard Rail Post; Greenhouse Gas; GHG; Galvanized Steel

1. Introduction

A highway department's selection of a guard rail system and its materials primarily is based on safety; however factors such as cost, aesthetics, and environmental acceptance play a role in decisions made. While most highway guard rails are made of W-beam galvanized steel, the supporting posts are mostly either preserved wood or galvanized steel; The feasibility of composite materials as guard rail posts, has been studied [1], but the current use does not represent a significant portion of the guard rail post market.

While wood products are susceptible to degradation when left untreated, wood preservative treatments can extend the useful life of a wood product by 20 to 40 times that of untreated wood [2] when used in weather exposed or wet environments subject to microbial or insect attack. Chromated copper arsenate (CCA) was introduced in the 1930s and subsequently adopted through-

out the United States for many exterior and marine uses. CCA has a long history of proven performance in transportation systems [3]. While alternative copper-based water-borne preservatives such as alkaline copper quaternary (ACQ) and copper azoles became popular in the early 2000s, CCA is approved for industrial uses [4] and remains the waterborne preservative of choice for many demanding, commercial applications, including guard rail systems. CCA is a mixture of chromic acid, cupric oxide, and arsenic pentoxide. Because CCA fixes strongly to wood, it provides wood with excellent protection from decay in a variety of environments. Wood post products fulfill the same function as galvanized steel posts and both products have advantages and disadvantages.

Consumer and regulatory agency concern about environmental impacts resulting from the manufacture, use, and disposal of infrastructure products, such as highway guard rail posts, has resulted in increased scrutiny during selection of transportation construction products. In

---

*Corresponding author.

many cases, products such as CCA-treated wood guard rail posts are replaced with galvanized steel guard rail posts based on perception rather than scientific consideration of potential environmental concerns. This study provides a basis for understanding the environmental impacts associated with the production, use, and final disposition of CCA-treated guard rail posts with comparison to galvanized steel posts.

2. Goal and Scope

The goal of this study is to provide a comprehensive, scientifically-based, fair, and accurate understanding of environmental burdens associated with the manufacture, use, and disposition of CCA-treated wood guard rail posts using primary data collected at U.S. treating plants. Other studies [5,6], discuss material performance. This study only includes performance as an estimate of service life.

The scope of this study includes investigation of cradle-to-grave life cycle environmental impacts for CCA-treated wood guard rail posts for highway applications, using life cycle assessment (LCA) methodologies. The results of the CCA-treated guard rail post LCA are compared to LCA findings for galvanized steel guard rail posts. LCA is the tool of choice for evaluating the environmental impacts of a product from cradle to grave, and determining the environmental benefits one product might offer over its alternative(s) [7].

3. Methodology

The LCA methodologies used in this study are consistent with the principles and guidance provided by the International Organization for Standardization (ISO) in standards ISO 14040 [8] and 14044 [9]. The study includes the four phases of an LCA: 1) Goal and scope definition; 2) Inventory analysis; 3) Impact assessment; and 4) Interpretation. The environmental impacts of CCA-treated and galvanized steel highway guard rail posts are assessed throughout their life cycles, from the extraction of the raw materials through processing, transport, primary service life, reuse, and recycling or disposal of the product.

This LCA assumes CCA-treated and galvanized steel guard rail posts can be used interchangeably. CCA-treated and galvanized steel guard rail posts are produced by many different manufacturers and variations exist. Therefore, a "typical product" has been estimated for both guard rail post products.

The LCA for galvanized steel guard rail posts does not include independently developed manufacturing inventory data (primary data). Such data might improve the detailed comparison of these products. However, the data that are available, including data on production of steel shapes [10], provide a basis for general comparison of

LCA impact indicators that is sufficient to understand how the guard rail post products compare.

4. Life Cycle Inventory Analysis

Life cycle inventory (LCI) data are collected at four main stages including raw material acquisition, manufacture, service life use, and disposition. LCI inputs and outputs are tallied and reported at a functional unit of one guard rail post per year of use.

4.1. CCA-Treated Guard Rail Post Inventory

LCI inputs and outputs for the CCA-treated wood guard rail post are quantified per 1000 cubic feet (Mcf). The cubic foot (cf) unit is a standard unit of measure for the U.S. guard rail post industry and is equivalent to 0.028 cubic meters (m^3). The cradle-to-grave life cycle stages considered in this LCI are illustrated in **Figure 1**.

This study builds on existing research for forestry resources and adds the treating (drying, CCA production, and pressure injection of preservative), service use, and disposition stages of CCA-treated wood highway guard rail posts. The previous studies, such as research conducted by the Consortium for Research on Renewable Industrial Materials (CORRIM), have investigated the environmental impacts of wood products. CORRIM's efforts build on a report issued under the auspices of the National Academy of Science regarding the energy consumption of renewable materials during production processes [11]. CORRIM's recent efforts [Johnson, *et al.* ([12-15]] have focused on an expanded list of environmental aspects necessary to bring wood products to market.

The main source of forest products LCI data used in this study are Johnson, *et al.* [12-14] and Milota, *et al*, [16]. Data include forestry practices applicable to rough cut southern pine softwood products grown on Southeastern U.S forest land with an average level of management intensity (*i.e.*, fertilization and thinning) and include the time frame from the sapling greenhouse (cradle) to the mill (gate). These data represent timber shipped to US wood preserving plants for treatment.

The data from Johnson *et al.* and Milota *et al.* are allocated for "typical" sawn and round guard rail posts. Sawn guard rail posts measure 5.5-inches (14 cm) wide by 7.25-inches (18 cm) deep by 6.0-feet (1.8 m) tall and have a volume of 1.66 cubic feet (ft^3) or 0.047 cubic meters (m^3). 1.0 Mcf of sawn timber posts is equivalent to 602 posts. Round posts measure 7.5-inches (19 cm) in diameter and are 6.0-feet tall and have a volume of 1.84 ft^3 (0.052 m^3). 1.0 Mcf of round posts equals 543 posts. Approximately 21% of guard rail posts are round and the rest are sawn rectangular shapes. Round posts are made of smaller diameter logs that only require peeling to remove bark and provide final shape and dimension. The

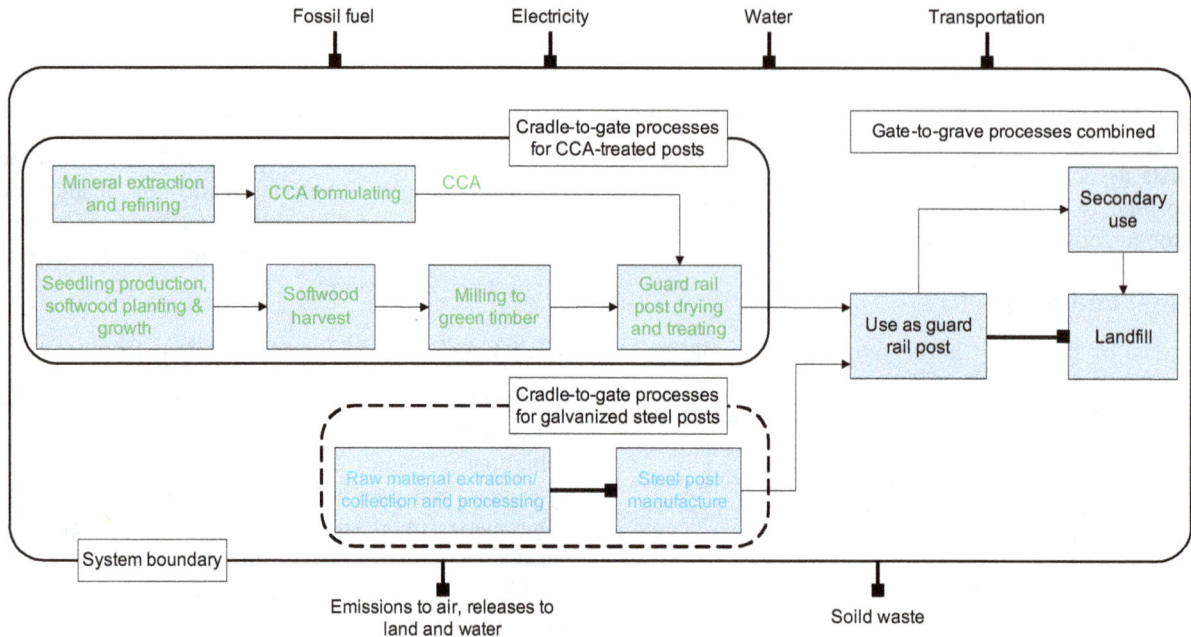

Figure 1. Life cycle stages of CCA-treated guard rail posts.

inputs and outputs of all guard rail posts are modeled in this LCA assuming that data for rough-cut, green lumber are applicable, acknowledging that rough cutting isn't required for round posts.

Six CCA treating plants in the U.S. provided the primary data responses covering operations at their respective treating plant in either 2007 or 2008. The total volume of CCA-treated guard rail posts reported in the surveys is approximately 0.8 million ft³ (800 Mcf) of product. Vlosky [17] estimates US industry total CCA highway construction material treatment in 2007 at approximately 2200 Mcf. Therefore, the primary data used in this study represents approximately 36% of the US highway construction material treating industry.

Southern pine species green timbers are calculated to have an average density of approximately 61.1 pounds per cubic foot (pcf), using USDA [18] wood property factors. Timber posts are dried prior to treatment by either air drying or heat applied processes, reducing the timber density to 39.7 pcf (25% moisture content). Surveyed treaters report that 35% of the total guard rail posts manufactured are dried with heat and 65% are air dried. Half of the respondents report using biomass for at least part of the heat energy needs.

CCA preservative is produced to meet the AWPA Standard for Waterborne Preservatives P5-09 [19]. CCA-C is the formulation currently in use in the U.S. and the preservative modeled in this study. The AWPA Standards specify CCA guard rail post preservative retentions for Use Category 4A (0.4 pcf outer 1.0-inch) and 4B/4C (0.6 pcf outer 1.0-inch) for sawn southern pine posts and 4A (0.4 pcf outer 1.0-inch) and 4B/4C (0.5 pcf outer

1.0-inch) for round posts [20]. A calculation of theoretical guard rail post retention is made using minimum retentions and assumes the "inner" retention in the zone from 1.0-inch deep to center is at 75% of the minimum retention level, acknowledging that the inner zone includes a heartwood section that accepts very little preservative. The calculated average retention for sawn timber posts (in their entirety) at UC4A is 0.35 pcf, for sawn timber posts at UC4B/4C is 0.53 pcf, and for round posts at UC4B/4C is 0.43 pcf. It is assumed that posts are treated at an average 15% over minimum AWPA standards to minimize retreating. The weighted average of these with 15% over-minimum treatment is 0.57 pcf. This theoretical guard rail post retention level compares well to the survey reported preservative use of 0.56 pcf.

Surveyed treaters report that wood treating facilities use a mix of both fossil and biogenic fuel for process heat necessary in facility processes such as kiln drying of posts. The survey respondents report approximately 2.2 tons of wood biomass and 8,500 cubic feet of natural gas per Mcf of guard rail post is used for kiln drying.

Posts are assumed to be installed with spacing of six-foot 3-inches on centers [21]. Service life is a function of quality and species of wood, quality and type of treatment, soil and climatic conditions at the installation location, and use factors. Often, posts are removed from service for other than quality reasons, such as for accident repair, road widening, or following repaving (so guards must be reinstalled higher). A 40-year average service life for CCA-treated guard rail posts is modeled in this LCI. Maintenance applications of preservative to an installed guard rail post are considered rare and are

Life Cycle Assessment of CCA-Treated Wood Highway Guard Rail Posts in the US with Comparisons to Galvanized
Steel Guard Rail Posts

61

not included in this LCA. Other components of a high-way guard rail installation, such as the rails and attachment hardware, are considered equivalent for use with wood and alternative post material and thus, not included within the system boundaries of this LCA.

At the end of useful life, this study assumes removal from service with 90% disposed in a solid waste landfill and 10% reused as fence posts or landscape timbers, or other applications that extend the use of the wood product.

Removed CCA-treated guard rail posts disposed in landfills are modeled as if decayed to a point where 17% of the carbon is released as carbon dioxide, 6% is released as methane, and 77% [21] of the wood carbon and 100% of the preservative remain in long-term storage in the landfill, following the primary phase of anaerobic degradation. Methane capture efficiencies are modeled based on landfill type. Of the captured methane, a portion is used to generate electricity, and applied as an energy credit, and the remainder is assumed to be destroyed by combustion (flaring), so that all the recovered methane is converted to carbon dioxide. Inputs and outputs related to landfill construction and closure are apportioned on a mass disposed basis using data from Menard *et al.* [22].

Transportation-related inputs and outputs are quantified for each life cycle process. Distances and transport modes for preservative supply to treaters, inbound untreated guard rail posts, and outbound treated guard rail posts are based on weighted averages of primary data.

4.2. Galvanized Steel Highway Guard Rail Posts Inventory

This LCA includes an LCI of galvanized steel guard rail posts. The "representative" galvanized steel guard rail post is an I-Beam (W6 × 8.5, W6 × 9, or W9 × 9) with a web width of approximately 6 inches (15 cm), a weight of approximately 8.5 or 9.0 pounds (3.9 to 4.1 kilograms) per foot and 6.0 feet in length and spaced at 6-foot 3-inches on centers [23]. The steel post is hot-dip galvanized to limit corrosion, assuming ASTM A123 standards of 2.0 ounces per 1 square foot of steel [24] or 1.7 pounds of zinc per guard rail post are met. Energy and resources needed to galvanize the steel I-Beams are modeled in the LCI.

Steel source is estimated as a mix of domestic and international sources. As with CCA-treated guard rail posts, transportation-related inputs and outputs are quantified for each life cycle process. Because there are fewer steel post manufacturing facilities than CCA-treating facilities, distances are assumed at least as great as the data received as part of surveyed CCA treaters; thus, the CCA-treated post and galvanized steel distribution distances are the same. Disposition transport to recycle sites

is included in the model.

The estimated average life of galvanized steel guard rail posts is assumed to be the same as for wood posts, acknowledging that some steel posts will be installed in regions of high corrosivity and some will be removed due to highway work. When removed from service, it is assumed that 100% are recycled as steel scrap.

New steel posts are assumed to be produced from typical blast furnaces using a combination of iron ore and approximately 29% recycled steel [26]. All steel posts are assumed to be recycled after service. The LCA allows for 5% loss in recycling [10]. Since the inputs needed to melt and shape the steel shapes cannot be recovered in recycling, the input of electric energy to melt and form steel in an electric arc mini-mill process is "taken back" from the recycle benefit. Thus, as steel recycling reaches 100% nationally, the lowest possible energy input for shapes from recycled steel is that required to process steel in an electric arc furnace since that is required in every cycle.

A summary of selected inventory inputs and outputs for CCA-treated and galvanized steel guard rail posts is provided in **Table 1**.

5. Life Cycle Impact Assessment

5.1. Selection of the Impact Indicators

The impact assessment phase of the LCA uses the LCI results to calculate impact indicators of interest. The LCIA environmental impact indicators are considered at "mid-point" rather than at "end-point". For example, the amount of greenhouse gas (GHG) emission in pounds of carbon dioxide equivalent (CO_2-eq) at mid-point is provided rather than estimating end-points of global temperature or sea level increases. The LCIA is performed using USEPA's Tool for the Reduction and Assessment of Chemical and Other Environmental Impacts (TRACI) [25,26] to assess GHG, acidification, ecotoxicity, eutrophication, and smog impacts potentially resulting from life cycle air emissions. Other indicators of interest also are tracked, such as fossil fuel use and water use.

5.2. Impact Indicators Considered But Not Presented

The TRACI model, a product of USEPA, and the USEtox model [27] a product of the Life Cycle Initiative (a joint program of the United Nations Environmental Program (UNEP) and the Society for Environmental Toxicology and Chemistry (SETAC)), offer several additional impact indicators that were considered during the development of the LCIA, including, but not limited to, human health impacts and impacts to various impact indicators from releases to soil and water. The decision was made

Table 1. CCA-Treated and galvanized steel highway Guard Rail (GR) post life cycle inventory summary (cradle-to-gate per post and cradle-to-grave per post).

Infrastructure process	Units	CCA-treated post (per post)		Galvanized steel post (per post)	
		Cradle-to-gate	Cradle-to-grave	Cradle-to-gate	Cradle-to-grave
Inputs from technosphere					
Electricity-avg. of US grid	kWh	8.5	18	0.071	65
Natural gas (feedstock)	ft3	19	36	0.11	118
Natural gas, combusted in boiler	ft3	26	27	11	18
Diesel fuel, at plant (feedstock)	gal	0	0	0	0
Diesel fuel, combusted in boiler	gal	0.010	0.015	0.00011	0.040
LPG, combusted in equipment	gal	0.00099	0.0010	0	0
Residual oil, processed (feedstock)	gal	0.0043	0.0043	0	0
Residual oil, combusted in boiler	gal	0.0084	0.0090	0.000050	0.0045
Diesel fuel, combusted in equipment	gal	0.14	0.14	0	0
Gasoline, combusted in equipment	gal	0.0051	0.0057	0.000046	0.0043
Hog fuel/biomass (50%MC)	lb	11	11	0.0016	1.8
Coal-bit. & sub. combusted in boiler	lb	0.0070	0.010	0.000016	0.018
Coal-feedstock	lb	0.0020	0.0020	0	0
Energy (unspecified)	Btu	77	77	0	0
Truck transport	ton-miles	56	59	0.042	25
Rail transport	ton-miles	3.4	5.1	7.1	19
Barge transport	ton-miles	0.20	0.41	0.0013	1.5
Ship transport	ton-miles	19	19	26	27
Diesel use for transportation	gal	0.59	0.63	0.018	0.31
Residual oil use for transportation	gal	0.036	0.037	0.048	0.048
Limestone from mine	lb	1.4	1.4	0	0
Rough, green timber from sawmill	ft3	2.0	2.0	0	0
Treated guard timber	ft3	0	0.15	0	0
Zinc	lb	0	0	1.7	1.7
Steel	lb	0	0	51	51
Landfill capacity	ton	0	0.033	0	0
Inputs from nature					
Water	gal	10	10	21	11
Bark from harvest	ft3	0.15	0.15	0	0
Unprocessed coal	lb	4.7	10	38	38
Unprocessed U_3O_8	lb	0.000012	0.000026	0.000025	0.000095
Unprocessed crude oil	gal	0.13	0.15	3.4	0.30
Unprocessed natural gas	ft3	23	23	54	2.6
Biomass/wood energy	Btu	0	0	0.0072	0.00034
Hydropower	Btu	2196	4687	3860	17,366
Other renewable energy	Btu	163	349	1.1	-
Biogenic carbon (from air)	lb	27	19	0	0
Other mined mineral resources	lb	0	0	75	3.6
Outputs to nature					
CO_2-fossil	lb	36	52	119	118
CO_2-non-fossil	lb	−102	−78	0.011	1.9
Carbon monoxide	lb	0.080	0.086	1.3	0.11
Ammonia	lb	0.00014	0.00016	0.000072	0.00012
Hydrochloric acid	lb	0.0038	0.0071	0.0011	0.022
Hydrofluoric acid	lb	0.00035	0.00075	0.00015	0.0028
Nitrogen oxides (NOx)	lb	0.15	0.19	0.19	0.14
Nitrous oxide (N_2O)	lb	0.0011	0.0012	0.000043	0.00022
Nitric oxide (NO)	lb	0.00063	0.00063	0	0
Sulfur dioxide	lb	0.12	0.22	0.10	0.71
Sulfur oxides	lb	0.014	0.016	0.22	0.026
Particulates (PM10)	lb	0.11	0.11	0.0049	0.0074
VOC	lb	0.031	0.032	0.013	0.012
Methane	lb	0.052	2.2	0.064	0.24
Acrolein	lb	0.00019	0.00019	0.00000021	0.0000053
Arsenic	lb	0.0000022	0.0000033	0.00000078	0.0000078
Cadmium	lb	0.00000043	0.00000060	0.00000021	0.0000012
Lead	lb	0.0000036	0.0000047	0.00000077	0.0000081
Mercury	lb	0.00000043	0.00000066	0.00000042	0
Arsenic	lb	0.0000094	0.015	0	0
Chromium	lb	0.000037	0.014	0	0
Copper	lb	0.000021	0.0070	0	0
Zinc	lb	0.00000054	0.00000054	0.0033	0.21
Solid wastes	lb	3.6	80	3.2	65
Process solid & hazardous waste	lb	0.010	0.010	0	0

to not include these impact indicators because of limited and/or insufficient data or concerns regarding misinterpretation. The LCI includes releases of chemicals associated with impacts (such as human health and land and water ecological impacts), but impact indicators for these categories are not calculated.

6. Life Cycle Interpretation

6.1. Findings

Impact indicator values are totaled at two stages for CCA-treated and galvanized steel guard rail post products: 1) the new guard rail post at the manufacturing facility after production, and 2) after service and final disposition. A summary of impact indicator values is provided in **Table 2**. Comparisons are made per post per year of service.

Impact indicator values are normalized to cradle-to-grave CCA-treated guard rail post values of one (1.0), with the galvanized steel guard rail post impact indicator values being a multiple of one (if larger) or a fraction of one (if smaller). The normalized results of **Table 2** are shown graphically in **Figure 2**, illustrating the comparative assertions about the life cycle impacts of CCA-treated guard rail posts and galvanized steel guard rail posts.

Table 2. Summary of impact indicator totals at life cycle stages for CCA-treated and galvanized steel guard rail posts (per post and per year of use assuming a 40-year service life).

Impact Indicators	Units	CCA-treated post (per post per year)		Galvanized steel post (per post per year)	
		Cradle-to-gate[a]	Cradle-to grave[b]	Cradle-to-gate[a]	Cradle-to grave[b]
Anthropogenic GHG	lb-CO2-eq	0.94	2.5	3.0	3.1
Net GHG	lb-CO2-eq	−1.6	0.52	3.0	3.1
Fossil fuel use	MMBTU	0.0063	0.0082	0.022	0.015
Total energy input	MMBTU	0.0089	0.011	0.023	0.017
Acidification	H+-mole-eq	0.33	0.48	0.61	1.1
Water use	gal	0.30	0.30	0.52	0.26
Smog	g NOx/m	0.0025	0.0029	0.0036	0.0037
Eutrophication	lb-N-eq	0.00017	0.00018	0.00021	0.00015
Ecotoxicity	lb-2,4-D-eq	0.0027	0.0041	0.0026	0.010

[a]Cradle-to-gate includes pre-treatment and treating stages for the CCA-treated guard rail post, where gate is defined as point the product leaves the treating facility. Cradle-to-gate includes steel acquisition (recycled and virgin), and steel post manufacture; [b]Cradle-to-grave includes cradle-to-gate, use, and final disposition.

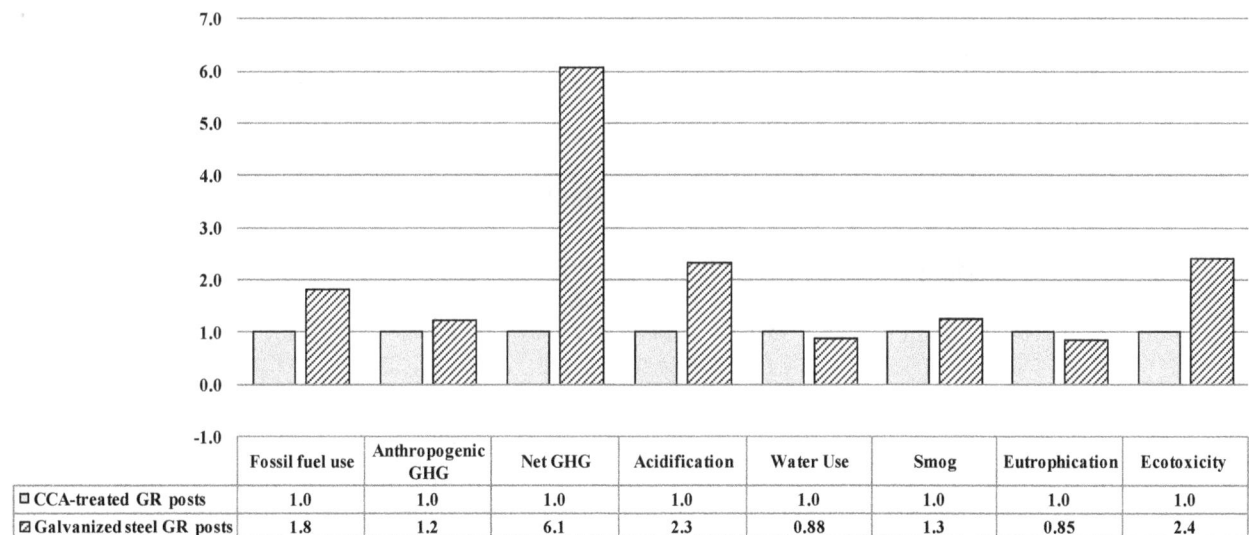

	Fossil fuel use	Anthropogenic GHG	Net GHG	Acidification	Water Use	Smog	Eutrophication	Ecotoxicity
CCA-treated GR posts	1.0	1.0	1.0	1.0	1.0	1.0	1.0	1.0
Galvanized steel GR posts	1.8	1.2	6.1	2.3	0.88	1.3	0.85	2.4

Figure 2. CCA-treated wood and galvanized steel guard rail posts normalized impact comparisons (values normalized to CCA-treated guard rail posts cradle-to-grave = 1.0).

National normalization can be used to provide a means to compare the impact indicator values for guard rail posts to total US annual impact values. Impacts associated with guard rail posts are very small for all indicators for both materials at less than 0.001% of U.S. national impacts. Since relative impacts are so small, further discussion is not included.

6.2. Data Quality Analyses

Data quality analyses per ISO 14044 include a gravity analysis, uncertainty analysis, and sensitivity analysis.

6.2.1. Gravity Analysis

The gravity analysis identifies the CCA-treated guard rail post manufacture, use, and disposition processes most significant to the impact indicator values. This gravity analysis only addresses CCA-treated guard rail posts. The gravity of impacts by life cycle stage is shown in **Figure 3**.

Anthropogenic GHG emissions most notably are impacted by decay of the posts in landfills (45%), landfill construction (16%), truck transport in all stages combined (14%), and electricity use at the treating plant (12%). Net GHG most significantly is impacted by tree growth (credit of 47%), decay of the posts in landfills (26%), emissions from fossil and non-fossil energy sources at the treating plant (11%), landfill construction (7%), and combined truck transport (6%).

Fossil fuel use most notably is impacted by fuel use related to landfill construction and disposal (23%), combined truck transport (23%), guard rail post production prior to treatment (11%), and electricity use (17%) and fuel use (19%) at the treating plant.

The potential to cause acidification is most notably impacted by landfill construction (29%), electricity use at the treating plant (23%), truck transport (20%), natural gas used for drying and facility energy (7%), and ship transport (6%).

Water use includes treatment of the post (38%), preservative manufacture (36%), kiln drying (14%), and tree growth (12%).

The potential to cause smog most notably is impacted by transportation in all stages of the life cycle (67%), landfill construction (10%), wood combustion and kiln drying at the treating plant (9%), and electricity use at the treating plant (7%).

The potential to cause eutrophication is most notably impacted by transportation in all stages of the life cycle (82%) and wood combustion at the treating plant (7%).

The potential to cause ecotoxicity most notably is impacted by landfill construction (35%), electricity use at the treating plant (27%), wood combustion at the treating plant (25%), and fossil fuel use at the treating plant (5%).

6.2.2. Uncertainty Analysis

Areas of uncertainty identified in this LCA include:

The CCA preservative producers did not provide detailed LCI input and output data for CCA production. This LCA relies on industry experts for CCA manufacture LCI data.

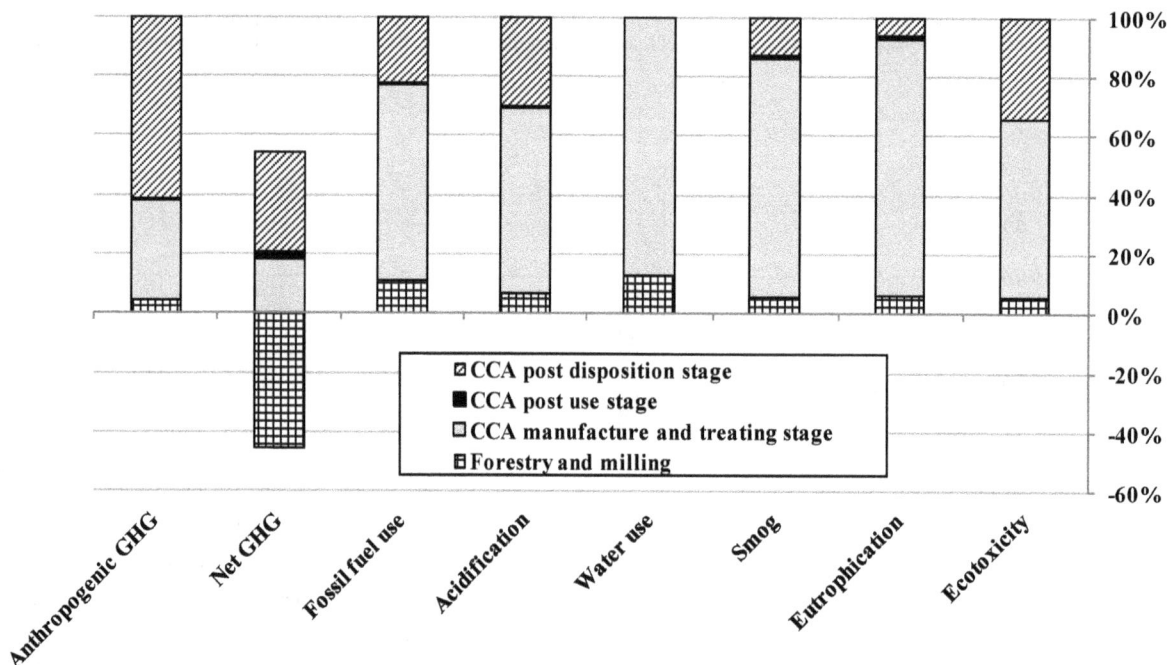

Figure 3. Contributions to impact indicators by life cycle stage of CCA-treated highway guard rail posts.

Landfill fate and release models are based on USEPA GHG emission inventory data [28], and modeled assumptions result in variability of impact indicator values, especially GHG. In this LCA, CCA-treated guard rail posts are conservatively assumed to degrade to the same degree and at the same rate as untreated round wood limbs disposed in a landfill.

The comparative analysis phase of this LCA includes the assembly of an LCI for galvanized steel highway guard rail posts. The cradle-to-grave LCI of galvanized steel posts includes data inputs that involve professional judgments, as no survey of manufacturers of the steel posts was done.

6.2.3. Sensitivity Analysis

Sensitivity analysis determines the magnitude of changes to impact indicators resulting from alternative assumptions. Certain items or categories stand out as most important in affecting the sensitivity of LCA impact indicator outcomes.

Copper source. Copper used in CCA preservative generally comes from recycled, off-specification sources. This LCA applies a fraction of the burdens associated with the production of market-grade copper to the use of recycled copper in CCA. LCI data for recycled copper was not found, so the baseline evaluation assumes one-third of the inputs and outputs associated with market-grade copper is representative as a surrogate for the recycled off-specification copper used in CCA. A sensitivity test assumes that inputs for copper are the same as if all was from primary production. This analysis results in impact indicator increases between 0% and 12%.

CCA preservative use. If CCA retention is increased to 125% of baseline, net GHG (19% increase) and water use (9% increase) impact indicators are most notably impacted. The sensitivity test did not change the comparative results with galvanized steel posts.

CCA-treated highway guard rail post service life. Altering the estimated average service life (40 years) of CCA-treated highway guard rail posts to either 20 or 60 years results in notable impact indicator value changes. Reducing the service life to 20 doubles all of the impact indicators. Similarly, increasing the service life to 60 years, decreases all impact indicators by 33%. Even with service life shortened to half that of galvanized steel, many of the impact indicators for CCA-treated guard rail posts, including net GHG, acidification, and ecotoxicity continue to compare favorably to steel posts.

Post-use disposition of CCA-treated guard rail posts and the impact. The baseline case assumes 10% of used guard rail posts have a secondary use application and 90% are disposed at a landfill. A sensitivity test considers 70% of posts being recycled for energy using combustion cogeneration facilities with appropriate air emis-

sion control devises and 20% being landfilled. Beneficial energy recovery at a cogeneration facility, instead of landfill disposal, reduces anthropogenic GHG (155%) and net GHG (596%), fossil fuel use (165%), acidification (212%), smog (53%), and ecotoxicity (253%) impact indicator values and increase eutrophication (9%) in comparison to the baseline values. Impact indicator reductions result from fossil fuel offsets generated with the use of the wood product for energy recovery and the absence of landfill construction and landfill emission impacts. Reductions of greater than 100% result in overall impact indicator credits.

Landfill decay models. Barlaz [29] reports that approximately 77% of the carbon in wood fiber of branches disposed in landfills is sequestered after primary decomposition has occurred. The presence of lignin (a major carbon-based component of wood) can interfere greatly with cellulose and hemicellulose degradation under the anaerobic conditions of landfills. Laboratory research shows lignin to be very resistant to decay in landfills because cellulose and hemicellulose are embedded in a matrix of lignin [30-32]. Preservative in disposed CCA-treated guard rail posts is expected to further increase carbon sequestration by retarding decay, but such effects are not considered in the baseline assumptions. To demonstrate the sensitivity of carbon storage, a test case assumes 90% wood fiber carbon storage. Increasing wood fiber storage to 90% reduces the anthropogenic GHG (24%) and net GHG (158%) impact indicators, and results in increases for most other impact indicators (most notably ecotoxicity (4%)) because less methane is collected to generate power. Comparisons of indicators between products do not change.

Galvanized steel guard rail post service life. Changes in service life affect all galvanized steel guard rail post impact indicators proportionally. Increasing service life 50% results in a of 33% decrease in impact indicator values.

7. Conclusions and Recommendations

7.1. Conclusions

CCA treated wood guardrail posts offer notably lower environmental impacts for fossil fuel use (almost half), net GHG emissions (one-sixth), acidification (approximately half), and ecotoxicity (approximately half) relative to galvanized steel posts. The other indicators are approximately the same; anthropogenic GHG, water use, smog, and eutrophication. See **Figure 2**.

The LCA process demonstrates the advantage of wood products in relation to GHG. Only wood products begin their life cycles by taking carbon out of the air. This is shown in **Figure 3** where the Net GHG value is negative for the forestry and milling stage. Even with wood posts disposed in landfills following use, the net full life GHG emissions of treated wood posts are one sixth that of galvanized steel posts.

The GHG advantage of wood is dramatically increased under a scenario in which most used wood guardrail posts are recycled for energy production. The LCI credit for energy from recycled wood offsets fossil energy inputs and impacts, resulting in negative impacts (benefits to the environment) for the following; fossil fuel use, anthropogenic and net GHG emissions, acidification, and ecotoxicity.

Recycling of steel has less benefit than expected because, at best, the electric energy input to an electric arc furnace is required for every cycle of use and recycle.

7.2. Recommendations

Production facilities of guard rail posts should continue to strive to reduce energy inputs through conservation and innovation, including sourcing materials from locations close to point of treatment and use. Also, the use of biomass as an alternate energy source can reduce some impact category values compared to the use of fossil fuel energy or electricity off the grid.

The treated wood industry and highway authorities should seek to find beneficial secondary use opportunities for out-of-service wood guard rail posts. Secondary use reduces disposal of wood products in landfills and includes opportunities for beneficial energy recovery in cogeneration or synthetic gasification systems or with reuse as agricultural fencing or landscaping applications. If disposed in a landfill, selection of a disposal facility with methane capture can reduce emissions of GHGs and can result in energy recovery through the capture and reuse of methane.

This study includes the comparison of CCA-treated highway guard rail posts to galvanized steel guard rail posts. The results conform with the ISO 14040 and ISO 14044 standards and are suitable for public disclosure. A detailed, peer-reviewed Procedures and Findings Report can be requested by contacting the TWC at www.treated-wood.org/contactus.html. This LCA covers one treated wood product in a series of LCAs commissioned by the Treated Wood Council (TWC). The other treated wood product LCAs are for alkaline copper quaternary (ACQ)-treated lumber [33], borate-treated lumber [34], pentachlorophenol-treated utility poles [35], creosote-treated railroad ties, and CCA-treated marine pilings.

8. Acknowledgements

The authors wish to thank the TWC for their funding of this project. The TWC members and its Executive Director, Mr. Jeffrey Miller, have been integral in its completion. We also thank the internal reviewers, James Clark, Craig McIntyre, and Maureen Puettmann, and the independent external reviewers, Mary Ann Curran, Paul Cooper, and Yurika Nishioka for their support, patience, and perseverance in seeing this project through to completion.

REFERENCES

[1]   A. Atahan, R. Bligh and H. Ross, "Evaluation of Recycled Content Guardrail Posts," *Journal of Transportation Engineering*, Vol. 128, No. 2, 2002, pp. 156-166.

[2]   J. Morrell, "Disposal of Treated Wood," In: *Environmental Impacts of Preservative Treated Wood*, National Pesticide Information Center, Orlando, 2004, p. 27.

[3]   J. Bigelow, S. Lebow, C. Clausen, L. Greiman and T. Wipf, "Preservation Treatment for Wood Bridge Application," *Transportation Research Record: Journal of the Transportation Research Board*, No. 2108, Transportation Research Board of the National Academies, Washington DC, 2009, pp. 77-85.

[4]   US Environmental Protection Agency (USEPA), "Reregistration Eligibility Decision for Chromated Arsenicals," Office of Prevention, Pesticides and Toxic Substances, Washington DC, 2008.

[5]   C. Plaxico, G. Patzner and M. Ray, "Finite Element Modeling of Guardrail Timber Posts and the Post-Soil Interaction," *Transportation Research Record: Journal of the Transportation Research Board*, Vol. 1647, No. 1998, 1998, pp. 139-146.

[6]   R. Faller, J. Reid, D. Kretschmann, J. Hascall and D. Sicking, "Midwest Guardrail System with Round Timber Posts," *Transportation Research Record: Journal of the Transportation Research Board*, Vol. 2120, No. 2009, 2009, pp. 47-59.

[7]   K. Andersson, M. Eide, U. Lundqvist and B. Mattsson, "The Feasibility of Including Sustainablility in LCA for Product Development," *Journal of Cleaner Production*, Vol. 6, No. 3-4, 1998, pp. 289-298.

[8]   International Organization for Standardization (ISO), "Environmental Management—Life Cycle Assessment—Principles and Framework," ISO 14040:2006, Geneva, 2006.

[9]   International Organization for Standardization (ISO), "Environmental Management—Life Cycle Assessment—Requirements and Guidelines," ISO 14044:2006, Geneva, 2006.

[10] International Iron and Steel Institute (IISI), "Life Cycle Inventory Data for Steel Products and Application of the IISI LCI Data to Recycling Scenarios," Brussels, 2006.

[11] C. Boyd, *et al.*, "Wood for Structural and Architectural Purposes," *Wood and Fiber Science*, Vol. 8, No. 1, 1976, pp. 3-72.

[12] L. Johnson, B. Lippke, J. Marshall and J. Comnick, "Forest Resources-Pacific Northwest and Southeast. CORRIM Phase I Final Report Module A: Life-Cycle Environmental Performance of Renewable Building Materials in the Context of Residential Building Construction," 2004.

http://www.corrim.org/reports

[13] L. Johnson, B. Lippke, J. Marshall and J. Comnick, "Life-Cycle Impacts of Forest Resource Activities in the Pacific Northwest and the Southeast United States," Wood Fiber Science, Vol. 37, No. 12, 2005, pp. 30-46.

[14] B. L. Johnson, E. Oneil, J. Comnick and L Mason, "Forest Resources-Inland West. Corrim Phase II Report Module A. Environmental Performance Measures for Renewable Building Materials with Alternatives for Improved Performance," Seattle, 2008.
http://www.corrim.org/pubs/reports/2005/phase1/

[15] E. Oneil, et al., "Life-Cycle Impacts of Inland Northwest and Northeast/North Central Forest Resources," Wood and Fiber Science, Vol. 42, Special Issue, 2010, pp. 29-51.

[16] M. Milota, C. West and I. Hartley, "Phase I Final Report, Module C, Softwood Lumber—Southeast Region," 2004.
http://www.corrim.org/pubs/index.asp#2004

[17] R. Vlosky, "Statistical Overview of the US Wood Preserving Industry: 2007," Louisiana Forest Products Development Center, Baton Rouge, 2009.

[18] USDA, Forest Products Service, Forest Products Laboratory, "Wood Handbook, Wood as an Engineering Material," General Technical Report FPL-GTR-190, USDA, Forest Products Service, Forest Products Laboratory Madison, 2010.

[19] American Wood Protection Association, "Standard P5-09 Standard for Waterborne Preservations," In: 2010 AWPA Book of Standards, American Wood Protection Association, Birmingham, 2010.

[20] American Wood Protection Association, "Standard U1-10 Use Category System: User Specification for Treated Wood," In: 2010 AWPA Book of Standards, American Wood Protection Association, Birmingham, 2010.

[21] US Environmental Protection Agency (USEPA), "Solid Waste Management and Greenhouse Gases: A Life Cycle Assessment of Emissions and Sinks," 3rd Edition, USEPA, Washington DC, 2006.

[22] J. Menard, et al., "Life Cycle Assessment of a Bioreactor and an Engineered Landfill for Municipal Solid Waste Treatment," 2003.
www.lcacenter.org/InLCA-LCM03/Menard-presentation.ppt

[23] US DOT, "Drawings," 1994.
http://flh.fhwa.dot.gov/resources/pse/standard/st61711.pdf

[24] American Galvanizers Association, "Galvanize It," 2002.
http://www.galvanizeit.org/images/uploads/publicationPDFs/SP_SGRS_02.pdf

[25] J. Bare, G. Norris, D. Pennington and T. McKone, "TRACI—The Tool for the Reduction and Assessment of Chemical and Other Environmental Impacts," Journal of Industrial Ecology, Vol. 6, No. 3-4, 2003, pp. 49-78.
http://mitpress.mit.edu/jie

[26] US Environmental Protection Agency (USEPA), "Tool for the Reduction and Assessment of Chemical and Other Environmental Impacts (TRACI)," 2009.
http://www.epa.gov/nrmrl/std/sab/traci/

[27] R. Rosenbaum, et al., "USEtox—The UNEP-SETAC Toxicity Model: Recommended Characterization Factors for Human Toxicity and Freshwater Ecotoxicity in Life Cycle Impact Assessment," The International Journal of Life Cycle Assessment, Vol. 13, No. 7, 2008, pp. 532-546.

[28] US Environmental Protection Agency (USEPA), "Inventory of US Greenhouse Gas Emissions and Sinks: 1990-2007," Report No: EPA 430-R-09-004, USEPA, Washington DC, 2009.

[29] M. A. Barlaz, "Carbon Storage during Biodegradation of Municipal Solid Waste Components in Laboratory-Scale Landfills," Global Biogeochemical Cycles, Vol. 12, No. 2, 1998, pp. 373-380.

[30] R. Ham, P. Fritschel and M. Norman, "Refuse Decomposition at a Large Landfill," Proceedings Sardinia 93, 4th International Landfill Symposium, Cagliari, 11-15 October 1993, pp. 1046-1054.

[31] R. Ham, P. Fritschel and M. Norman, "Chemical Characterization of Fresh Kills Landfill Refuse and Extracts," Journal of Environmental Engineering, Vol. 119, No. 6, 1993, pp. 1176-1195.

[32] Y.-S. Wang, C. S. Byrd and M. A. Barlaz, "Anaerobic Biodegradability of Cellulose and Hemicelluloses in Excavated Refuse Samples Using a Biochemical Methane Potential Assay," Journal of Industrial Microbiology, Vol. 13, No. 3, 1994, pp. 147-153.

[33] C. A. Bolin and S. Smith, "Life Cycle Assessment of ACQ-Treated Lumber with Comparison to Wood Plastic Composite Decking," Journal of Cleaner Production, Vol. 19, No. 6-7, 2011, pp. 620-629.

[34] C. A. Bolin and S. Smith, "Life Cycle Assessment of Borate-Treated Lumber with Comparison to Galvanized Steel Framing," Journal of Cleaner Production, Vol. 19, No. 6-7, 2011, pp. 630-639.

[35] C. A. Bolin and S. Smith, "Life Cycle Assessment of Pentachlorophenol-Treated Wooden Utility Poles with Comparisons to Steel and Concrete Utility Poles," Renewable and Sustainable Energy Reviews, Vol. 15, No. 5, 2011, pp. 2475-2486.

8

Forecasting Baltic Dirty Tanker Index by Applying Wavelet Neural Networks

Shuangrui Fan[1], Tingyun Ji[1], Wilmsmeier Gordon[2,3], Bergqvist Rickard[1]

[1]Logistics and Transport Research Group, Department of Business Administration, School of Business, Economics and Law at University of Gothenburg, Göteborg, Sweden

[2]Economic Commission for Latin America and the Caribbean (ECLAC), Santiago, Chile

[3]Transport Research Institute (TRI), Edinburgh Napier University, Edinburgh, UK

ABSTRACT

Baltic Exchange Dirty Tanker Index (BDTI) is an important assessment index in world dirty tanker shipping industry. Actors in the industry sector can gain numerous benefits from accurate forecasting of the BDTI. However, limitations exist in traditional stochastic and econometric explanation modeling techniques used in freight rate forecasting. At the same time research in shipping index forecasting e.g. BDTI applying artificial intelligent techniques is scarce. This analyses the possibilities to forecast the BDTI by applying Wavelet Neural Networks (WNN). Firstly, the characteristics of traditional and artificial intelligent forecasting techniques are discussed and rationales for choosing WNN are explained. Secondly, the components and features of BDTI will be explicated. After that, the authors delve the determinants and influencing factors behind fluctuations of the BDTI in order to set inputs for WNN forecasting model. The paper examines non-linearity and non-stationary features of the BDTI and elaborates WNN model building procedures. Finally, the comparison of forecasting performance between WNN and ARIMA time series models show that WNN has better forecasting accuracy than traditionally used modeling techniques.

Keywords: BDTI; Tanker Freight Rates; Forecasting; Wavelets; Neural Networks; Shipping Finance

1. Introduction

Research on tanker shipping generally focus on freight rates, fleet arrangements, ship chartering decisions, shipping strategies, operation optimization, etc. [1-6], the freight rate complexity in connection to forecasting techniques is of particular interest in this paper. An inherent feature of shipping freight rates are fluctuations as a source of market risks for all market participants, including not only tanker shipping companies, but also hedge funds, commodity and even energy producers [7]. In essence, forecasting is about attaining and analyzing the right information of the present [cf.8]. Decision makers in the shipping industry often utilize information on historical freight rates to make strategic decisions [cf.8], thus appropriate forecasting techniques may enable the actors in shipping business to make better decisions.

Due to the high volatility in the tanker shipping market accurate prediction of future tanker freight rates is challenging. Currently, the crude oil tanker freight rate freight forward agreements (FFA) are mainly referred to the routes included in the Worldscale [9] and Baltic Dirty Tanker Index [10]. Due to the short history of the BDTI research concerning forecasting of crude oil tanker indexes and rates has only emerged in the last decade. This paper offers new thoughts on the application of hybrid artificial intelligent techniques—Wavelet Neural Networks, beyond traditional modeling and forecasting methods in the shipping sector.

2. Literature Review

Prevailing research determining and forecasting shipping freight rates is based on stochastic and econometric explanation modeling techniques. Stochastic modeling is related to probability theory in the modeling of phenomena in technology and natural sciences. There are numerous examples in bulk and tanker shipping research using stochastic modeling methods. Cullinane [11] firstly developed a short-term adaptive forecasting model for The Baltic International Freight Futures Exchange speculation, through a Box-Jenkins approach, which revolves around what is referred to as ARIMA (p, d, q) modeling. Cullinane measured the predictive power of his model and compared it with alternative forecasting models at that time.

By utilizing Autoregressive Conditional Heteroscedasticity (GARCH) model, Kavussanos [12] determined that smaller size tankers show more flexibility compared to larger tanker ships in terms of their business pricing and operations. Jonnala & Bessler [13] also used GARCH for the specification and estimation of an ocean grain rate equation. Based on sample freight rates data, Veenstra & Franses [14] demonstrated the series of freight rates to be non-stationary, utilizing freight rate data in a vector autoregressive model to forecast dry bulk shipping freight rates. Presenting a stochastic optimal control problem the property of the equilibrium of tanker shipping freight rates is discussed to be close to that of the standard geometric mean reversion process [15]. Adland & Cullinane [5] utilized a general non-parametric Markov diffusion method to investigate the dynamics of tanker spot freight rates, arguing that non-linear stochastic models can best describe these dynamics. However, due to the high volatility in spot freight rates, the difficulties in detecting slow-speed mean reversion in high frequency data became a constraint for the model [5]. Batchelor et al. [10] examined the performance of popular time series models in forecasting spot and forward rates on major seaborne freight routes, and they found the vector equilibrium correction model (VECM), to deliver the best in-sample-fit, however, the predictive ability of the VECM in reality is poor. Batchelor et al. [10] suggested ARIMA and VAR model as better models for forecast. Goulielmos & Psifia [16] demonstrated the non-normality and non-linearity in trip and time charter freight rate indices by using the BDS test, thus arguing that linear and other traditional models are not suitable for modeling the distributions of the indices [16]. Sødal et al. [17] explored the market switching of different ship types based on numerical experiments with a mean-reverting model. They concluded that new combination carriers may enter the market in the near future [17].

The non-linearity, non-stationary and complex inherent characteristics of shipping freight rates make traditional stochastic modeling methods hard to achieve a good balance between forecasting accuracy and theoretical feasibility of the models.

The earliest econometric explanation modeling application in tanker freight rate research can be traced back to 1930s. Koopmans [18] studied the determinants of tanker freight rates by modeling the supply and demand for tanker shipping services. Since that time, numerous tanker freight rate studies have been following Koopmans ideas. Zannetos [19] argued that static expectations in tanker ship supply and demand necessarily imply a future price level equal to current prices, thus the future price level of tanker freight rates under static conditions can be derived from objective data before any changes in present prices occur. Evans [20] analyzed the matching of supply and demand for bulk shipping in the short and long run. Even though initially Evans' research was not

designed for tanker shipping, the use of static models shows is of similar nature to that of Zanneto's in analyzing the supply and demand in shipping by econometric explanation modeling methods. In contrast to static models, dynamic econometric models emphasize the interrelationship and interactions among the variables. It has been demonstrated that mathematical analysis of dynamic econometric models is a valuable tool for predicting future time tracks of certain economic variables [21]. In tanker shipping freight rate mechanism studies, recent research exerts the stochastic extension of traditional partial shipping supply and demand equilibrium models of the VLCC spot freight market, and simulates the probability distribution of future spot freight rates and fleet size based on present market situations collective with stochastic demand by using a dynamic model [22].

Econometric modeling methods are able to incorporate causality. Econometric explanation modeling methods are widely adopted in the shipping freight rate related research, despite its known limitations. In comparison to stochastic modeling methods, econometric modeling methods are generally less capable in modeling uncertainties. Additionally, even if the rationales or causalities behind the fluctuations in shipping freight rate may be explained by econometric models, in order to model the causalities, bundles of constraints and assumptions about the data and variables are requested by the econometric model, thus the used causalities are often found to be deficient in the modeling process.

Because of the limitations of traditional stochastic and econometric explanation models, the combination of wavelets and neural networks may provide an interesting opportunity. The Wavelet Neural Networks (WNN) concept is based on wavelet transformation theory and was first proposed by Zhang & Benveniste [23], as an alternative to feed forward neural networks for approximating arbitrary nonlinear functions [23]. The main feature of WNN is that it combines the time frequency localization properties and adaptive learning nature of neural networks [24,25], thus making WNN a useful tool for analyzing and forecasting time series, especially when the data series is non-linear and non-stationary.

3. Modeling

3.1. Determinants of BDTI

The Baltic Exchange Dirty Tanker Index (BDTI) indicates the cost of shipping unrefined petroleum oil, on a basis of the average costs of 17 routes. The average rate (AV_i) of each route is calculated by the Baltic Exchange in cooperation with a panel of major ship broking companies. The panelists calculate weighting factors of each route. The BDTI is defined as the sum of the multiplications of the average rates (AV_i) of each route with the weighing factor (WF_i) of that particular route. The calculation can be illustrated by the following function:

$$BDTI = BDTI = \sum_{i}^{n}\left(AV_i \cdot WF_i\right) \quad [26].$$

The supply of ships balanced against cargo demand, the balance of fleet capacity and cargo volumes are considered to be the most significant assessment factors and direct indicators for the market [27]. The *BDTI* is calculated on the basis of its internal determinants, which are route, haul, cost & revenue and vessel specification:

The route: due to changes in the geography and balance of fronthaul and backhaul routes of oil trade only a limited number of trade routes can become the standard routes in the BDTI index. The supply of ships balanced against cargo demand determines routes [27].

Load and discharge ports: The load and/or discharge port may be fixed outside the route definitions the difference between the ports is assessed on the basis of market significance. This normally includes the factors of port costs, extra steaming time or savings and other added value due the geographical difference [27].

Vessel specifications: include deadweight, overall length, draft, capacity, service speed and bunker consumption [27]. Due to economies of scale the average vessel capacity determines patterns and trends of the *BDTI*.

In this paper, the set of internal determinants are extended further by including the external determinants of oil demand, supply price, world economy and fleet capacity. Correspondingly, six indices are introduced to represent the external determinants and applied in the WNN model for forecasting BDTI trends (see **Figure 1**).

3.2. The Method

A Wavelet is a "small wave", which grows and decays within a limited time period. A Wavelet is in clearly contrary to "big wave", which swing up and down for a very long, even infinite time period, for instance, the sine wave is a typical "big wave"

$$\left(\sin\left(u\right), \text{where } u \in \left(-\infty, +\infty\right)\right).$$

By defining a real value function $\psi(\cdot)$ over a real axis $\left(-\infty, +\infty\right)$, the quantifying notion of a wavelet can be expressed as follows [cf.28]:

$$\int_{-\infty}^{+\infty} \psi\left(\mu\right)d\mu = 0 \tag{1}$$

$$\int_{-\infty}^{+\infty} \psi^2\left(\mu\right)d\mu = 1 \tag{2}$$

Therefore, in a given interval $\left[-T, +T\right]$, for any

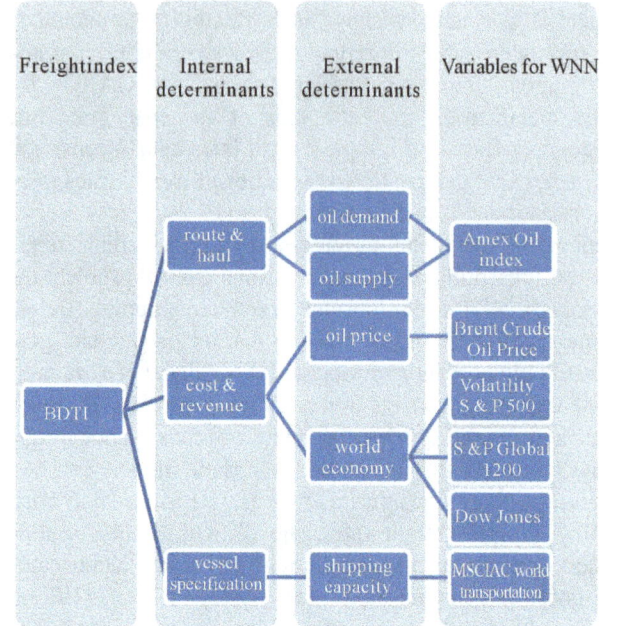

Source: Authors

Figure 1. Determinants of BDTI.

$0 < \epsilon < 1$, there must be

$$\int_{-T}^{+T} \psi^2\left(\mu\right)d\mu > 1 - \epsilon \tag{3}$$

Equation (2) describes function $\psi(\cdot)$ having a movement away from zero, while Equation (1) describes that movements above zero must be canceled out by movements below zero. Since the interval $\left[-T, +T\right]$ is utmost small compared to the infinite long real axis $\left(-\infty, +\infty\right)$, the non-zero movement of $\psi(\cdot)$ therefore can be regard as limited to a small time interval.

In order to utilize wavelets, the following common additional condition defined as admissibility condition needs to be considered:

$$C_\psi \equiv \int_0^\infty \frac{\left|\psi\left(f\right)\right|^2}{f} df \text{ where } 0 < C_\psi < \infty \tag{4}$$

Under this condition, a wavelet $\psi(\cdot)$ is admissible, thus allowing reconstruction of a function from its wavelet transform.

The application of wavelets allows the adapting the representations of data to the nature of the information. Wavelets have many advantages over traditional Fourier analysis methods in analyzing conditions where discontinuities and sharp spikes exist in the signal [29]. Traditional Fourier analysis, which expresses a signal as a sum of a series of sine and cosine ("big waves"), is good for studying stationary data; however, since many changes in a market will happen within a transient time, Fourier analysis is not well suited for studying data with transient events that can be hardly predicted from the historical data [30]. Wavelets fill this gap by the notion of small wave processing, which is designed with a high level of non-stationary data capturing ability.

Wavelets are well-suited for approximating data with sharp discontinuities [29]. The BDTI can be regard as a kind of signal, with sharp spikes all along the line (**Figure 2**).

Using wavelets analysis allows isolating and processing specific types of patterns concealed in the masses of data. Wavelet transformations allow time-frequency localization for the signal or data [30]. This enables to not only capture repeating background signals in the *BDTI*, but also individual, localized, specific variations in the background. Therefore, the authors use wavelets to approximate *BDTI* data, thus cutting the data into different frequency components each with its own solution (pattern) matched to its scale, and then process these components with neural networks.

Generally speaking, wavelet analysis explores the signals with short duration and finite energy. Wavelet analysis transforms the signal under investigation into another presentation, which expresses the signal in a more useful pattern [30]. Wavelet analysis procedure adopts a wavelet prototype function, often called, mother wavelet, through temporal analysis, which is performed with a contracted, high-frequency version of the prototype function, and frequency analysis, which is performed using dilated, low-frequency version of the prototype wavelet, so to expand and shift wavelet of varying scales and positions to approximate the original signal or data [28,29]. The Morlet wavelet function $\psi(\cdot)$ is often referred to as the mother wavelet (6) [cf.31]:

$$\psi_{\mu,s}(t) = \left(\frac{1}{\sqrt{s}}\right)\psi\left(\frac{(t-\mu)}{s}\right) \quad (6)$$

where $s > 0, \mu \in$ real number. s is the scaling parameter, μ is the translation parameter, t is the time, and $\psi_{\mu,s}(t)$ is the basis function of the wavelet $\psi(t)$.

For the purpose of this paper, the Morlet wavelet is adopted as the mother wavelet since it offers a good formulation of time frequency methods and a resolution for smoothly changing time series [28,32]. Hence, the Morlet wavelets are used for wavelet transformations and selected as the hidden layer nodes in the future neural network construction.

Each element of the wavelet set is a scaled (dilated or compressed) and translated (shifted) Morlet mother wavelet, that can be formulated as follows:

$$\psi(t) = \cos(1.75t)\exp\left(-\frac{t^2}{2}\right) \quad (7)$$

Certain functions of the selected wavelet family can then be presented without any loss of information as a linear combination in the discrete case, or as an integral in the continuous case using wavelet transformation [33].

The wavelet functions constitute the wavelet transformation to analyze the signal in its time, scale, and frequency content [29]. Wavelet transformation can be localized in both time and frequency domains, by using a scaling and translating wavelet function [34]. The continuous wavelet transformation (CWT) and discrete wavelet transformation (DWT) are main classes for different wavelets [28].

The continuous wavelet transformation is designed to deal with the time series defined over a continuous real axis:

$$W(\lambda,t) \equiv \int_{-\infty}^{+\infty} x(\mu)\psi_{\lambda,\tau}(\mu)\,d\mu, \text{where } \psi_{\lambda,t}(\mu)$$
$$\equiv \frac{1}{\sqrt{\lambda}}\psi\left(\frac{\mu-\tau}{\lambda}\right) \quad (8)$$

where $\lambda > 0$ and $-\infty < t < +\infty$, $x(\mu)$ is the function or signal. The idea behind CWT is to calculate the amplitude coefficient thus making $\psi_{\lambda,\tau}$ best fit the signal $x(\mu)$, given λ dilation factor, and τ translation factor. The adjustments in λ enables to see how the wavelet fits the signal from dilation, and changes in τ reveal the nature of signal changes over time. CWT preserves all the information from the original data series or signal- $x(\mu)$. If this $\psi(\cdot)$ and signal $x(\mu)$ meet the given conditions in Equations (1)-(4), then the wavelet can be reconstructed following inverse transformation:

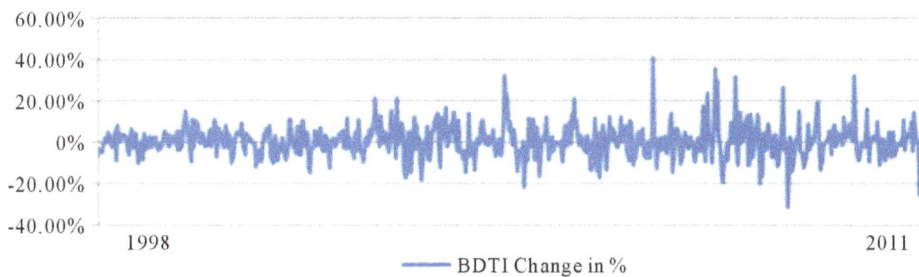

Source: Authors, based on BDTI

Figure 2. BDTI change in percentage, 1998 to 2011.

$$x(t)=\frac{1}{C_\psi}\int_0^\infty\left[\int_{-\infty}^{+\infty}(x,\psi_{\lambda,\tau})\psi_{\lambda,\tau}(t)d\mu\right]\frac{d\lambda}{\lambda^2}\quad(9)$$

where C_ψ is defined in Equation (4).

Thus, the CWT and signal $x(\mu)$ represent the same entity in contents, but CWT shows the information in a new manner [28]. Since the Morlet wavelet is a typical continuous wavelet the authors utilize CWT to gain new insights in BDTI data series.

Discrete wavelet transform (DWT) works with time series defined over a specific range of integers [28], and it is the operation that generates a data structure containing $\log_2 n$ components of a variety of lengths, then filling and transforming data structure into a different data vector of length 2^n [29].

As presented in previous section, CWT consists a function of C_ψ and $\psi(\cdot)$, therefore, leading to a result of over abundant information when analyzed. By implementing DWT, we are able to retain the critical features of CWT and view DWT as a discrete sample of the CWT.

$$W(a,b)=\int_t f(t)\frac{1}{\sqrt{|a|}}\psi\left(\frac{t-b}{a}\right)dt\quad(10)$$

This discrete critical sampling of CWT is obtained through $a=2^{-j},b=k2^{-j}$ in which discrete translation and dilation are presented by integers j and k. The substitution can be described as following:

$$\int_t f(t)2^{j/2}\psi(2^j t-k)dt\quad(11)$$

The function of j and k is donated as W(j, k), thus, a critical sampling of CWT defines the resolution of DWT from both the time and frequency perspective [30]. The wavelet coefficients can be found by wavelets that follow the values given by:

$$\psi_{j,k}(t)=2^{j/2}\psi(2^j t-k)\quad(12)$$

After knowing the relationship between CTW and DTW, the discrete wavelet is defined as:

$$\psi_{j,k}(t)=\frac{1}{\sqrt{\lambda_0^j}}\psi\left(\frac{t-k\lambda_0^j\tau_0}{\lambda_0^j}\right)\quad(13)$$

where k and j are integers, $\lambda_0>1$, is the fixed dilation factor, τ_0 is the translation factor and depends on the dilation factor. Then the scaling function is defined as [30]:

$$\varphi(2^j t)=\sum_{i=1}^k h_{j+1}(k)\varphi(2^{j+1}t-k)\quad(14)$$

The wavelet translation function is defined as:

$$\psi(2^j t)=\sum_{i=1}^k g_j+1(k)\psi(2^{j+1}t-k)\quad(15)$$

So, a signal or a series of data can be expressed as a combination of (14) and (15):

$$f(t)=\sum_{i=1}^k \varphi(2^{j-1}t)+\psi(2^{j-1}t)\quad(16)$$

The additive decomposition is known as a "multiresolution analysis", and the generalization of DWT is known as the "wavelet packet" [28]. The process of DWT is achieved by Matlab programming. Therefore, by applying decomposition to the original BDTI series, the sample variance of the original series of BDTI is decomposed based on scale and the amount of redundant information in the data is filtered by this sub-sampling processes.

An example of a 3-layer feed-forward artificial neural network with n inputs, n hidden nodes, and n outputs is illustrated in **Figure 3**.

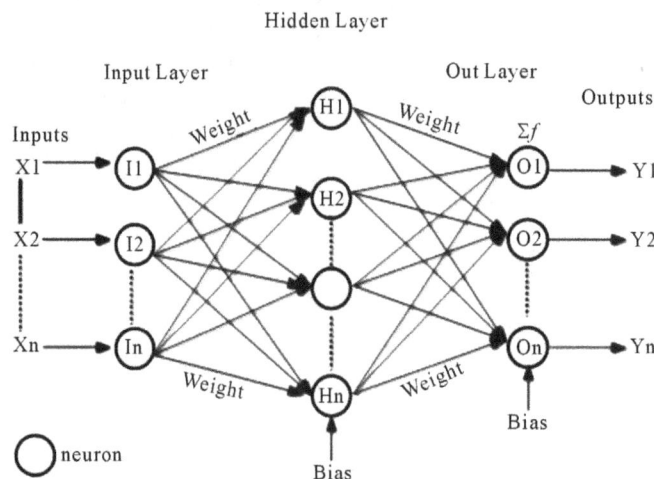

Source: Authors.

Figure 3. An example of neural network.

A Neuron is a linear or nonlinear, parameterized, bounded function. The variables of the neuron are called inputs of the neural while its corresponding value is its outputs [35]. The parallel structure of the system means the neurons functions can carry out the computation at the same time. Input layer is not consistent with a fully functional neuron, instead, it is more like a data recorder, which receives the values of a data series from outside the neural networks and then constitutes inputs to next layer. The hidden layer is the processing layer where inputs and outputs remain within neural networks. Output layer sends the output results out of neural networks.

The weights $\{w_j\}$ and parameters are assigned to the inputs $\{x_i\}$ of the neurons, thus making a neuron a nonlinear combination of function

$$y = f\left(x_1, x_2, \cdots, x_n; w_1, w_2, \cdots, w_n\right).$$

Bias is an additional constant that is added to the combination of inputs and weights. The most frequently used linear combination v is expressed as a weighted sum of inputs, with bias:

$$v = w_0 + \sum_{i=1}^{n-1} w_i x_i \qquad (17)$$

The function f is termed as activation function, which is applied to the weighted sum of inputs of a neuron in order to produce the output under a certain threshold, in another words, the activation function manages the amplitudes of the output. Currently, the majority of NN's use sigmoid functions as activation function such as the tan$h$ function or tangent function as inverse-of the output y of a neuron with inputs $\{x_i\}$ is produced by

$$y = \tan h\left[w_0 + \sum_{i=1}^{n-1} w_i x_i\right] \quad [35].$$

The weights are non-linearly assigned to the neurons. The *function f* to a great extent defines the assigns of weights. For example, the output of Radical Basis Function (RBF) is given by:

$$y = \exp\left[\frac{-\sum_{i=1}^{n}\left(x_i - w_i\right)^2}{2w_{n+1}^2}\right] \qquad (18)$$

where w_i is weight, w_{n+1} is standard deviation.

The authors utilize the Morlet wavelet, as the function to assign the weights and to produce the neuron outputs for forecasting. As explained above, a neuron is a nonlinear, and weighted function of its input variables, thus, the neural network is the composition of the nonlinear functions of two or more neurons [35]. Generally, the neural network can be divided into two classes: feedforward neural networks and recurrent/feedback neural networks.

A feedforward neural network is *a nonlinear function of its inputs, which is the composition of the functions of its neurons* [35]. In this kind of network, only neurons in subsequent layers can be connected, and the information strictly flows from the input layer to the output layer. Each layer receives information only to and from its connected layers. The data processing extends over multiple layers in feedforward direction and no feedback connections exist in the neural network [36].

In a recurrent neural network, there exists at least one path leading back to the neuron. Therefore, in comparison to feedforward neural networks, the recurrent neural network contains feedback connections [37]. Recurrent neural networks are sensitive and can be adapted to the past inputs [38,39]. The recurrent neural network may be effective in this BDTI forecasting research, however, as suggested by Pearlmutter [40], it is more sensible to begin with trying multi-layered feedforward neural networks to solve the problem before applying recurrent neural network [40]. In this paper the authors only utilize feedforward neural networks for forecasting BDTI forecasting.

The weights in the networks are obtained through training. Training is the algorithmic procedure, by which the weights of such a network are estimated, for a given family of functions [35]. The neural network is trained by learning rules then determines suitable weights by itself. Generally, there are two types of learning, supervised and unsupervised learning [41].

This research uses backpropagation, a type of supervised learning that is widely used and a computationally economical method to train neural networks [35,42,43]. Backpropagation is realized by iteratively processing a set of training samples, comparing the outputs of network for each sample with actual known values. The mean square error between real outputs and actual known values is minimized by adjusting the weights for each sample, and this adjust is conducted in a backward direction, from the output layer to the hidden layer [44].

Metrics of backpropagation can be generally described as simple and local nature [45]. Moreover, there are two distinctive advantages of backpropagation training [46]; firstly, it is mathematically calculated to minimize the mean squared aggregate error in all training samples. Secondly, it is a supervised training since the input information vector can be compared with a desired output or target vector. Therefore, it can be implemented on parallel neural networks for BDTI forecasting. Next section elaborates on the aspects of backpropagation training algorithms and WNN.

3.3. The WNN Model

There are two types of WNN. In first type of WNN, input

data are merely preliminary treated through the wavelet function, and thus the wavelet and the neural network processing are conducted separately. The second type of WNN replaces the neurons by wavelets, in this situation, weights and thresholds of the neural network are modified by the dilation and translation of wavelets [23]. Theoretically, the three-layer neural networks structure is adequate to solve any arbitrary function approximation [47]. For the BDTI forecasting the latter WNN type is used and a three-layer WNN model is constructed (**Figure 4**).

In a first step the weights, thresholds of neural network and the dilation and translation parameters of wavelet function are initialized and the dilation parameter as a_j, the translation parameter as b_j, neural network connections weights as w_{jk} and w_{ij}, the learning rate as η, and a momentum factor as λ and initial values for these parameters are defined. The sampling data calculator set as 1. The learning rate determines how fast the network can learn, that is to say how much the link weights and node biases can be modified based on the change in direction and change rate. A momentum rate allows the network to potentially avoid through local minima.

The second step includes training input data and the corresponding expected output d_p^i, and calculating the outputs of hidden and output layer; where the output of the hidden layer:

$$O_j^p = h\left(\frac{\sum_{k=1}^{K} w_{jk} x_k^p}{a_j}\right) \qquad (19)$$

and where the output of the output layer:

$$y_i^p = h\left(\sum_{i=1}^{N} \omega_{ij} O_j^p\right) \qquad (20)$$

In the Equations (19) and (20), x_k^p is the input of the output layer, O_j^p is the output of the hidden layer, $h(\cdot)$ is the Morlet wavelet (see (6) and (7)).

The third step calculates the error and gradient vectors using the following equations:

$$E = \frac{1}{2}\sum_{i=1}^{N}\left(d_i^p - y_i^p\right)^2 \qquad (21)$$

$$\delta_i^j = \frac{\partial E_i^p}{\partial \omega_{ij}} = \left(d_i^p - y_i^p\right)y_i^p\left(1 - y_i^p\right) \qquad (22)$$

$$\delta_{jk} = \frac{\partial E_i^p}{\partial \omega_{jk}} = \sum_{i=1}^{N}\left(\delta_{ij}\omega_{ij}\right)\frac{\partial O_j^p}{\partial a_j}x_k^p \qquad (23)$$

$$\delta_{aj} = \frac{\partial E_i^p}{\partial \alpha_j} = \sum_{i=1}^{N}\left(\delta_{ij}\omega_{ij}\right)\frac{\partial O_j^p}{\partial \alpha_j} \qquad (24)$$

$$\delta_{bj} = \frac{\partial E_i^p}{\partial b_k} = \sum_{i=1}^{N}\left(\delta_{ij}\omega_{ij}\right)\frac{\partial O_j^p}{\partial b_j} \qquad (25)$$

In the fourth step the errors will be back propagated through the network, and the weights are adjusted correspondingly as given below:

$$\omega_{ij}^{new} = \omega_{ij}^{old} + \eta\sum_{k=1}^{p}\delta_{ij} + \lambda\Delta_1\omega_{ij}^{old} \qquad (26)$$

$$\omega_{jk}^{new} = \omega_{jk}^{old} + \eta\sum_{k=1}^{p}\delta_{kj} + \lambda\Delta_1\omega_{kj}^{old} \qquad (27)$$

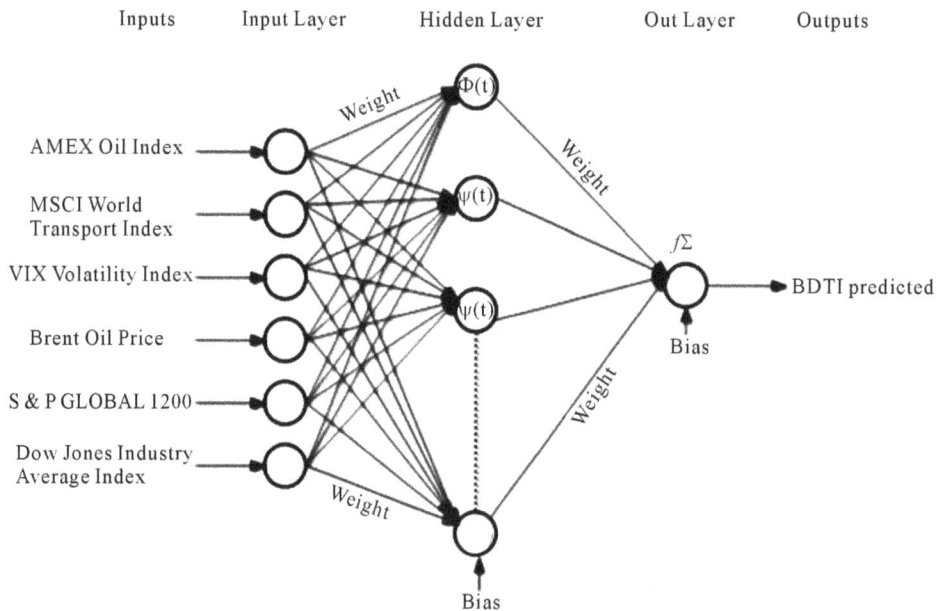

Figure 4. Forecasting BDTI by WNN Model.

Source: Authors.

$$a_j^{\text{new}} = a_j^{\text{old}} + \eta \sum_{k=1}^{p} \delta_{aj} + \lambda \Delta_1 a_j^{\text{old}} \qquad (28)$$

$$b_j^{\text{new}} = b_j^{\text{old}} + \eta \sum_{k=1}^{p} \delta_{bj} + \lambda \Delta_1 b_j^{\text{old}} \qquad (29)$$

The fifth steps inputs the next sample date, so $p = p+1$. If the expected error $E < \varepsilon$, where ε is the predefined accuracy specification value $(\varepsilon > 0)$, the training of the WNN will be ceased, otherwise, the p calculator will be reset to 1 and the training will restart from step 2.

3.4. ARIMA Model for Forecasting Comparison

ARIMA (p,d,q) time series is a record of a random variable realized from a stochastic process [48]. The model transforms the non-stationary time series into stationary ones by differencing and logging of original data. In the ARIMA model d, p, q are non-negative integers, and refer to autoregressive terms, non-seasonal differences and moving averages respectively. The mathematic expression of ARIMA is as follows:

An ARIMA (p,q) model is given by:

$$\left(1 - \sum_{i=1}^{p} \alpha_i L^i\right) X_t = \left(1 + \sum_{i=1}^{p} \theta_i L^i\right) \varepsilon_t \qquad (30)$$

where X_t is the time series data, and t is time. L is the lag operator and α_i is the autoregressive term. ε_t refers to the error term.

If the linear polynomial $\left(1 - \sum_{i=1}^{p} \alpha_i L^i\right)$ has a unitary root of multiplicity d, then

$$\left(1 - \sum_{i=1}^{p} \alpha_i L^i\right) = \left(1 + \sum_{i=1}^{p-d} \varnothing_i L^i\right) \varepsilon_t$$

thus an ARIMA (p, d, q) procedure expresses these polynomial factorization characteristics, as given by [49]:

$$\left(1 - \sum_{i=1}^{p} \varnothing_i L^i\right)(1 - L)^d X_t = \left(1 + \sum_{i=1}^{q} \theta_i L^i\right) \varepsilon_t \qquad (31)$$

The means absolute error (MAE), root mean square deviation (RMSD) and mean absolute percentage error (MAPE) are used to examine the forecasting accuracy and are defined as:

$$MAE = \frac{1}{n} \sum_{t=1}^{n} |x_t - \hat{x}_t| \qquad (32)$$

where $\hat{x}_t$ is the predicted value.

$$RMSD = \sqrt{MSE(\hat{x}_t)} = \sqrt{\frac{1}{n} \sum_{t=1}^{n} (x_t - \hat{x}_t)^2} \qquad (33)$$

$$MAPE = \frac{1}{n} \sum_{t=1}^{n} \left|\frac{A_t - F_t}{A_t}\right| \qquad (34)$$

4. Modeling Results

4.1. Data

The data includes the daily BDTI value between August 3rd 1998 and February 25th 2011, which is equal to 3147 trading days. The data were supplied by Baltic Exchange London, Ltd.

For the same period, the data of the Brent Oil Price Index, CBOE SPX Volatility Index, and S&P Global 1200 Index were collected from Bloomberg Financial Laboratory; the Amex Oil Index and Dow Jones Industry Average Index data were collected through Yahoo Finance [50], with quotation codes ^XOI and ^DJI.

Following Ripley [51], the data is cataloged into three sets: the training set, validation set and test set. The training set refers to a set of example data used for learning, in order to identify the weights. The validation set is a set of examples used to select the number of hidden nodes in a neural network. This set is combined with test data since the hidden nodes will be set through experience. The test set is related to assess the performance of networks. Data sets for 4 weeks (20 days [52]), 12 weeks (60 days), 24 weeks (120 days) and 48 weeks (240 days) are used to examine the performance of WNN and ARIMA.

The data were cleaned in order to make these data valid for the same point in time. The time points of the other six indices data are strictly in line with those of the BDTI, gap values for certain trading days for the six indices are filled using the value of its nearest previous trading day, and redundant values are deleted.

All the data are normalized for WNN to recognize and process them. This is because every Morlet wavelet node's signal is restricted to a 0 to 1 range and training targets therefore should be normalized between 0 and 1 (see (2), (4), (7), (12) and (15)). The data were normalized as follows:

$$\overline{x}_t = \frac{x_t - x_{\min}}{x_{\max} - x_{\min}} \qquad (35)$$

4.2. Forecasting BDTI Using ARIMA

While the WNN has no pre-request in properties of input series data, the ARIMA model can process only stationary series. Therefore, it is necessary to examine whether the BDTI is stationary through auto-correlation analysis using SPSS 19 software.

The autocorrelation decreases as the time lag increases (see Appendix, **Table A1**), confirming that the BDTI series is a non-stationary time series. For ARIMA forecasting, the BDTI data series needs to be transformed into a stationary time series by order difference. The Partial ACF approaches values close to zero after three lags the value of coefficient approaches zero, therefore, the p value is set as 3 (**Figure A1**).

According to experience, the d, trend value, is usually set by 0, 1, or 2. 0 means no trend exists in the time series. When time series are differenced by one, $d = 1$, the linear trend is removed. When $d = 2$, both linear and seasonal trends are removed. Usually, d values of 1 or 2 are adequate to make the mean stationary [48]. Here d *is set as* 2 for the ARIMA model.

The moving average q is often used to smooth short-term fluctuations, thus highlighting longer-term trends [48]. According to the BDTI research above, no significant long-term trend has been found, therefore, p is set as 0 for the ARIMA model.

The ARIMA model is set at (3,2,0) for a 120 days forecast, the ARIMA model is only used as a reference model, therefore, the other possible combinations of value (p, d, q) for ARIMA model may be detected by export modeler function of SPSS are ignored.

The prediction accuracy of the ARIMA model seems to be acceptable for periods of 40 trading days. The MAPE value is relatively low which means average errors between actual and forecasting values are within $\pm 8\%$ in average (**Figure A2**, **Table 3**).

When increasing the forecasting period to 60 days the prediction accuracy of ARIMA model decreases significantly. Given the linear nature of ARIMA, it can hardly catch any fluctuation in this period.

The forecasting results for 120 days are not acceptable, since the errors terms are too significant (**Figure A4**, **Table A3**), which indicates the poor forecasting accuracy of ARIMA model in mid or long term (more than three months) non-linear forecast. In **Figure A3**, it can be observed that ARIMA forecasts a downturn of the BDTI, however, in reality the BDTI experienced a significant increase during that period.

The results for 240 days forecasts (**Figure A5**, **Table A3**) reveal the short coming of linear forecasting when forecasting long term non-linear time series.

The results show that the ARIMA model has relatively low predictive accuracy in forecasting BDTI, especially in longer terms. The only acceptable result is for a 20 days ahead forecasting situation, with an 8% average forecasting error, and this result is in line with El-Shaarawi & Piegorsch [53].

In addition, the forecasting results by ARIMA are presented in form of linear regression, which can hardly match the non-linear, non-stationary, and volatile nature of BDTI. In 120 and 240 days ahead forecasting, the ARIMA model offers obviously contrary trends of the BDTI contradicting the actual development of the index, the results thus may lead to miserable decision making errors.

4.3. BDTI Forecasting Using WNN

The structure of the WNN model for BDTI forecasting is presented in **Figure 4**. Six indices are employed that may externally affect the BDTI development, in addition one day BDTI time delay is used as input, to forecast the BDTI. These variables will interact in the "black box" of WNN, and BDTI forecasting value will be used as an output.

The number of hidden notes are assigned by experiences and adjusted according to the time frame of forecasting. The weights are initialed by default random value of the network. The time delay is defined at 1, which means the BDTI index can be influenced by its previous day's closing value. The learning rate is set at 0.01. The number of iterations for training is defined at a maximum of 800. The margin of error tolerance for training is set at less than 0.001.

Since the WNN forecasting model contains much information, hidden characteristics of the BDTI need to be made more obvious for the six variables to catch. Figuratively speaking, the authors intend to make specific trees become more visible in the forest; therefore, the DWT philosophy and wavelet reconstruction methods are employed. The wavelet transform is used to decompose the time series into varying scales of temporal resolution [54]. The latter provides a sensible decomposition of the data so that the underlying temporal structures of the original time series become more traceable.

The BDTI data is decomposed and then a new BDTI series is reconstructed. The noise and redundant information in the original BDTI data series is filtrated in this process (**Figure 5**).

The results of WNN for 20 trading days forecasting show that the WNN model in does not well catch the trend of the actual movement of BDTI (**Figure 6**). However, the WNN model converges really fast, after around 80 training iterations (**Figure A6**). The rapid convergence demonstrates the efficiency of wavelet neuron algorisms in neural networks.

The relative errors indicate the difference between the actual and forecasting values of BDTI. The relative error transforms errors into percentage expressions. The forecasting errors stay within a range of $\pm 100\%$ (**Figure A7**). The relative error is between 20% and -10% (**Figure 7**).

In 60 days ahead forecasting, the WNN captured the upward trend in the first 20 days, but it underestimated the extent of the drop in the BDTI in the following days (**Figure 8**). The network started to converge after 100 trainings (**Figure A8**).

In the 120 days ahead situation the WNN is able to approximate major movements of the BDTI, however, after 110 days, the forecasting values surged much faster than the actual values (**Figure 10**).

In long term forecasting, the non-linear movements of the BDTI are well approximated by the WNN the model. The WNN model captured the three peaks of the BDTI,

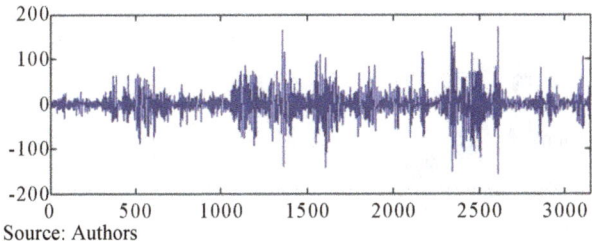

Source: Authors

Figure 5. Filtrated noise—filtrated high frequency components after wavelet reconstruction.

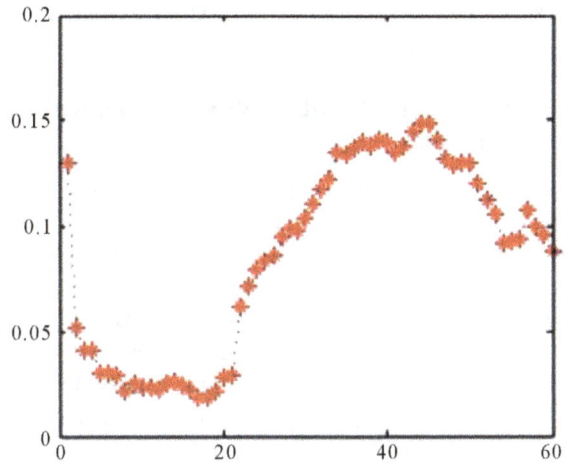

Source: Authors

Figure 6. 20 Days ahead BDTI forecasting by WNN (January 31st 2011-February 25th 2011).

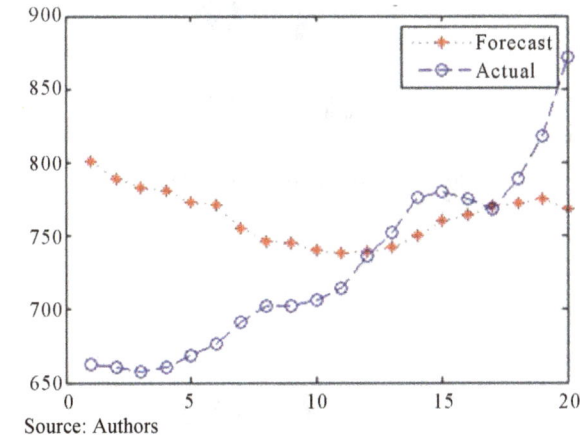

Source: Authors

Figure 7. Relative error, 20 days ahead.

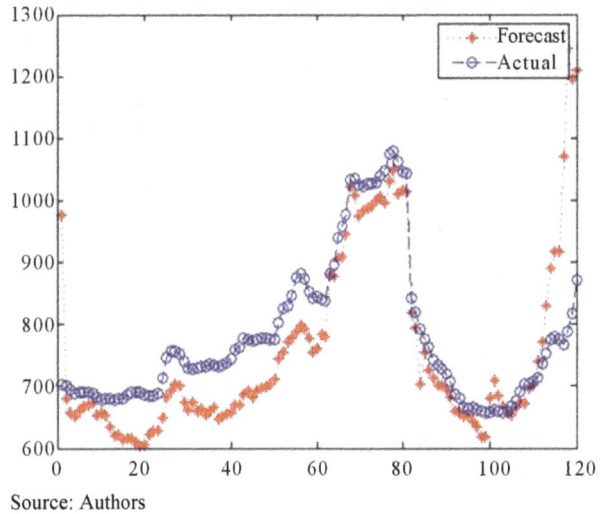

Source: Authors

Figure 8. 60 Days ahead BDTI forecasting by WNN (November 26th 2010-February 25th 2011).

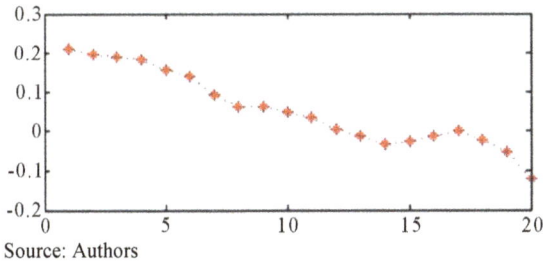

Source: Authors

Figure 9. Relative error, 60 days ahead.

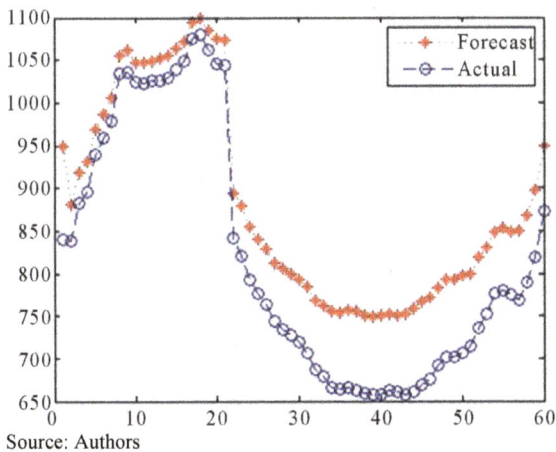

Source: Authors

Figure 10. 120 Days ahead BDTI forecasting by WNN (September 3rd 2010-February 25th 2011).

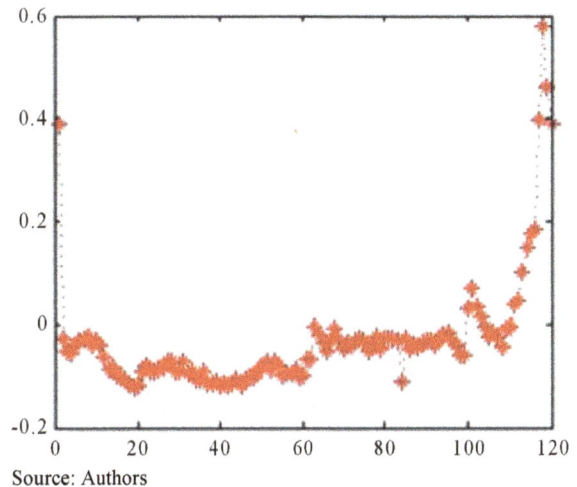

Source: Authors

Figure 11. Relative error, 120 days ahead.

however, it failed to precisely predict the troughs in their full extent (**Figure 12**).

4.4. Forecasting Results of WNN in Challenging Situations

Based on the significant results in comparison to the ARIMA model this section tests the performance of the WNN forecasting model in two very specific and "challenging" time periods, see **Figure 14**.

In September 2004 to February 2005, BDTI surged to its historical high within three months, and then plummeted in the following three months. **Figure 15** shows that the WNN model was able to appropriate the movement of the BDTI during this time period.

Although the WNN model did not adequately predict the climax of BDTI, it provides useful information about the movement trends of BDTI (**Figure 16**). The current financial crisis started at the end of 2007, and the world economies have still not fully recovered from the crisis. The tanker shipping freight rates and BDTI fluctuated significantly during the global financial crisis.

From **Figure 17**, it can be seen that the forecasted movements of the BDTI by WNN fluctuated more vocatively than the actual BDTI during this one year time period. The forecasting error maximum can be 100% more than actual value (**Figure 18**). During the first 100 trading days the WNN presented a relatively good performance, but in later period the WNN seemed to be more sensitive in forecasting, and volatilization became significant.

5. Conclusions

Concluding, there seems to be no significant difference

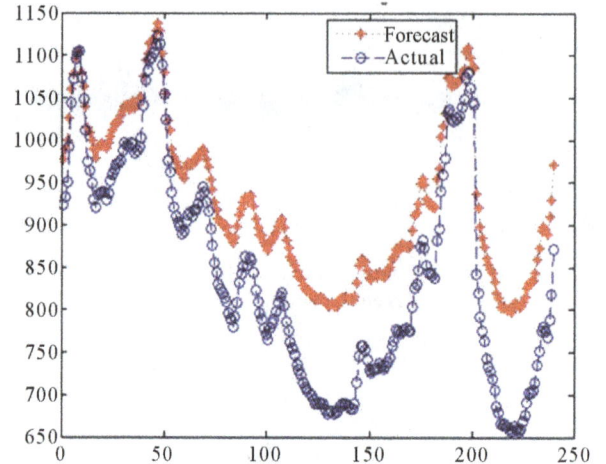
Source: Authors

Figure 12. 240 days ahead by WNN (March 12th 2010-February 25th 2011).

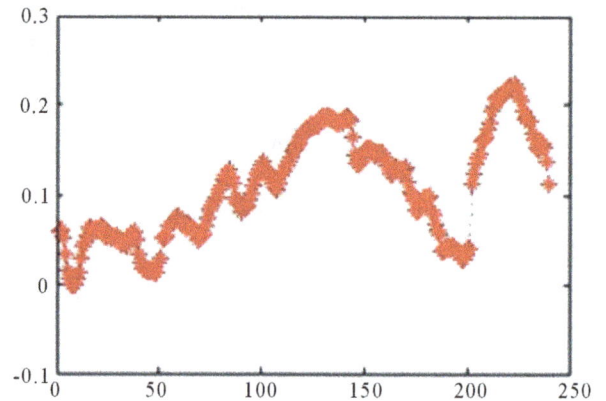
Source: Authors

Figure 13. Relative error, 120 days ahead.

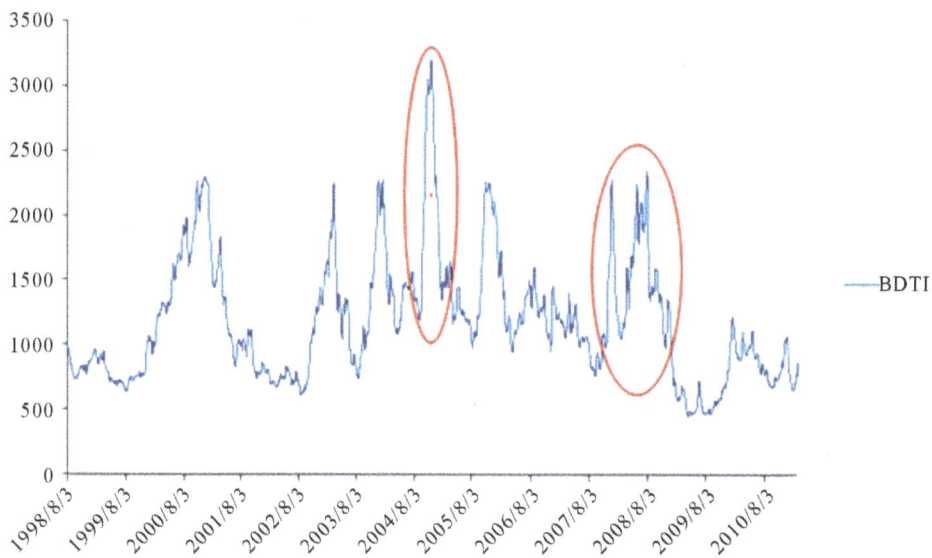
Source: Authors

Figure 14. BDTI and "challenging situations".

in performance between WNN and ARIMA model in BDTI short term forecasting. However, for longer periods, the WNN model shows some superiority over ARIMA, offering reasonable non-linear forecasts about the BDTI movements. The forecasting accuracy of WNN decreases as forecasting times increase (**Tables A3-A5**). Although the WNN model did not perfectly fulfill the forecasting tasks in "challenging situations", it was able to predict and capture of useful information about trends and movements of the BDTI index.

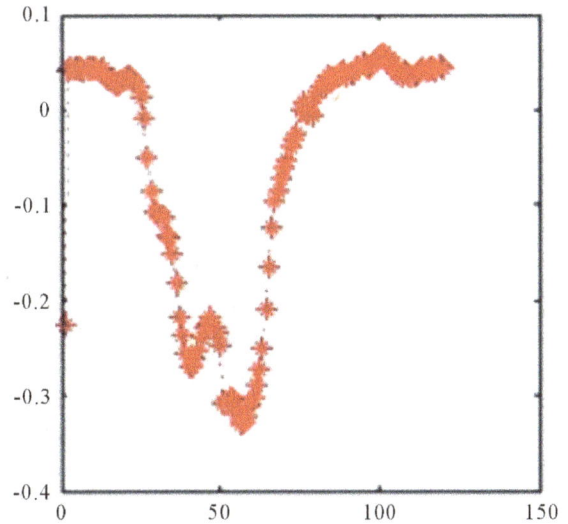

Source: Authors

Figure 15. "Big wave" forecasting by WNN (September 2004-February 2005).

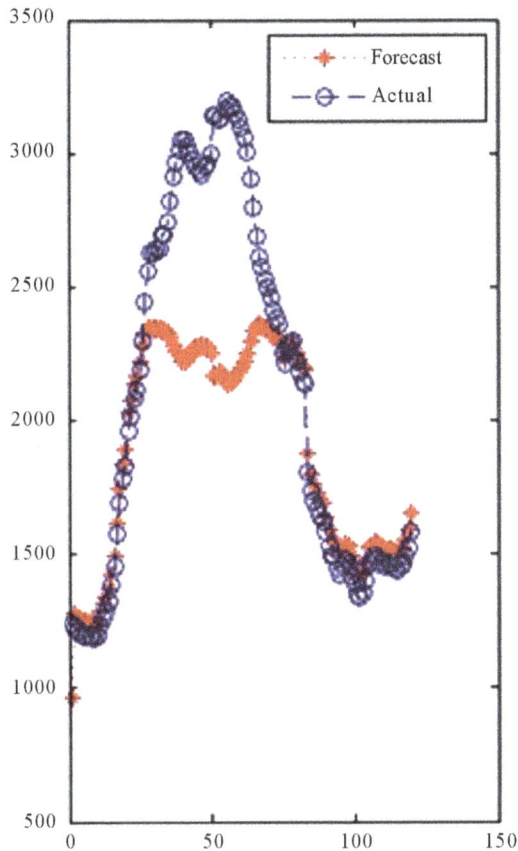

Source: Authors

Figure 16. Relative error, "big wave".

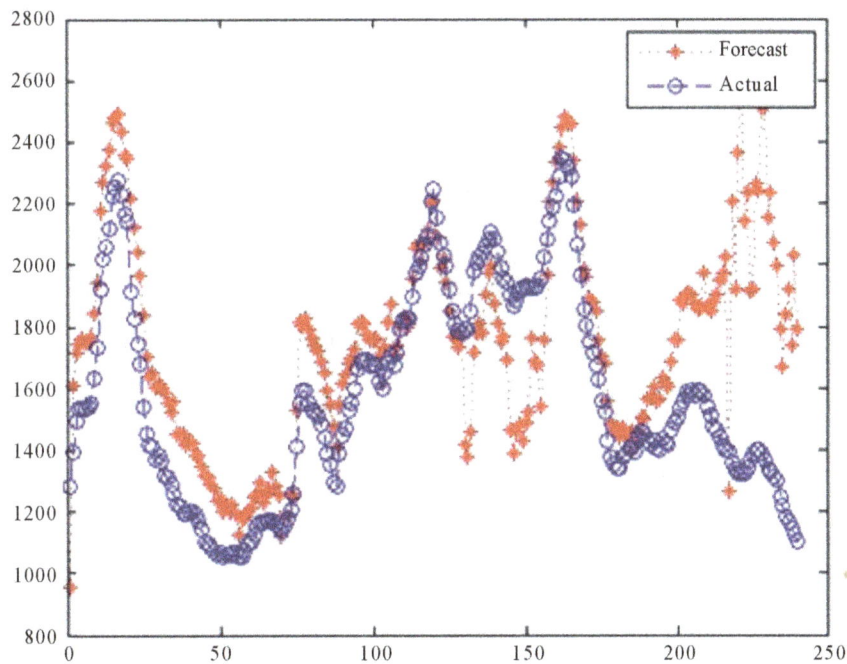

Source: Authors

Figure 17. BDTI forecasting during financial crisis (November 2007-November 2008).

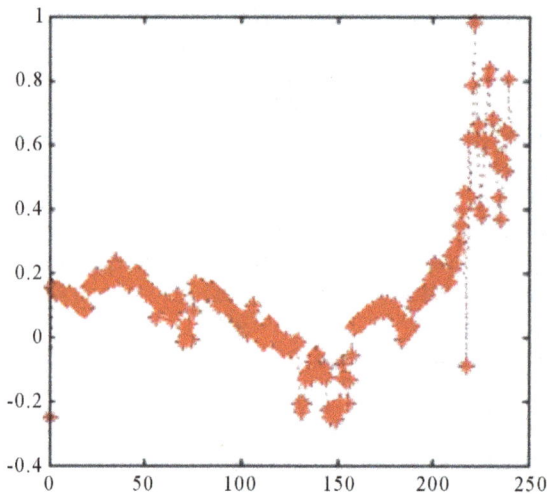

Figure 18. Relative error, "big wave".

This paper illustrates that artificial intelligent methods can constitute powerful problem solving tools in engineering and natural science, but also have great application potential in shipping research. Traditional stochastic and econometric explanation models are significantly different from machine learning and artificial intelligent methods in nature. Generally, comparing with traditional stochastic and econometric explanation methods, machine learning methods regard to the nature of data mechanism as unknown and complex and allow models to learn from and adapt to their circumstances. Wavelet neural network is a type of hybrid neural network. WNN combines the time frequency localization properties and adaptive learning nature of neural networks thus making it a potential tool for forecasting in complex circumstance.

Examining BDTI forecasting performance, the authors identify that traditional ARIMA forecasting method is weak in forecasting this non-linear and highly fluctuating shipping index, especially for longer time periods. In contrast, the WNN model adopts the artificial intelligent algorisms and combines the wavelet and neural networks, with adequate and appropriate inputs, network design and training, WNN can be a very effective method in forecasting the non-linear and non-stationary shipping index, such as BDTI.

Since WNN can be applied to forecast BDTI, this method can probably also be applied for other analysis of other shipping sectors. WNN offers a good prediction of future trends of the BDTI, which can be used as a relevant tool in market intelligence, business negotiation, decision making and financial budgeting. Shipping companies require rational decision making and market knowledge to optimize their fleet to balance demand and supply in the whole tanker market, which is of particular difficulty and importance in volatile markets.

One short coming in the WNN forecasting model is that the initial weights for neural network are randomly defined by computing software. If the initial weights are far from suitable values, then the network may have to iterate many more times than usual and thus the model may have difficulty in converging. This may lead to poor predicting accuracy and unstable forecast performance. However, this can be partly overcome by applying genetic algorithm optimization methods [55] and/or particle swarm optimization [56] algorithm.

In addition, the number of hidden nodes, training times and lags are assigned according to researcher's experience. The selection of appropriate number of hidden nodes is a very difficult and tricky task for the design of WNN

REFERENCES

[1] D. Hawdon, "Tanker Freight Rates in the Short and Long Run," *Applied Economics*, Vol. 10, No. 3, 1978, pp. 203-218.

[2] D. Glen, M. Owen and R. Meer, "Spot and Time Charter Rates for Tankers, 1970-77," *Journal of Transport Economics and Policy*, Vol. 15, No. 1, 1981, pp. 45-58

[3] M. Beenstock and A. Vergottis, "An Econometric Model of the World Tanker Market," *Journal of Transport Economics and Policy*, Vol. 23, No. 3, 1989, pp. 263-280.

[4] A. Perakis and W. Bremer, "An Operational Tanker Scheduling Optimization System: Background, Current Practice and Model Formulation," *Maritime Policy and Management*, Vol. 19, No. 3, 1992, pp. 177-187.

[5] R. Adland and K. Cullinane, "The Non-Linear Dynamics of Spot Freight Rates in Tanker Markets," *Transportation Research Part E: Logistics and Transportation Review*, Vol. 42, No. 3, 2006, pp. 211-224.

[6] R. Laulajainen, "Operative Strategy in Tanker (Dirty) Shipping," *Maritime Policy and Management*, Vol. 35, No. 3, 2008, pp. 315-341.

[7] T. Angelidis and S. G. Skiadopoulos, "Measuring the Market Risk of Freight Rates: A Value-at-Risk Approach," *International Journal of Theoretical and Applied Finance*, Vol. 11, No. 5, 2008, pp. 447-469.

[8] M. Stopford, "Maritime Economics," 3rd Edition, Routledge, New York, 2009.

[9] Worldscale Association, "Introduction to Worldscale Freight Rate Schedules," 2011. http://www.worldscale.co.uk/company%5Ccompany.htm

[10] R. Batchelor, A. Alizadeh and I. Visvikis, "Forecasting Spot and Forward Prices in the International Freight Market," *International Journal of Forecasting*, Vol. 23, No. 1, 2007, pp. 107-114.

[11] K. Cullinane, "A Short-Term Adaptive Forecasting Model

for BIFFEX Speculation: A Box—Jenkins Approach," *Maritime Policy and Management: The Flagship Journal of International Shipping and Port Research*, Vol. 19, No. 2, 1992, pp. 91-114.

[12] M. Kavussanos, "Price Risk Modelling of Different Sized Vessels in Tanker Industry Using Autoregressive Conditional Heteroscedasticity GARCH Models," *Transportation Research Part E: Logistics and Transportation Review*, Vol. 32, No. 2, 1996, pp. 161-176.

[13] F. Jonnala, S. Fuller and D. Bessler, "A GARCH Approach to Modelling Ocean Grain Freight Rates," *International Journal of Maritime Economics*, Vol. 4, No. 2, 2002, pp. 103-125.

[14] A. W. Veenstra and P. H. Franses, "A Co-Integration Approach to Forecasting Freight Rates in the Dry Bulk Shipping Sector," *Transportation Research Part A: Policy & Practice*, Vol. 31, No. 6, 1997, pp. 447-458.

[15] J. Tvedt, "Shipping Market Models and the Specification of Freight Rate Processes," *Maritime Economics and Logistics*, Vol. 5, No. 4, 2003, pp. 327-346.

[16] A. M. Goulielmos and M. Psifia, "A Study of Trip and Time Charter Freight Rate Indices: 1968-2003," *Maritime Policy and Management*, Vol. 34, No. 1, 2007, pp. 55-67.

[17] S. Sødal, S. Koekebakkera and R. Adland, "Market Switching in Shipping—A Real Option Model Applied to the Valuation of Combination Carriers," *Review of Financial Economics*, Vol. 17, No. 3, 2008, pp. 183-203.

[18] T. Koopmans, "Tanker Freight Rates and Tankship Building," *The Economic Journal*, Vol. 49, No. 196, 1939, pp. 760-762.

[19] Z. S. Zannetos, "The Theory of Oil Tankship Rates: An Economic Analysis of Tankship Operations," MIT—Massachusetts Institute of Technology, Cambridge, 1964, pp. 60-64.

[20] J. J. Evans, "An Analysis of Efficiency of the Bulk Shipping Markets," *Maritime Policy and Management: The Flagship Journal of International Shipping and Port Research*, Vol. 21, No. 4, 1994, pp. 311-329.

[21] S. Reutlinger, "Analysis of a Dynamic Model, with Particular Emphasis on Long-Run Projections," *Journal of Farm Economics*, Vol. 48, No. 1, 1966, pp. 88-106.

[22] R. Adland and S. P. Strandenes, "A Discrete-Time Stochastic Partial Equilibrium Model of the Spot Freight Market," *Journal of Transport Economics and Policy (JTEP)*, Vol. 41, No. 2, 2007, pp. 189-218.

[23] Q. Zhang and A. Benveniste, "Wavelet Networks," *IEEE Transactions on Neural Networks*, Vol. 3, No. 6, 1992, pp. 889-898.

[24] Z. Wang and Y. Tan, "Research of Wavelet Neural Network Based Host Intrusion Detection Systems," *Proceedings of the International Computer Conference 2006 on Wavelet Active*, Chongqing, 29-31 August 2006, pp 1007-1012.

[25] K. K. Minu, M. C. Lineesh and C. J. John, "Wavelet Neural Networks for Nonlinear Time Series Analysis," *Applied Mathematical Sciences*, Vol. 4, No. 50, 2010, pp. 2485-2495.

[26] K. G. Goulias, "Transport Science and Technology," Elsevier Ltd., Amsterdam, 2007.

[27] The Baltic Exchange, "Manual for Panelists—A Guide to Freight Reporting and Index Production," Unpublished Manuscript, The Baltic Exchange, London, 2011.

[28] D. B. Percival and A. T. Walden "Wavelet Methods for Time Series Analysis," Cambridge University Press, Cambridge, 2006.

[29] A. Graps, "An Introduction to Wavelets," *IEEE Computational Sciences and Engineering*, Vol. 2, No. 2, 1995, pp. 50-61.

[30] K. P. Soman, K. I. Ramachandran and N. G. Resmi, "Insight into Wavelets," 3rd Edition, PHI Learning Pvt. Ltd., Coimbatore, 2010.

[31] L. Debnath, "Wavelet Transforms and Their Applications," Springer, Boston, 2002.

[32] J. Lewalle, "Wavelets without Lemmas on Applications of Continuous Waveletsto Data Analysis," Syracuse University, Syracuse, 1998. http://www.ecs.syr.edu/faculty/lewalle/papers/vki1.pdf

[33] D. Veitch, "Wavelet Neural Networks and Their Application in the Study of Dynamical Systems," *Networks*, Vol. 1, No. 8, 2005, pp. 313-320.

[34] I. Daubechies, "The Wavelet Transform, Time-frequency Localization and Signal Analysis," *IEEE Transactions on Information Theory Society*, Vol. 36, No. 5, 1990, pp. 961-1005.

[35] G. Dreyfus, "Neural Networks: Methodology and Applications," Springer-Verlag, Berlin, Heidelberg, New York, 2005.

[36] A. Abraham, "Artificial Neural Networks. Handbook of Measuring System Design," John Wiley and Sons Ltd., Hoboken, 2005.

[37] D. P. Mandic and J. A. Chambers, "Recurrent Neural Networks for Prediction: Learning, Algorithms, Architectures and Stability," John Wiley and Sons Ltd., Hoboken, 2001.

[38] M. Casey, "The Dynamics of Discrete-Time Computation, with Application to Recurrent Neural Networks and Finite State Machine Extraction," *Neural Computation*, Vol. 8, No. 6, 1996, pp. 1135-1178.

[39] G. Dematos, M. S. Boyd, B. Kermanshahi, N. Kohzadi and I. Kaastra, "Feedforward versus Recurrent Neural Networks for Forecasting Monthly Japanese Yen Exchange Rates," *Asia-Pacific Financial Markets*, Vol. 3, No. 1, 1996, pp. 59-75.

[40] B. A. Pearlmutter, "Dynamic Recurrent Neural Networks," School of Computer Science Carnegie Mellon University, Defense Advanced Research Projects Agency, Information Science and Technology Office, 1990. http://www.bcl.hamilton.ie/~barak/papers/CMU-CS-90-196.pdf

[41] K. Cannons and V. Cheung, "An Introduction to Neural Networks," Iowa State University, Ames, 2002. http://www2.econ.iastate.edu/tesfatsi/NeuralNetworks.Ch eungCannonNotes.pdf

[42] R. Hecht-Nielsen, "Theory of the Backpropagation Neural Network," *International Joint Conference on Neural Networks (IJCNN)*, Vol. 1, Washington DC, 18-22 June 1989, pp. 593-605.

[43] P. J. Werbos, "Backpropagation Through Time: What It Does and How To Do It," *Proceedings of the IEEE*, Vol. 78, No. 10, 1990, pp. 1550-1560.

[44] J. Han and M. Kamber, "Data Mining: Concepts and Techniques," Morgan Kaufmann, San Francisco, 2006.

[45] T. Hastie, R. Tibshirani and J. Friedman, "The Elements of Statistical Learning: Data Mining, Inference, and Prediction," Springer, New York, 2009.

[46] P. K. Simpson, "Neural Networks Theory, Technology, and Applications, Institute of Electrical and Electronics Engineers, Technical Activities Board," University of Michigan, Ann Arbor, 1996.

[47] R. M. Golden, "Mathematical Methods for Neural Network Analysis and Design," The MIT Press, Cambridge, 1996.

[48] K. M. Vu, "The ARIMA and VARIMA Time Series: Their Modelings, Analyses and Applications," AuLac Technologies Inc., Ottawa, 2007.

[49] T. C. Mills, "Time Series Techniques for Economists," Cambridge University Press, Cambridge, 1990.

[50] Yahoo, "Get Quotes, Historical Prices," 2012. http://finance.yahoo.com/

[51] B. D. Ripley, "Pattern Recognition and Neural Networks," Cambridge University Press, Cambridge, 1996.

[52] The 20 trading days is based on the following assumption: 1 week = 5 trading days, 1 month = 4 week.

[53] A. H. El-Shaarawi and W. W. Piegorsch, "Encyclopedia of Environmetrics," Wiley, Hoboken, 2002.

[54] A. Aussem and F. Murtagh, "Combining Neural Network Forecasts on Wavelet-Transformed Time Series," *Connection Science*, Vol. 9, No. 1, 1997, pp. 113-122.

[55] J. J. Merelo, M. Patón, A. Cañas, A. Prieto and F. Morán, "Optimization of a Competitive Learning Neural Network by Genetic Algorithms," *New Trends in Neural Computation*, Vol. 686, 1993, pp. 185-192.

[56] J. Kennedy and R. Eberhart, "Particle Swarm Optimization," *Proceedings of IEEE International Conference on Neural Networks*, Vol. 4, Perth, 27 November-1 December 1995, pp. 1942-1948.

Appendix (Figures)

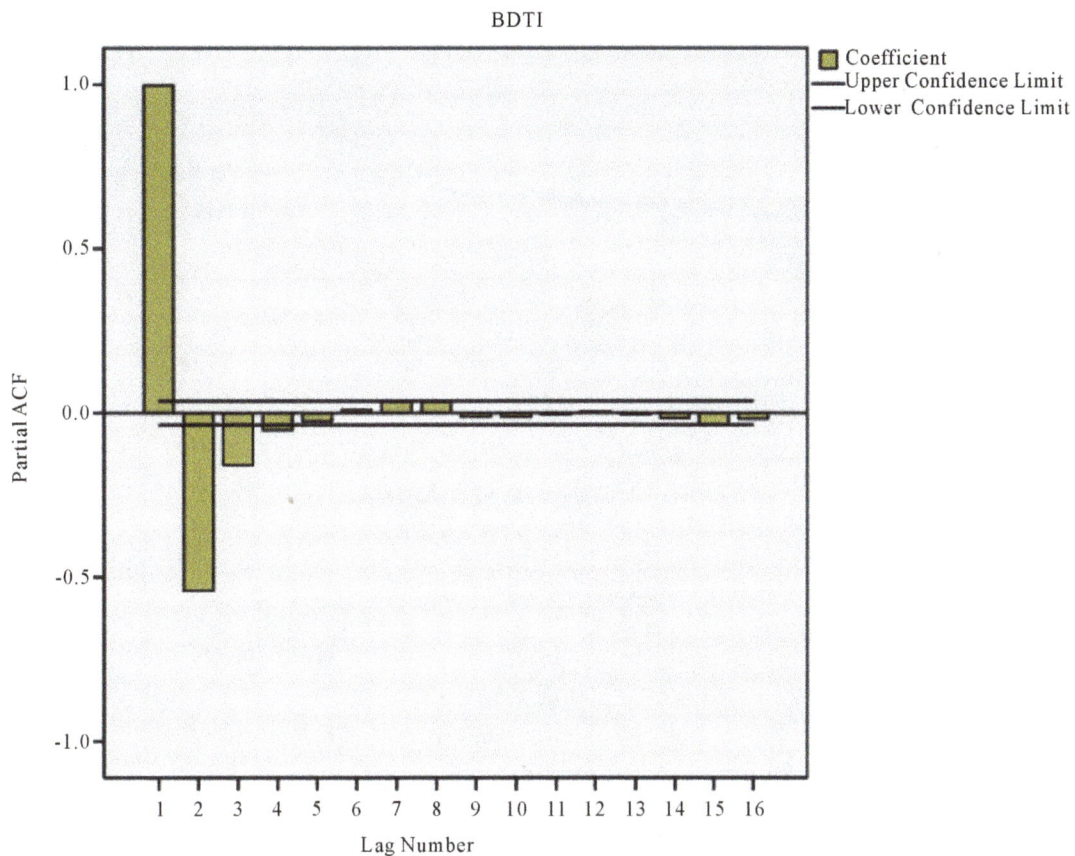

Source: Authors.

Figure A1. Partial ACF of BDTI.

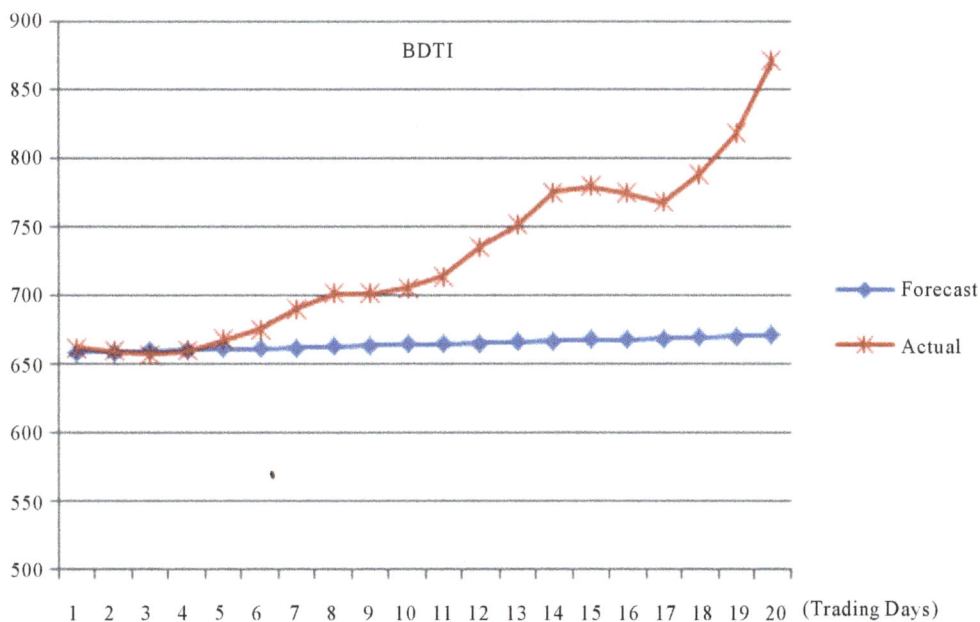

Source: Authors.

Figure A2. 20 days ahead by ARIMA.

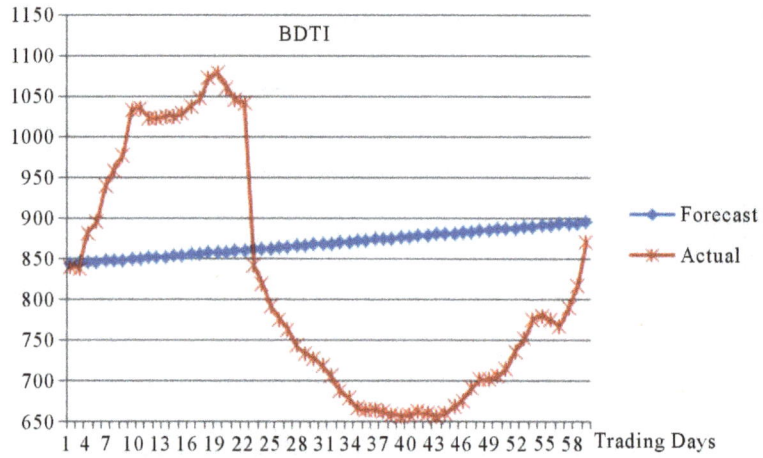

Source: Authors.

Figure A3. 60 days ahead by ARIMA.

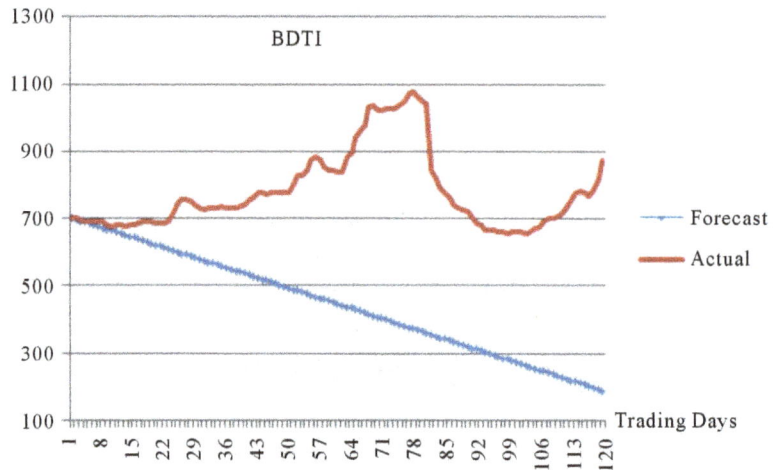

Source Authors.

Figure A4. 120 days ahead by ARIMA.

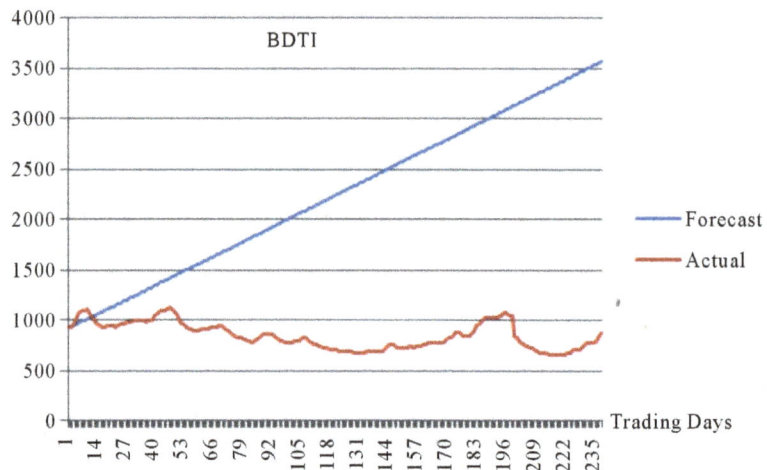

Source: Authors.

Figure A5. 240 days ahead by ARIMA.

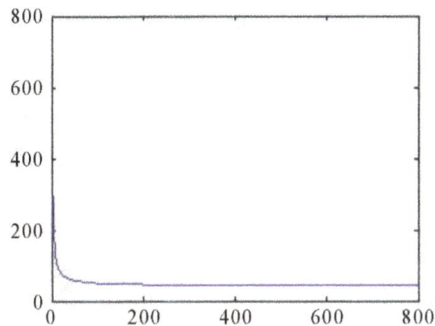
Source: Authors.

Figure A6. Error changes with training times—convergence speed, 20 days ahead.

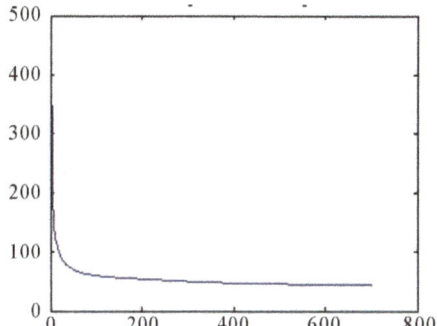
Source: Authors.

Figure A7. Errors, 20 days ahead.

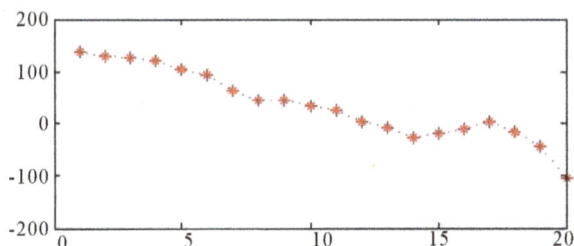
Source: Authors.

Figure A8. Error changes with training times—convergence speed, 60 days ahead.

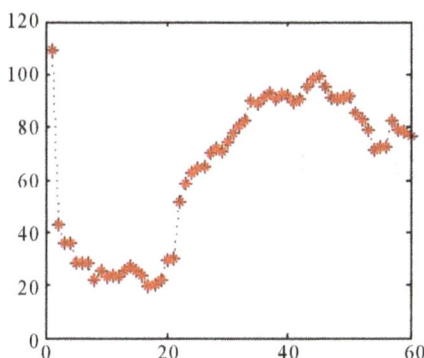
Source: Authors.

Figure A9. Errors, 60 days ahead.

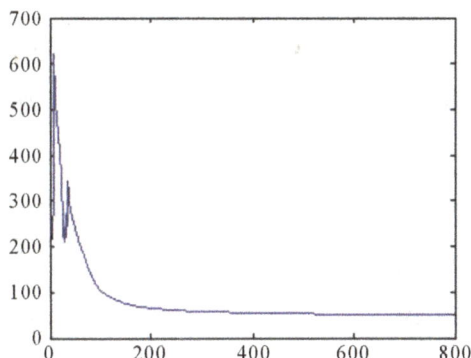
Source: Authors.

Figure A10. Error changes with training times—convergence speed, 120 days ahead.

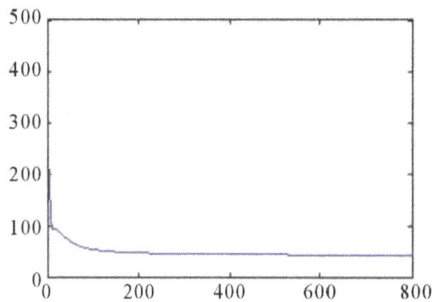
Source: Authors.

Figure A11. Errors, 120 days ahead.

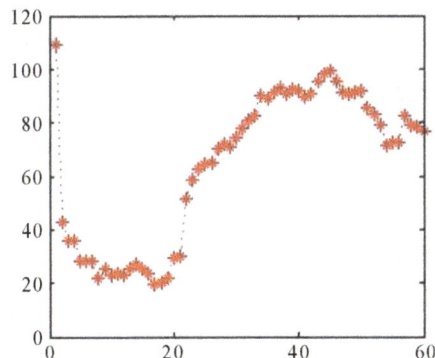
Source: Authors.

Figure A12. Error changes with training times—convergence speed, 240 days ahead.

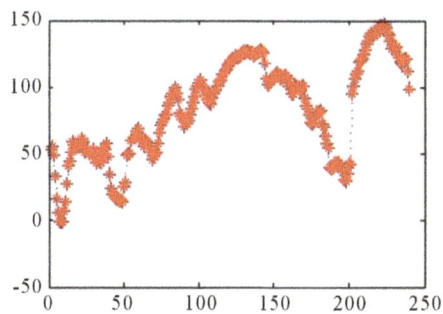
Source: Authors.

Figure A13. Errors, 240 days ahead.

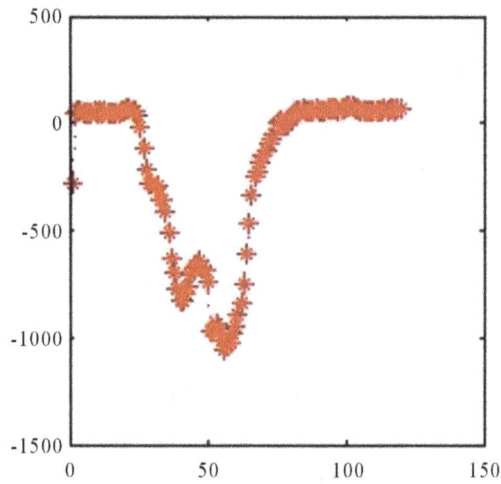

Source: Authors.

Figure A14. Errors, "big wave".

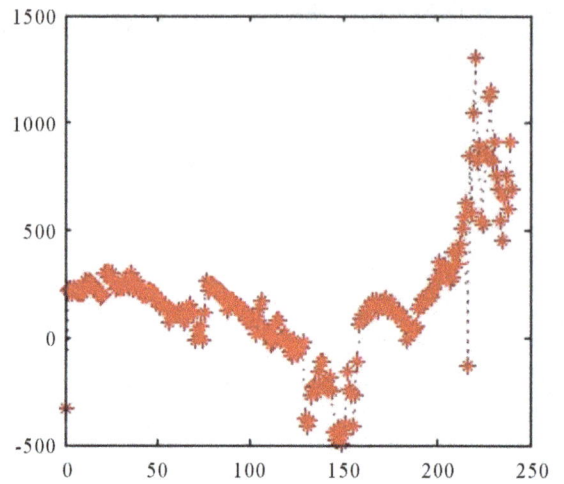

Source: Authors.

Figure A15. Errors, "the crisis".

Appendix (Tables)

Table A1. Autocorrelations of BDTI series.

Lag	Autocorrelation	Std. Error[a]	Box-Ljung Statistic		
			Value	df	Sig.[b]
1	0.998	0.018	3017.257	1	0.000
2	0.994	0.018	6009.101	2	0.000
3	0.987	0.018	8965.531	3	0.000
4	0.980	0.018	11879.276	4	0.000
5	0.972	0.018	14745.009	5	0.000
6	0.963	0.018	17559.256	6	0.000
7	0.954	0.018	20320.352	7	0.000
8	0.944	0.018	23028.042	8	0.000
9	0.935	0.018	25682.434	9	0.000
10	0.925	0.018	28283.883	10	0.000
11	0.916	0.018	30832.950	11	0.000
12	0.906	0.018	33330.448	12	0.000
13	0.897	0.018	35777.179	13	0.000
14	0.887	0.018	38173.726	14	0.000
15	0.878	0.018	40520.129	15	0.000
16	0.868	0.018	42816.286	16	0.000

Source: Authors

Table A2. Autocorrelations of BDTI after 3 differences.

Lag	Autocorrelation	Std. Error[a]	Box-Ljung Statistic		
			Value	df	Sig.[b]
1	−0.598	0.018	1083.278	1	0.000
2	0.095	0.018	1110.649	2	0.000
3	−0.012	0.018	1111.076	3	0.000
4	0.014	0.018	1111.710	4	0.000
5	0.007	0.018	1111.843	5	0.000
6	0.011	0.018	1112.183	6	0.000
7	−0.037	0.018	1116.380	7	0.000
8	0.017	0.018	1117.251	8	0.000
9	0.005	0.018	1117.334	9	0.000
10	0.002	0.018	1117.346	10	0.000
11	−0.004	0.018	1117.384	11	0.000
12	0.004	0.018	1117.424	12	0.000
13	−0.020	0.018	1118.641	13	0.000
14	0.013	0.018	1119.149	14	0.000
15	0.031	0.018	1122.134	15	0.000
16	−0.054	0.018	1130.905	16	0.000

Source: Authors

Table A3. BDTI forecasting performance statistics of ARIMA.

Statistics \ (Days) Ahead	20	60	120	240
MAE	63.63	152.20	332.79	1392.60
RMSD	84.27	164.10	391.17	1625.48
MAPE	0.082	0.19	0.41	1.78

Source: Authors

Table A4. BDTI forecasting performance statistics of WNN.

Statistics \ (Days) Ahead	20	60	120	240
MAE	58.00	63.80	201.31	83.05
RMSD	67.33	28.35	238.00	36.55
MAPE	0.083	0.086	0.09	0.11

Source: Authors

Table A5. BDTI forecasting performance statistics of WNN, in challenging situations.

Statistics \ (Days) Ahead	Big Wave	The Crisis
MAE	63.25	242.83
RMSD	89.61	283.86

Source: Authors

9

Simulation Based Evaluation of Highway Road Scenario between DSRC/802.11p MAC Protocol and STDMA for Vehicle-to-Vehicle Communication

Vaishali D. Khairnar, Srikhant N. Pradhan
Computer Department, Institute of Technology Nirma University, Ahmadabad, India

ABSTRACT

In this paper the DSRC/IEEE 802.11p Medium Access Control (MAC) method of the vehicular communication has been simulated on highway road scenario with periodic broadcast of packets in a vehicle-to-vehicle situation. IEEE 802.11p MAC method is basically based on carrier sense multiple accesses (CSMA) where nodes listen to the wireless channel before sending the packets. If the channel is busy, the vehicle node must defer its access and during high utilization periods this could lead to unbounded delays. This well-known property of CSMA is undesirable for critical communications scenarios. The simulation results reveal that a specific vehicle is forced to drop over 80% of its packets/messages because no channel access was possible before the next message/packet was generated. To overcome this problem, we propose to use self-organizing time division multiple access (STDMA) for real-time data traffic between vehicles. Our initial results indicate that STDMA outperforms CSMA for time-critical traffic safety applications in *ad-hoc* vehicular networks.

Keywords: STDMA; CSMA; MAC; DSRC; IEEE 802.11p; Vehicle-to-Vehicle Etc

1. Introduction

Vehicular Ad-hoc Network area has attracted a lot of attention during the last few years due to the range of new applications enabled by emerging wireless communication technologies. Existing vehicle-to-vehicle safety systems together with new cooperative systems use wireless data communication between vehicles which can potentially decrease the number of accidents on the highway road in India. A tremendous interest in cooperating safety systems for vehicles has been notified through the extensive range of project activities around all over the world. Lane departure warning messages merge assistance and emergency vehicle routing are all examples of applications [1].

These new traffic safety systems implies increased requirements on the wireless communication and the challenge is not only to overcome the behavior of the unpredictable wireless channel but also to cope with rapid network topology changes together with strict timing and reliability requirements. The timing requirements can be deduced from the fact that it is only relevant to communicate about an upcoming dangerous situation before the situation is a fact and perhaps can be avoided (e.g., communicate a probable collision *before* the vehicles are

colliding) [2]. One thing in this respect is how the shared communication channel should be divided in a fair and predictable way among the participating users. This is done through the medium access control (MAC). Much attention within the standardization of vehicle communication systems has been devoted to enhancing the MAC by introducing different quality of service (QoS) classes for data traffic with different priorities [3]. The MAC layer in a traffic safety application is unlikely to need many different service classes or transfer rates. Instead, to guarantee that time-critical communication tasks meet their deadlines, the MAC layer must first of all provide a finite worst case access time to the channel. Once channel access is a fact, different coding strategies, diversity techniques and retransmission schemes can be used to achieve the required correctness and robustness against the impairments of the unpredictable wireless channel. Information that is delivered after the deadline in a critical real-time communication system is not only useless, but implies severe consequences for the traffic safety system. This problem has also been pointed out in [4].

The IEEE 802.11p, also known as Dedicated Short-Range Communication (DSRC), is an upcoming WLAN standard intended for future traffic safety systems. Currently this is the only standard with support for direct

Simulation Based Evaluation of Highway Road Scenario between DSRC/802.11p MAC Protocol and STDMA for Vehicle-to-Vehicle Communication

89

vehicle-to-vehicle (V2V) communication [5]. The original DSRC standards, which are found in Europe, Japan and Korea, are more application-specific standards containing the whole protocol stack with a physical (PHY), a MAC and an application layer. They are intended for hot spot communication such as electronic toll collection systems. The PHY in 802.11p and its capabilities have been treated in several articles [6-8]. The PHY mainly affects the reliability (error probability) of the system; however, if we do not get channel access the benefits of the PHY cannot be exploited.

The 802.11p MAC layer is based on carrier sense multiple accesses (CSMA), where nodes listen to the wireless channel before sending the packet. If the channel is busy, the node must defer its access and during high utilization periods this could lead to unbounded delays. Evaluations and enhancements of CSMA have been proposed in [9-13]. In [9] an investigation of 802.11p is made using real-world application data traffic, collected from 1200 vehicles communicating with each other on a Mumbai-Pune highway road in Maharashtra India. However, this scenario does not show the scalability problem of the MAC protocol. All MAC layer will function well as long as they are not loaded in terms of nodes and data traffic. Hence, a worst case analysis of a vehicular communication system is needed. The performance of 802.11p is evaluated analytically and through simulation in [10]. It is concluded that 802.11p cannot ensure time-critical message dissemination due to the amount of data that needs to be sent. The solution proposed in [10] is to decrease the amount of data traffic. The suggested enhancements of 802.11p include trying to avoid packet collisions by using a polling scheme [11] or by decreasing the amount of data traffic, [12,13]. However, none of these papers clearly point out that the MAC layer lacks the real-time properties required by traffic safety systems. This paper evaluates the requirements on the MAC protocol when used in *ad hoc* vehicle communication systems for low-delay traffic safety applications. Next two different MAC methods are evaluated by means of computer simulations: the MAC method in 802.11p, CSMA, and a solution potentially better suited for decentralized real-time systems, namely self-organizing time division multiple access (STDMA).

2. IEEE 802.11p/DSRC Protocol

The IEEE 802.11p standard (WAVE) emerges from the allocation of the (DSRC) Dedicated Short Range Communications spectrum band in the United States and the work done to define the technology to be used in this band. There are two types of channels in DSRC, all of them with a 10 MHz width: the control channel (CCH) and the service channel (SCH). The CCH is restricted to safety communications only, and the SCHs are available both for safety and non-safety use. Applications for vehicular communications can be placed in three main categories—traffic safety, traffic efficiency and value-added services (e.g. infotainment/business) [14-16]. In 1999, the US Federal Communication Commission (FCC) allocated these 75 MHz of spectrum at 5850 - 5925 GHz to be used exclusively for vehicle-to-vehicle and infrastructure-to-vehicle communications. The main objective is to enable public safety applications in vehicular environments to prevent accidents (traffic safety) and improve traffic flow (traffic efficiency).

In Europe, the spectrum allocated by the ETSI for cooperative safety communications has a range 5875 - 5925 GHz. It is divided into traffic safety (30 MHz) and traffic efficiency (20 MHz). In the traffic safety spectrum, two SCHs and one CCH are allocated **Figure 1**. In the traffic efficiency two SCHs are allocated [16]. As stated before, WAVE has its origins in the standardization of DSRC as a radio technology. WAVE is fully intended to serve as an international standard, which is meant to: describe the functions and services required by WAVE stations to operate in VANETs, and define the WAVE signaling technique and interface functions that are controlled by the IEEE 802.11 MAC.

WAVE is an amendment to the Wireless Fidelity (WiFi) standard IEEE 802.11 [17]. It is inside the scope of IEEE 802.11a, which is strictly a PHY and MAC level standard. In other words, IEEE 802.11p is an adaptation of the IEEE 802.11a protocol to vehicular situations, such as: rapidly changing environment. With a short time frame transactions required, and without having to join a Basic Service Set (BSS) (Peer-to-Peer (P2P) and ad-hoc networks). WAVE is only a part of a group of standards related to all layers of protocols for DSRC-based operations

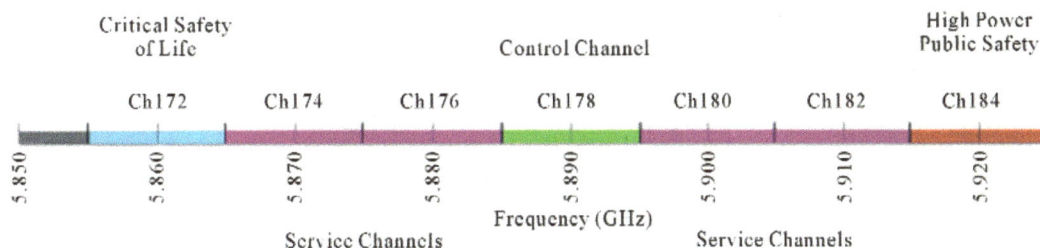

Figure 1. DSRC spectrum band and channels.

as can be seen in **Figure 2**. In this paper, we are only going to focus on IEEE 802.11p.

ETSI has defined two central types of messages: CAMs, defined before, and Decentralized Environmental Notification Messages (DENM). IEEE 802.11p has adopted these two types of messages. As a reminder, CAMs are broadcast packets sent periodically at a concrete heartbeat rate. A CAM packet contains information about the stating vehicle speed, position and driving direction of the transmitter [18]. DENMs are event-driven and application specific messages, which are sent on emergency cases. They are triggered in case of a hazard and are continuously broadcasted until this hazard disappears [18].

Every node in the network manages the information received by remote nodes, as well as the data generated by the own vehicle. All data is contained in a database called Local Dynamic Map (LDM). CAMs and DENMS are used to update the LDM of each vehicle with the information gathered from the rest of nodes of the network.

2.1. Physical Layer Architecture

The physical layer defined in the standard IEEE 802.11 consists of two sub-layers: Physical Medium Dependent (PMD) sub-layer and Physical Layer Convergence Protocol (PLCP) sub-layer. The PMD sub-layer defines the parameters to build up the signal to be sent, such as modulation, demodulation and channel coding. This sub-layer interfaces directly with the wireless medium, RF in the air. The PLCP sub-layer, on the other hand, is in charge of dealing with interferences among different PHY layers and makes sure that the MAC layer receives the data in a common format, independently from the particular PMD sub-layer. The PLCP communicates with the PMD sub-layer and MAC layer through the correspondent Service Access Points (SAP). This can be seen in **Figure 3**.

2.2. Data Frame Format

A procedure carried out by the PLCP sub-layer is the convergence procedure, in which it converts the actual data frame being sent, named PLCP Service Data Unit (PSDU) into the PLCP Protocol Data Unit (PPDU). In this procedure, the preamble and header are appended to the PSDU to obtain the PPDU. The preamble consists of 12 training symbols, 10 of which are short and are used for establishing automatic gain control, diversity selection and the coarse frequency offset estimate of the carrier signal. The receiver uses 2 long training symbols for channel and fine frequency offset estimation. It takes up to 16 ms to train the receiver after first detecting a signal on the RF medium. The header, also called the SIGNAL field of the PPDU frame, is always transmitted at 6 Mbps using BPSK modulation. It contains information about the transmission data rate and type of modulation (BPSK, QPSK, 16QAM or 64 QAM) in the RATE field and the length in number of octets of the PSDU that the MAC is currently requesting to transmit in the LENGTH field; as well as a parity bit (Parity field), based on the first 17 bits, and a Tail field with all bits set to 0. The PSDU itself is pre-pended with the Service field, with the first 7 bits as zeros to synchronize the descrambler in the receiver and the remaining 9 bits reserved for future use and set to all 0s, and appended with the Tail field and Pad Bits field, which are the number of bits that make the DATA field a multiple of the number of coded bits in an OFDM symbol (48, 96, 192, or 288). The Service field, PSDU, Tail field and Pad Bits field form the DATA field of PPDU frame. The IEEE 802.11 PPDU frame format is shown in **Figure 4** [19,20].

2.3. OFDM

The transmission of data is based on Orthogonal Frequency Division Multiplexing (OFDM) technique.

OFDM divides the available band into K sub-bands or sub-carriers, which are separated a frequency bandwidth F. From this perspective, OFDM is similar to Frequency Division Multiple Access (FDMA). However, in FDMA all subcarriers require spectral guard intervals in order to prevent interferences between closely allocated subcarriers.

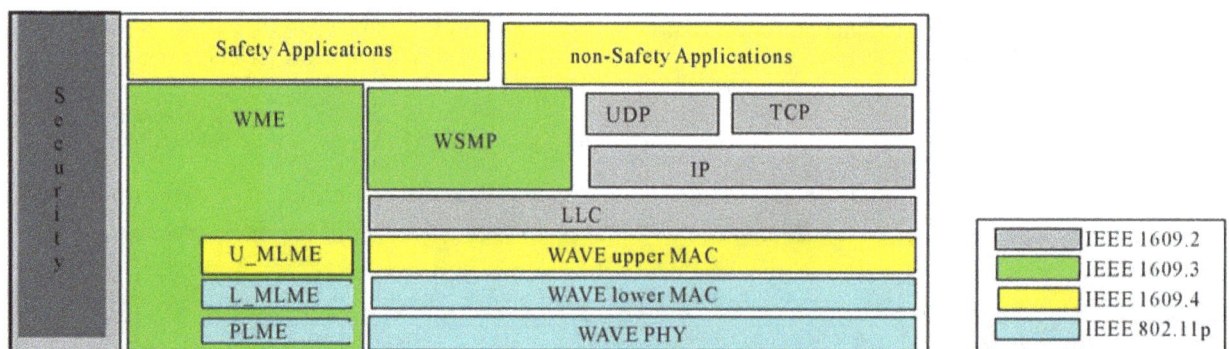

Figure 2. DSRC standards and communication stack.

Simulation Based Evaluation of Highway Road Scenario between DSRC/802.11p MAC Protocol and STDMA for
Vehicle-to-Vehicle Communication

91

OFDM uses the spectrum much more efficiently than FDMA since it makes all the subcarriers orthogonal to each other. This way, it is possible to have the subcarriers all together as close as possible and prevent any interference amongst them. Orthogonality of the subcarriers means that an integer multiple of cycles is contained in each symbol interval in every different subcarrier. Thus the spectrum of each subcarrier has a null at the central frequency of each of the other subcarriers. This attenuates the problems of overhead carrier spacing and guard interval allocation required in FDMA. In IEEE 802.11p there are defined 64 subcarriers, but only the 52 inner subcarriers are used. 48 out of these 52 actually contain the data and 4 of them, called pilot subcarriers, transmit a fixed pattern used to mitigate frequency and phase offsets at the receiver side.

Each of these 48 data subcarriers can be modulated, as explained before, with BPSK, QPSK, 16QAM or 64QAM. In combination with different coding rates, this leads to a nominal data rate from 6 to 54 Mbps if full clocked mode with 20 MHz bandwidth is used [21]. However, a change has been done in terms of sampling rate, for the adaptation of IEEE 802.11a to IEEE 802.11p: in IEEE 802.11p a channel of 10 MHz bandwidth is used. This way, the guard interval is long enough to prevent Inter-Symbol Interference (ISI) caused by multipath channel during the transmission and hence it fits the high-speed vehicular environment that characterizes the VANETs. The parameters in the time domain are doubled, compared to the parameters in IEEE 802.11a

[17]. In **Table 1** some of these parameters are shown.

2.4. The Transmitter

The binary data that is to be sent over the wireless medium, which is the PSDU, is encoded and modulated. The resulting coded data string is constantly being assigned to a certain complex number in a signal constellation and groups of 48 of these complex numbers are mapped to OFDM subcarriers. The operation in the assembler block is, mainly, to insert 4 pilot subcarriers among the 48 data subcarriers and form the OFDM symbol. In the next block, the OFDM sub-carriers are converted to the time domain using the Inverse Fast Fourier Transformation (IFFT) and prep ends a time

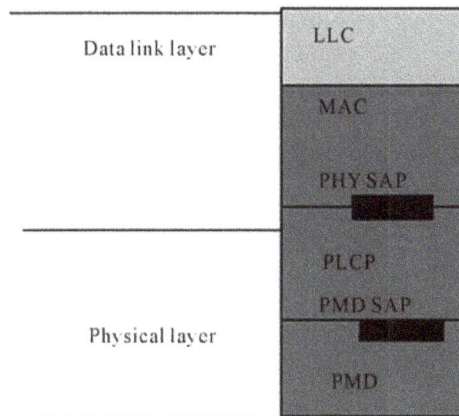

Figure 3. Physical layer and data link layer.

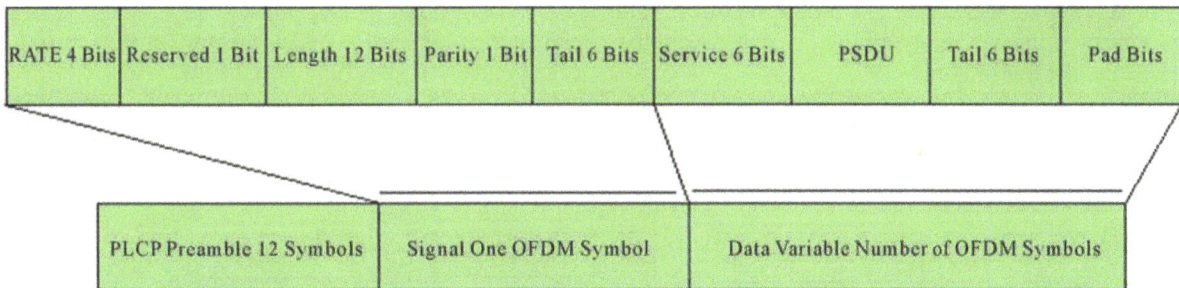

Figure 4. The IEEE 802.11 PPDU frame format.

Table 1. Comparison of PHY parameters in IEEE 802.11a and IEEE 802.11p.

Parameters	IEEE 802.11a	IEEE 802.11p	Changes
Channel bandwidth	20 MHz	10 MHz	Half
Bit rate (Mbps)	6, 9, 12, 18, 24, 36, 48, 54	3, 4.5, 6, 9, 12, 18, 24, 27	Half
Modulation Mode	BPSK, QPSK, 16QAM, 64QAM	BPSK, QPSK, 16QAM, 64QAM	No change
Number of subcarriers	52	52	No change
Symbol duration	4 µs	8 µs	Double
Guard Interval Time	0.8 µs	1.6 µs	Double

domain signal with circular extension of itself to generate the cyclic prefix. In the last stage in the transmitter, all the OFDM symbols are appended one after the other to form the PSDU and appended again with the PLCP preamble, the PLCP header (SERVICE field of the PPDU) and the fields Service, Tail and Pad Bits. This way the PPDU is obtained and is ready for transmission [23]. The block diagram of the transmitter is depicted on **Figure 5**.

2.5. The Channel

All types of wireless communications, the medium is the radio channel between transmitter and receiver. The signal is propagated through different paths, that can be either Line-Of-Sight paths (LOS) or Non-Line-Of-Sight (NLOS) between transmitter and receiver. In each of these paths, the signal can suffer from reflections, scattering and diffractions by different objects during its itinerary. These are just few of the conditions that can affect the multipath communications in this medium, and have a big impact on the propagation of the VANETs. Most times it is very complicated to take into account all of the adversities found this medium, therefore simplified model channels are used. For VANETs, these models must take into account the existence of multiple propagation paths and the high relative velocities among nodes.

Two key parameters that are directly affected by the channel conditions are:

Signal-to-Noise Ratio (SNR) is broadly defined as the ratio of the desired signal power to the noise power. This ratio indicates the reliability of the link between the receiver and the transmitter.

$$SNR[\text{dB}] = \text{Power Rcvd}[\text{dB}] - 10 * \log 10 (\text{noise}) \quad (1)$$

Signal-to-Interference-to-Noise Ratio (SINR) is de-

fined as the ratio of the desired power to the noise power plus the interferences generated by other transmitters close to the analyzed one, which are also considered as noise for the receiver.

$$SINR[\text{dB}] = \text{Power Rcvd}[\text{dB}]$$
$$-10 * \log 10 \left(\sum (\text{Power Int}) + \text{noise} \right) \quad (2)$$

2.6. The Receiver

In the receiver part, for the adaptation of IEEE 802.11a to IEEE 802.11p, some required improved performances have been introduced in the receiver to avoid cross channel interferences from adjacent channels [16]. The first block in the receiver is the Serial-to-Parallel (S/P), in which the signal is divided in blocks of samples and the DATA field is separated from the Preamble and SIGNAL fields of the PPDU. Both DATA and Preamble are demodulated with the Fast Fourier Transform (FFT) algorithm. After that, the channel coefficients are estimated and based on them, the equalizer compensates the fading effects introduced by the channel and transmits the samples to the decoder. Finally the received and decoded binary data stream is compared to the transmitted one, in order to calculate the error ratio statistics [21]. The block diagram of the transmitter is depicted on **Figure 6**.

2.7. MAC Layer

For the adaptation of IEEE 802.11a to IEEE 802.11p, no changes in the MAC layer have been done. The MACprotocol used in 802.11p is the same as in 802.11a, the Enhanced Distributed Channel Access (EDCA), which is an enhanced version of the basic access mechanism in IEEE 802.11 using Quality of Service (QoS).

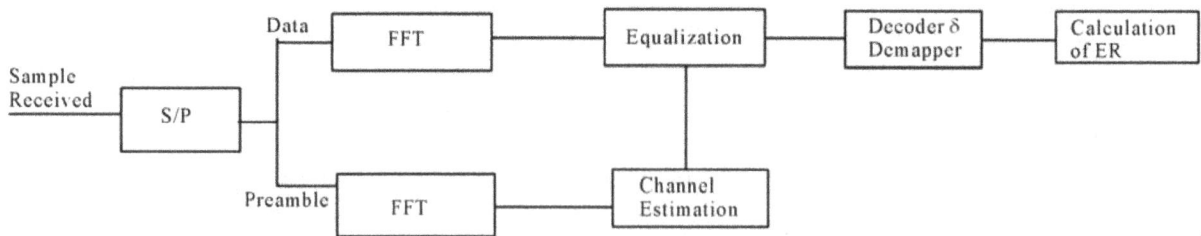

Figure 5. Block diagram of the transmitter.

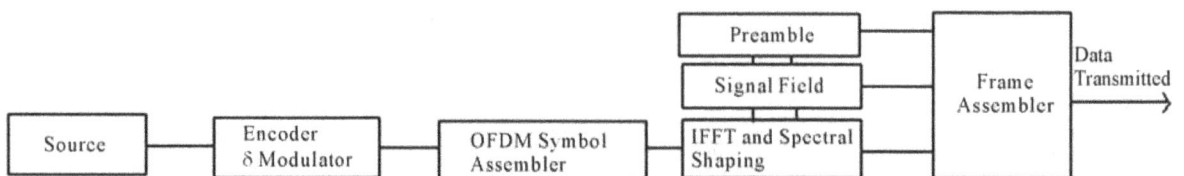

Figure 6. Block diagram of the receiver.

Simulation Based Evaluation of Highway Road Scenario between DSRC/802.11p MAC Protocol and STDMA for
Vehicle-to-Vehicle Communication

93

2.8. Overview of MAC Services

2.8.1. Data Services

This service provides peer entities in the LLC (Local Link Control) MAC sub-layer with the ability of exchanging MSDUs (MAC Service Data Units) using the underlying PHY-layer services. This delivery of MSDUs is performed in an asynchronous way, on a connectionless basis. By default, MSDU transport is based on best-effort. However, the QoS facility uses a Traffic Identifier (TID) to specify differentiated services on a per-MSDU basis. There are no guarantees that the MSDUs will be received successfully. Broadcast and multicast transport is part of the asynchronous data service provided by the MAC layer. Due to the characteristics of the wireless medium, broadcast and multicast MSDUs may experience a lower QoS, compared to that of unicast MSDUs. In our simulations, only broadcast MSDUs are sent and received and no acknowledgement is used and the vehicles, also called stations, are nQSTAs (non-QoS STAtions), this means that no QoS is used since all the transmitted messages are the same type (CAMs) [22]. In **Figure 7** it is shown the encapsulation of a MSDU inside a MPDU (MAC Protocol Data Unit), which becomes the PDSU when pro-

cessed at a PHY layer level.

2.8.2. MSDU Ordering

In nQSTAs, the ones simulated in this thesis, there are two service classes within the data service. By selecting the desired service class, each LLC entity initiating the transfer of MSDUs is able to control whether MAC entities are or are not allowed to reorder those MSDUs at reception. In an nQSTA, the MAC does not intentionally reorder MSDUs. If a reordering happens, the sole effect of this (if any), for the set of MSDUs received at the MAC service interface of any single STA, is a change in the delivery order of broadcast and multicast MSDUs originating from a single source STA address. If a higher layer protocol using the data service cannot tolerate this possible reordering, the optional Strictly Ordered service class should be used [22]. No reordering of MSDUs takes place in our simulations.

2.8.3. MAC Sub-Layer Functional Description

2.8.3.1. MAC Architecture

The MAC architecture can be described as shown in **Figure 8** as providing the Point Coordination Function (PCF) and Hybrid Coordination Function (HCF) through

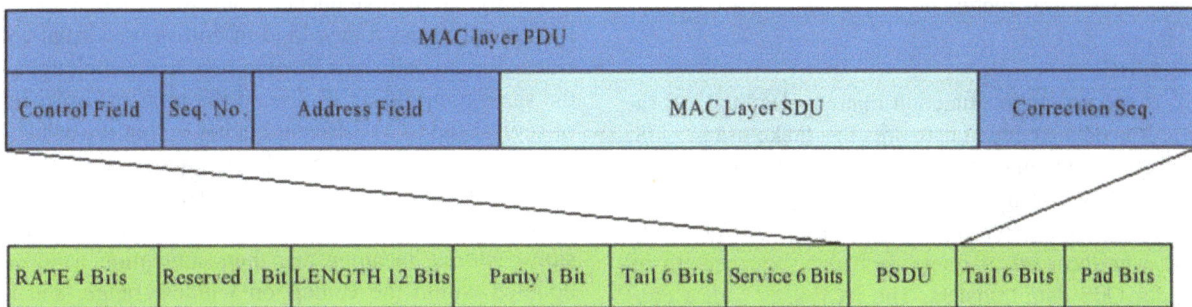

Figure 7. The IEEE 802.11 MPDU frame format.

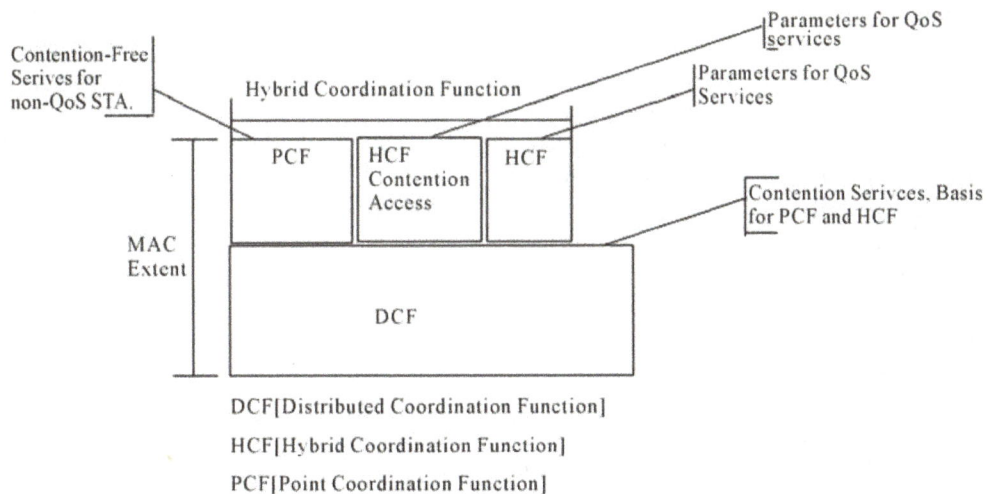

Figure 8. MAC architecture.

the services of the Distributed Coordination Function (DCF). The HCF is composed by the HCF contention-based channel access also called Enhanced Distributed Channel Access (EDCA), the HCF Controlled Channel Access (HCCA) and the Point Coordination Function (PCF) [21,22].

2.8.3.2. DCF
DCF is the fundamental MAC technique in the IEEE 802.11 standard. It employs an access function performed by the CSMA/CA algorithm and a collision management function carried out by the binary exponential back-off procedure.

2.8.3.3. PCF
The original IEEE 802.11 standard defines another coordination function in the MAC layer. It is only available in structure mode networks, where the nodes are intercomnected through at least one AP in the network. This mode is optional and only very few APs or Wi-Fi adapters actually implement it. The coordinator block is called Point Coordinator (PC). In the scope of this thesis, PCF is not used because we are not simulating an infrastructure network (with Access Points (APs)), but a VANET, where all nodes are peers: not only the vehicles but also the road-side infrastructure behave as peers in a VANET.

2.8.3.4. HCF
HCF is a coordination function that enables the QoS facility. It is only usable in networks that make use of QoS, so it is only implemented in the QSTAs. The HCF combines functions from the DCF and PCF with some enhanced, QoS-specific mechanisms and frame subtypes to allow a uniform set of frame exchange sequences to be used for QoS data transfers. The HCF uses both a controlled channel access mechanism, HCCA, for contention-free transfer and a contention-based channel access method mechanism, EDCA.

2.8.3.5. HCCA
HCCA works similarly to PCF. It uses a QoS-aware centralized coordinator, called a Hybrid Coordinator (HC), and operates under rules that are different from the PC of the PCF. HCCA is generally considered the most advanced (and complex) coordination function. With the HCCA, QoS can be configured with great precision. QSTAs have the ability to request specific transmission parameters which allow advanced applications to work more effectively on a Wi-Fi network.

2.8.3.6. The EDCA Channel Access Control
Every priority queue, also called Access Category (AC), has different values of Arbitrary Inter Frame Space (AIFS) and back-off range. The contention window lim-

its CWmin and CWmax, from which the random back-off is computed are variable depending on the AC. The highest the priority, the lowest the value of AIFS and the limits of the contention window [23,24]. The table with the different ACs and the values assigned to each one are shown in **Table 2**.

The duration AIFS(AC) is a duration derived from the value AIFSN(AC) by the relation:

$$AIFS(AC) = AIFSN(AC) * Slot\ Time + SIFS \qquad (3)$$

where the SIFS is the abbreviation for Short Inter-Frame Space period. The SIFS is the small time interval between the data frame and its acknowledgment. These values in IEEE 802.11 are defined to be the smallest of all inter frame spaces (IFSs) periods. A SIFS duration is a constant value and it depends on the amendments to the IEEE 802.11 standard. The shortest AIFS possible value in IEEE 802.11p is AIFS = 58 μs and this is the value used in our simulations. The slot time is derived from the PHY layer in use: in IEEE 802.11p, Slot Time = 13 μs. The back-off duration is calculated as:

$$\begin{aligned} Back\ off\ Duration \\ = Random\ Back\ off\ Value * Slot\ Time \end{aligned} \qquad (4)$$

Apart from real collisions (physical collisions on the medium) that involve queues from two different stations, EDCA introduces a new kind of collisions: virtual collisions. Virtual collisions involve two queues belonging to the same transmitting station. If the back-off procedures of several (up to 4) different queues within the same station finish at the same time slot, the queue with the highest priority has the right to be the first to try to access the medium, while the others will behave as if a real collision occurred, meaning that their contention window is doubled within the contention window range, and that will possibly delay its next trial to access the medium. In [24] a proposal solution to that is described.

In **Table 3**, default parameter settings for the different queues in 802.11p are found together with the CW setting.

2.8.4. Operation of the CSMA/CA Algorithm
Since we are not dealing with different type of messages, all the packets sent by the nodes have the same priority and the QoS enhancements explained before that EDCA

Table 2. Default EDCA parameters for each AC.

AC	CWmin	CWmax	AIFSN
AC VO	7	15	2
AC VI	15	31	2
AC BE	31	1023	3
AC BK	31	1023	7

Table 3. Default parameter setting in 802.11p for the EDCA mechanism.

	Queue-1	Queue-2	Queue-3	Queue-4
Priority	Highest	-	-	Lowest
AIFS	58 μs	58 μs	71 μs	123 μs
CWmin	3	7	15	15
CWmax	511	1023	1023	1023

adds are not needed. We give to these packets the highest priority and for that reason, we use AIFS = 58 μs and CW = CWmin = 3. Furthermore, we will not suffer in our simulations from virtual collisions, but only from real collisions [22]. In addition, all the messages sent are broadcasted and because of that we do not make use of the SIFS concept neither. We are dealing with nQSTAs, so HCF is not present in our simulations. What it is really of interest in this thesis from the IEEE 802.11p MAC layer are the CSMA/CA algorithm and the exponential back off procedure found in DCF. The CSMA/CA procedure according to IEEE 802.11p, it is, in the broadcast situation with periodic data traffic (CAM packets), is presented in **Figure 9**.

The transmitter node starts by listening to the channel activity during an AIFS amount of time (which in our simulations is 58 μs). If after this time, the channel is sensed free, the packet is transmitted. After that, the node checks if a new packet from the upper layers is ready to be transmitted, and when there is one, it performs the same action to transmit the new packet. If during AIFS, the channel is busy or becomes busy, then the node gets a random back off value, generated from an exponential distribution, by multiplying the integer from [0..CW] with the slot time 13 μs obtaining 0, 13, 26 or 39 μs. This value will be decreasing every time the node waits for an AIFS and senses the channel free. When the back off value gets to 0, then the packet can be transmitted. While the node is getting its back off value decreased, it keeps on checking constantly if a new packet was generated in the upper layers and is ready to be transmitted. When that happens, the old packet is dropped, and the node starts again with the whole transmission protocol.

Advantages and Drawbacks of CSMA/CA
CSMA/CA has some advantages and some drawbacks regarding to scheduling safety applications data. The advantages are:

In low-loaded networks, it performs in a quite optimal way: all nodes access in a fair way to the channel with a low access delay and with few transmission collisions.

On the other hand, the drawbacks are:

In high-loaded networks, it becomes an unfair and un-predictable MAC algorithm with a poor scalability, since a lot of nodes want to access to the channel at the same time and it generates many blocked transmitter nodes.

It turns out to be unfair because it may happen that some nodes that want to transmit a packet, always sense the channel busy and, other nodes may be luckier and be able to transmit their messages more often because they access the free channel.

It becomes unpredictable because it is not known how long it can take to a node to be able to transmit. If it senses the channel busy, then it performs a random back off and can be waiting more or less depending on how many times the channel has been sensed idle. So the channel access delay cannot be predicted.

It turns out to have a poor scalability, since the higher the number of nodes in the network, the worse the performance of this algorithm. There is a high number of blocked nodes waiting for an unpredictable time to get to the channel. It means that there are a high number of nodes suffering from starvation, and it implies that more packets are constantly dropped within the same node.

Since what we intend is to send safety messages, that should always arrive to the receivers in a short period of time, it is clear that the CSMA/CA algorithm is not suitable for safety purposes when performing in a high-loaded network. It takes us to think that another MAC algorithm is needed to handle this type of traffic. Some previous studies [22-24] have proposed an alternative MAC algorithm that outperforms CSMA/CA [24].

2.8.5. Self-Organizing Time Division Multiple Access (STDMA) MAC Layer Algorithm

The STDMA algorithm, invented in [14,15], is already used in commercial applications for surveillance, *i.e.*, the Automatic Identification System (AIS) used by ships and the VHF data link (VDL) mode 4 system used by the avionics industry. Traditional surveillance applications for airplanes and ships are based on ground infrastructure with radar support. Radar has shortcomings such as the inability to see behind large obstacles or incorrect radar images due to bad weather conditions. By adding data communication based on STDMA, more reliable information can be obtained about other ships and airplanes in the vicinity and thereby accidents can be avoided. Since STDMA is so successful in these systems, it is interesting to investigate if it can manage a more dynamic setting such as a vehicular network. STDMA is a decentralized MAC scheme where the network members themselves are responsible for sharing the communication channel. Nodes utilizing this algorithm, will broadcast periodic data messages containing information about their position. The algorithm relies on the nodes being equipped with GPS receivers. Time is divided into frames as in a TDMA system and all stations are striving

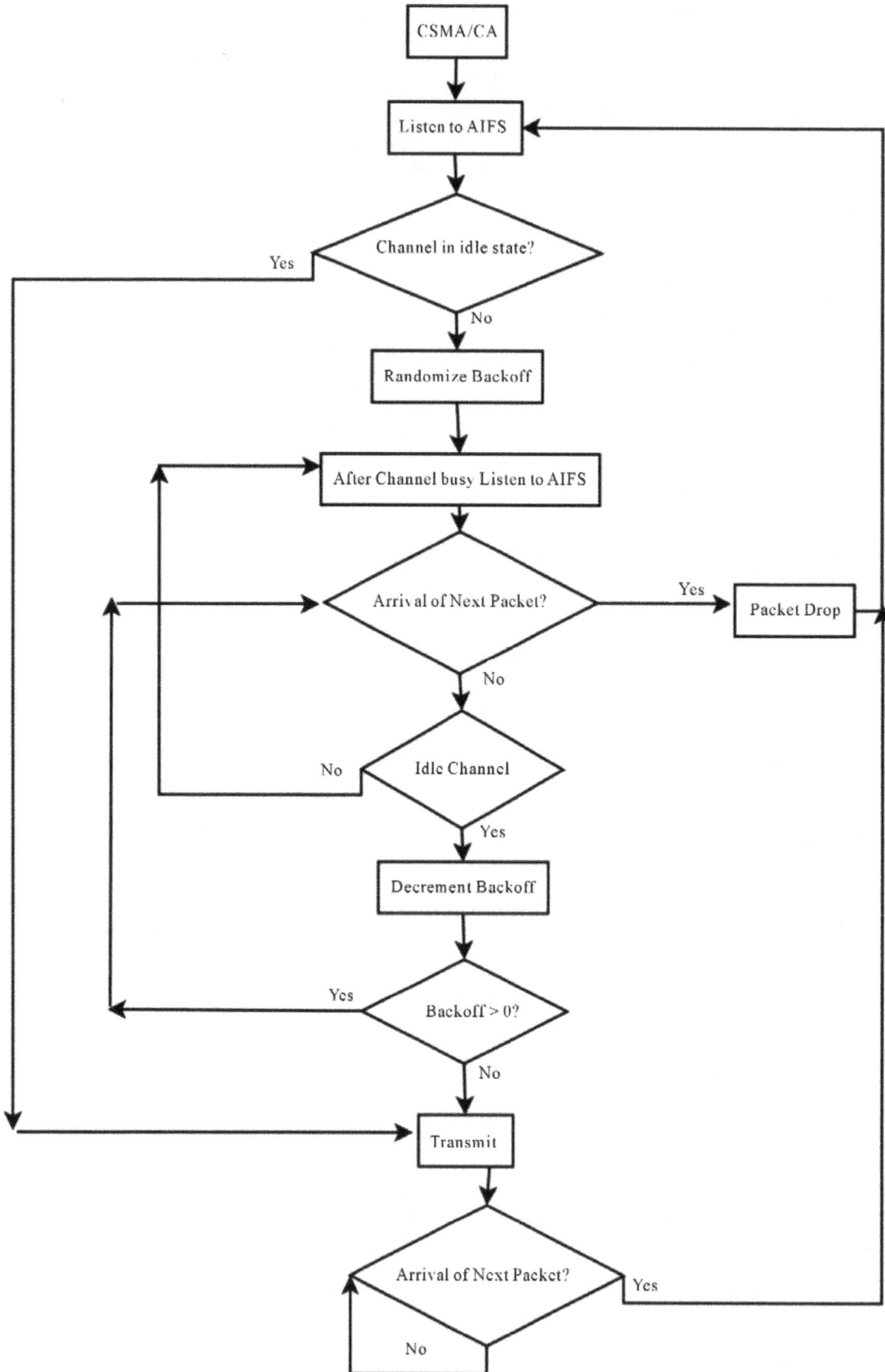

Figure 9. The CSMA/CA procedure according to 802.11p.

Simulation Based Evaluation of Highway Road Scenario between DSRC/802.11p MAC Protocol and STDMA for Vehicle-to-Vehicle Communication

97

for a common frame start. These frames are further divided into slots, which typically corresponds to one packet duration. The frame of AIS and VDL mode 4 is one minute long and is divided into 2250 slots of approximately 26 ms each. All network members start by determining a report rate, *i.e.*, how many position messages that will be sent during one frame. Then follows four different phases; *initialization*, *network entry*, *first frame*, and *continuous operation*. During the *initialization*, a node will listen to the channel activity during one frame length to determine the slot assignments. In the *network entry* phase, the node determines its own transmission slots within each frame according to the following rules: (*i*) calculate a nominal increment (*NI*) by dividing the number of slots with the report rate, (*ii*) randomly select a nominal start slot (NSS) drawn from the current slot up to *NI*, (*iii*) determine a selection interval

(SI) of slots as 20% of *NI* and put this around the NSS according to **Figure 1**, (*iv*) now the first actual transmission slot is determined by picking a slot randomly within SI and this will be the nominal transmission slot (NTS). If the chosen NTS is occupied, then the closest free slot within SI is chosen. If all slots within the SI are occupied, the slot used by a node furthest away from oneself will be chosen. When the first NTS is reached in the super frame, the node will enter the third phase called the *first frame*. Here a nominal slot (NS) is decided for the next slot transmission within a frame and the procedure of determining the next NTS will start over again. This procedure will be repeated as many times as decided by the report rate (*i.e.*, the number of slots each node uses within each frame) (**Figure 10**).

After the first frame phase (which lasts for one frame) when all NTS were decided, the station will enter the *continuous operation* phase, using the NTSs decided during the *first frame* phase for transmission. During the *first frame* phase, the node draws a random integer $n \in \{3, \cdots, 8\}$ for each NTS. After the NTS has been used for *n* frames, a new NTS will be allocated in the same SI as the original NTS. This procedure of changing slot after a certain number of frames is to cater for network changes, e.g., two nodes using the same NTS which were not in radio range of each other when the NTS was chosen could have come closer and will then interfere.

The STDMA relies on the position information sent by other network members and it will not work without this.

2.8.6. Continuous Operation Phase

The last phase is called continuous operation phase. Here, a new concept is introduced, the n reuse factor. Every message in a slot has an n value related to it, which decreases within every transmission. When n gets to 0, then the message has to be reallocated in a new slot within the same SI as the former slot. If all of the slots are busy, then the procedure is the same as in the second phase. Apart from a reallocation, a new n factor is assigned to the new NTS location. This factor is used to cater with changes in the network topology. When a node enters the same transmission range of another node, and both of them have a message allocated in the same slot within the frame, it will cause a co-located transmission and in case they are close to each other packets from both co-located transmitters might be lost by the receiving nodes. Without the use of the n reuse factor, they would be suffering a collision every time until they get out of the same range of transmission. The situation changes when one of them gets its n reuse factor value to 0, so its message has to be reallocated to a new slot avoiding from that moment, suffering a collision with the other node. The n reuse factor adds flexibility to STDMA, very important since we are dealing with VANETs, whose nodes are constantly moving. The continuous operation phase is depicted as a flow diagram in **Figure 11**.

Advantages and Drawbacks of STDMA

The advantages of STDMA compared to CSMA/CA are: STDMA is considered a fair algorithm because all the nodes that have packets to transmit are able to send them, and there are no distinctions in the performance among lightly crowded or heavily crowded channels,

The allocation of messages in the slots even if all of them are busy, allows all the nodes to transmit every time they have a packet and hence, there are no packet drops at all. Furthermore, it allows knowing the maximum packet access channel delay that will be a value delimited by the SI. It makes STDMA to be a predictable algorithm, and It is also considered a potentially scalable algorithm because the higher the number of nodes in the network, this algorithm will keep on performing in a fair and a predictable way. On the other hand, the number of collisions will increase but it is a normal fact since there will be more nodes transmitting in co-located positions.

Allocating the messages on the slot occupied by the furthest node allows diminishing the interference level and the SINR value will not drop so severely as if the two co-located transmitters were close to each other.

It is a best solution than dropping the packet straight, as happened with CSMA/CA. The main drawback of

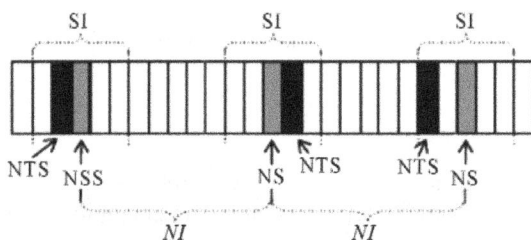

Figure 10. The STDMA algorithm in the first frame phase.

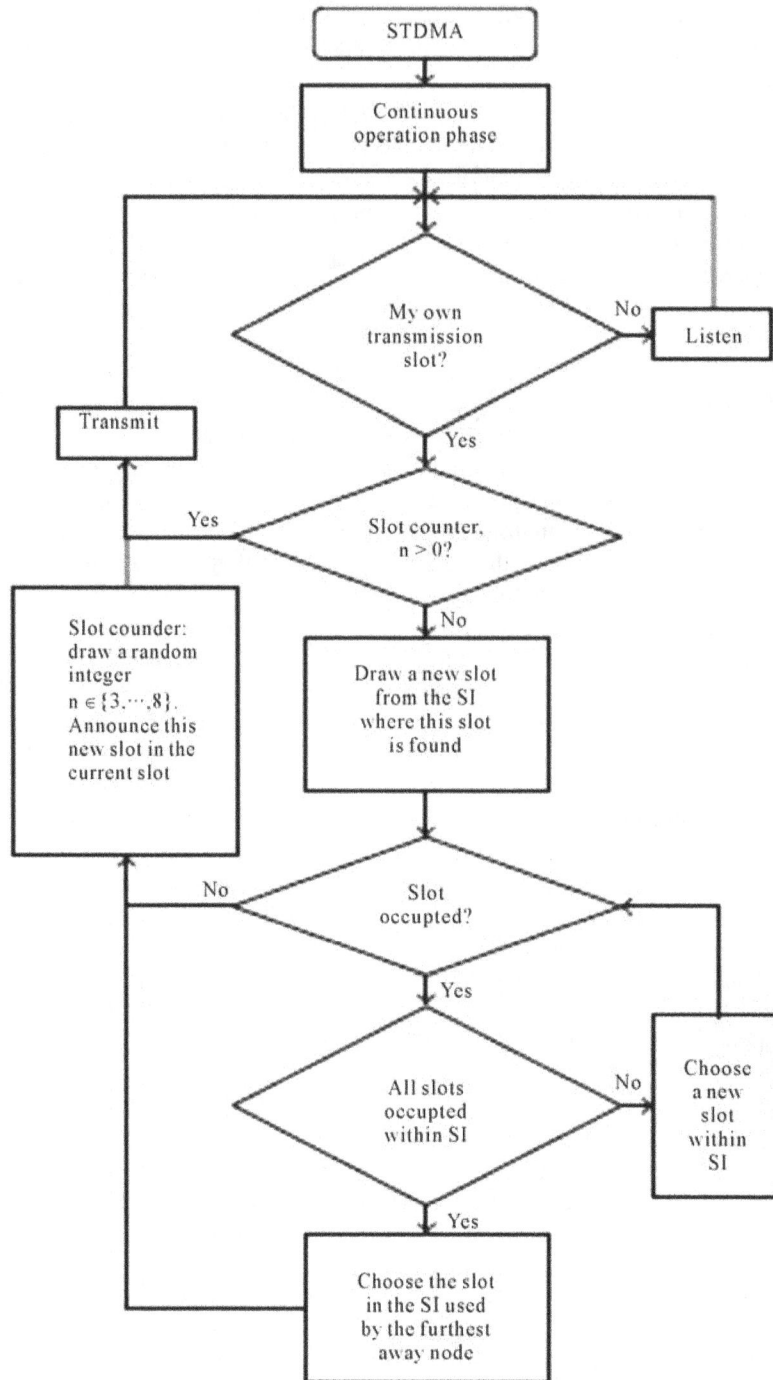

Figure 11. The continuous operation phase of STDMA.

STDMA is the need of every STDMA node of relying on a positioning system. It adds a second actor for safety applications and another point susceptible to fail.

For low-loaded networks, it has been shown in previous simulations [21,23,24] that both algorithms perform similarly, and even in some performance metrics such as the distance between concurrently transmitting nodes, STDMA outperforms CSMA/CA. For high-loaded or saturated networks, the advantages of STDMA become clearer.

3. Simulations

Many traffic safety systems will rely on vehicles periodically broadcasting messages containing their current state (e.g., current location, speed, average speed, distances travelled, total distance etc). We have developed a

Simulation Based Evaluation of Highway Road Scenario between DSRC/802.11p MAC Protocol and STDMA for Vehicle-to-Vehicle Communication

99

simulator using Open Street Map, eWorld, SUMO version 0.10.3 (traffic simulator), NS-2 version 2.34 (Network Simulator) and TraNs version 1.2 (Intermediate simulator between SUMO and NS2) also we require Gnu plot/Xgraph/Excel to plot the graphics presentation (**Figure 12** for simulation architecture and **Figure 13** for Simulation Flow diagram) where each vehicle sends a location message according to a predetermined range of 5 or 10 Hz. Simulations has been conducted both for the CSMA of 802.11p as well as for the proposed STDMA algorithm. The vehicle traffic scenario is a Mumbai-Pune Highway Road of 120 kilometer (km) with 3 lanes in each direction (*i.e.* total 6

lanes including both the directions) (**Figure 14**).

The Mumbai-Pune Highway Road scenario is chosen because here the highest relative speeds (*i.e.* min 80 km/h to max 120 or above km/h) in vehicular environments are found and hence it should constitute the biggest challenge for the MAC layer. The vehicles are entering each lane of the highway road according to a Poisson process with a mean inter-arrival time of 3 seconds (consistent with the 3-second-rule used in Sweden, which recommends drivers to maintain a 3-second spacing between vehicles). The speed of each vehicle is modeled as a Gaussian random variable with different mean values for each lane; 83 km/h, 108 km/h and 130 km/h, and a standard

Figure 12. Simulation architecture.

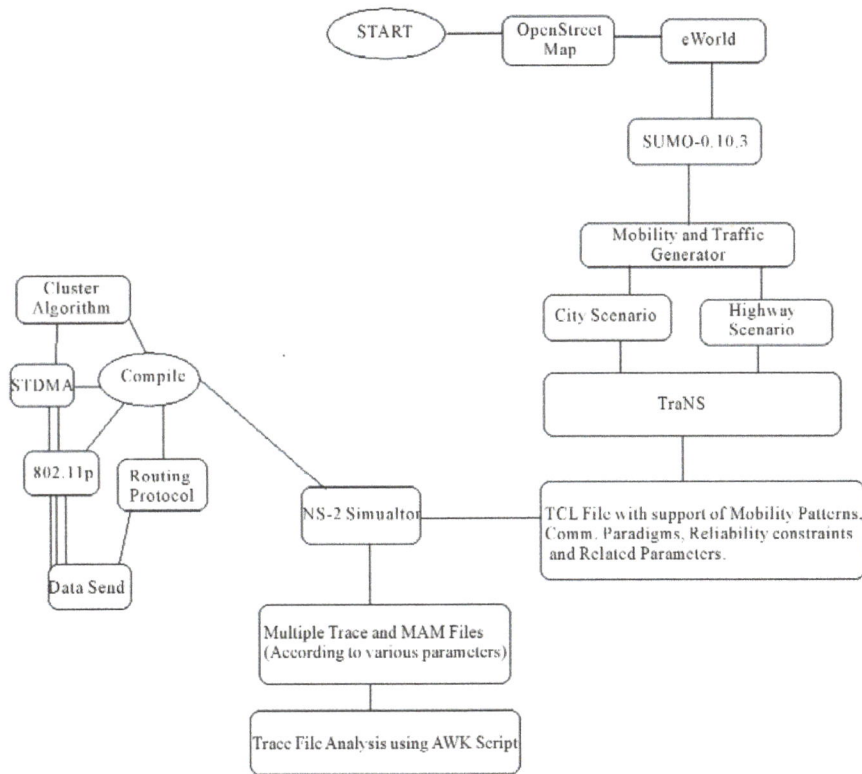

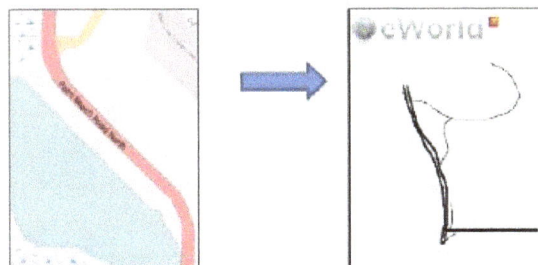

Figure 13. Simulation flow diagram.

deviation of 1 m/s. For simplicity we assume that no overtaking is possible and vehicles always remain in the same lane. There is no other data traffic in addition to the broadcast messages.

The channel model is a circular transmission model where all vehicles within a certain Sensing range will sense and receive packets perfectly. The simulated sensing ranges are 250 m and 500 m. We have tried to focus on how the two MAC methods perform in terms of time between channel access requests until actual channel access within each vehicle node. Three different packet lengths have been considered: 100, 300 and 500 byte. The shortest packet length is just long enough to distribute the location, direction and speed, but due to security overhead, the packets are likely longer. The transfer rate is chosen to be the lowest rate supported by 802.11p, namely 3 Mbps. Since all vehicles in the simulation are

broadcasting, no ACKs are used. **Table 4** contains a summary of the simulation parameter settings.

Figure 14. Scenario of mumbai-pune highway road.

Table 4. Simulation parameter setting for Mumbai-Pune Highway Road scenario simulation.

Parameter	Value
Start-point of Highway Road	Panvel
End-Point of Highway Road	Pune
Simulation Time	1 hour 30 mins (In-time 6.30 am & Out-time 8.00 am)
Length of Highway Road	120 Km
Traffic Direction	2 ways
Number of Lanes in Each Direction	3 lanes
Vehicle Type	Cars, Private vehicles, Buses, Trucks etc.
Number of Vehicle Nodes on Highway	1200
Speed of Vehicle Nodes	40 km/h, 60 km/h, 80 km/h to 130 km/h
Communication Protocol	802.11p and STDMA
Traffic Type	UDP
Packet Sending Frequency	5 Hz, 10 Hz
Packet Length	100 bytes (Ratio 30% vehicles), 300 bytes (Ratio 40% vehicles) and 500 bytes (Ratio 30% vehicles).
Transfer Rate	3 Mbps
Slot time, T_{slot}	9 μs
SIFS, T_{SIFS}	16 μs
CWmin	3
CWmax	Not used
Communication Range	250 meter, 500 meter
Backoff Time, $T_{Backoff}$	0, 9, 18, 27 μs
AIFS (Listening Time before Sending) CSMA Parameter	34 μs (highest priority)
STDMA Frame Size	1 s
No of Slots in the STDMA Frame	3076 slots (100 byte packets), 1165 slots (300 byte), 718 slots (500 byte)

4. Implementation and Results

Aim is to implement using simulators which will analyze the properties of 802.11p MAC protocols and behavior on a typical highway road. Issue here is how the MAC method will influence the capability of each sending node to delivers data packets within the clusters and within the deadlines. Two clusters are integrated with group of vehicles move in opposite directions on highway roads, merge and finally separate from each other. Such situation takes place in real life which happens on multilane highway roads, intersections in highways with two or more levels where vehicles go on a fly over's etc. vehicle in each cluster will transmit their packets without having a big number of packets entering the back off mode as the density of the vehicles per cluster is not so high. When the two clusters will merge, a sudden increase on the number of vehicles wanting to transmit occurs, so the channel is found most all the time busy by the transmitter vehicles. It makes that a high number of packets get into the back off mode and thus, a higher number or packets are dropped. This will last until both clusters totally surpass each other, so the number of transmitters willing to send their packets in the same portion of time will decrease. Highway has highest relative speed and this causes network topology to change more frequently. If a traffic accident occurs, N number of vehicles could be quickly being gathered in a small geographical area by sharing the information through wireless communication. Vehicles broadcast data packets at two different rates i.e. 5 hz & 10 hz. Assume there is no other data traffic in addition to these messages. Highway road is about 120 km long with five intersection points connected to it and contains 3 lanes in each direction as shown in **Figure 15**. Vehicles are entering in each lane of highway from different point of intersection. Vehicles are entering each lane of highway road accordingly with a mean of inter-arrival time of 3 seconds, speed of each vehicle modeled as a Gaussian random variable with different mean values for each lane, 80 km/h, 60 km/h, 40 km/h. The different speeds are chosen with the speed regulations of Indian National highway road in mind. Every node within the sensing area receives the message perfectly. Nodes could be exposed to two concurrent transmissions, where transmitters $TX1$ and $TX2$ are broadcasting messages at same time since transmitters cannot hear each other while receivers $RX1, RX2, \cdots, RXn$ will then experience the collisions of the two ongoing transmissions, unless some sort of power control is used. Simulation is to characterize the MAC channel access delay T_{acc}. Nodes enter the highway they will start to transmit after initial random delay between 0 to 120 mins. Simulation has been carried out with three different packet lengths $N = 100, 300$ and 500 bytes and different

sensing ranges as shown in table above. All vehicles use 802.11p described above, a broadcast packet will experience at most one backoff procedure due to lack of ACKs in a broadcast system. The contention window will never be doubled since at most one failed channel access attempt can occur. Since all data traffic in our highway scenario has same priority, only the highest priority AIFS and CWmin have been used. The backoff product of the slot time, Tslot, and a random integer uniformly distributed in the interval of backoff time, the CSMA procedure will broadcast with periodic message from every node next arrival packet will test if the new position message has arrived from MAC layer or not, if so old packet will be dropped.

STDMA Simulation, the vehicles will go through three phases, initialization, network entry and first frame before it enters in continuous operation. Vehicle stays in the continuous phase after it has been through the other three. STDMA always guarantees channel access even when all slots are occupied within SI, in which case a slot belonging to the node located furthest away will be selected. Time parameters are selected by the PHY specification of 802.11p. The CSMA transmission time, T_{CSMA} consists of AIFS period T_{AIFS} of 34 μs, 20 μs preamble $T_{preamble}$ and the actual packet transmission T_{packet}. STDMA transmission time T_{STDMA} which is same as the slot time, consists of two guard time $T_{GT} = 3$ μs each, $T_{preamble}$, T_{packet}, and two SIFS periods $T_{SIFS} = 16$ μs each derived from PHY layer. Total transmission time for CSMA is

$$T_{CSMA} = T_{AIFS+} T_{preamble} + T_{packet} \qquad (5)$$

and the total transmission time for STDMA is

$$T_{STDMA} = 2T_{GT} + 2T_{SIFS} + T_{preamble} + T_{packet} \qquad (6)$$

The delay that takes to a packet sent from the transmitting vehicle until it is decoded by the receiving vehicle at the MAC layer level. This measure shows not only the delay, but also the reliability of the messages since it takes into account the interference at the MAC level caused by other vehicles.

This delay is expressed as:

$$T_{MM} = Tca + Tp + Tdec \qquad (7)$$

At the receiver side, to be a packet candidate to be decoded and sent to higher layers, it should have arrived within 100 ms, which is the maximum allowed delay at the receiver vehicle for CAM messages to be considered.

The values analyzed from this performance indicator are the mean values of TMM for a concrete message transmitted by a concrete vehicle to all of the receiver vehicles.

The simulated has vehicle density of approximately one vehicle for every 10 meters in each lane. Vehicle density is chosen to examine the scaling performance of the two MAC layers. Simulation has been carried out in SUMO and NS2 simulator with the parameters settings shown in the table above for the highway scenario. When 500 bytes long packets are sent 10 times per second and the nodes have a sensing range in 1 km since this corresponds to the largest bandwidth requirements per unit area. MAC can handle 70 nodes that are in communication range of each other without packet collisions. Simulation contains situations that are overloaded and a node has 210 neighbors within communication range when the range is 1 km, and consequently some packet drops takes place.

Cumulative distribution functions (CDFs) for the channel access delay is

$$F_{T_{acc}}\left(x\right) = \Pr\left\{T_{acc} < x\right\} \qquad (8)$$

for CSMA as in **Figures 16 (a)** and **(b)** for different sensing ranges. Simulation statistics were collected from middle of the highway with the vehicle traffic. Dropped packets are considered to have infinite delays. Three plots in the figure represent CDF for the node performance in best average and worst case for different sensing range. In best case only 5% of generated and send packets are dropped while in worst case 65% to 70% packets are dropped for sensing range of 500 meters and 50% to 55% packets are dropped in average case for sensing range of 1000 meters. Lose of many consecutive packets which will make the node invisible to the surrounding vehicles for a period of time. CDF for number of consecutive packet drops is in **Figure 17**. In worst case a node can drop 100 consecutive packets, implying invisibility for over 10 seconds.

STDMA algorithm grants packets channel access since slots are reused if all slots are currently occupied within selection interval of the node. Node will choose the slot that is located furthest away hence there will be no packets drops at sending side when using STDMA and channel delay is small. **Figure 18** the CDF channel delay for STDMA for all nodes will choose a slot for transmission during selection interval therefore CDF for T_{acc} in STDMA is sending at unity after a finite delay compared to CDF for T_{acc} in CSMA in **Figure 16**.

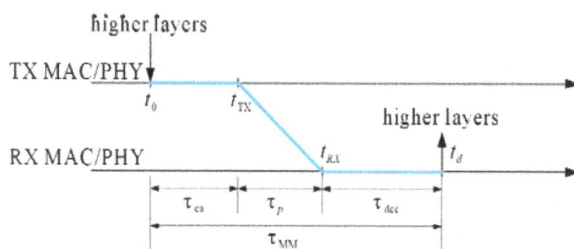

Figure 15. Mac-to-mac (E2E) delay.

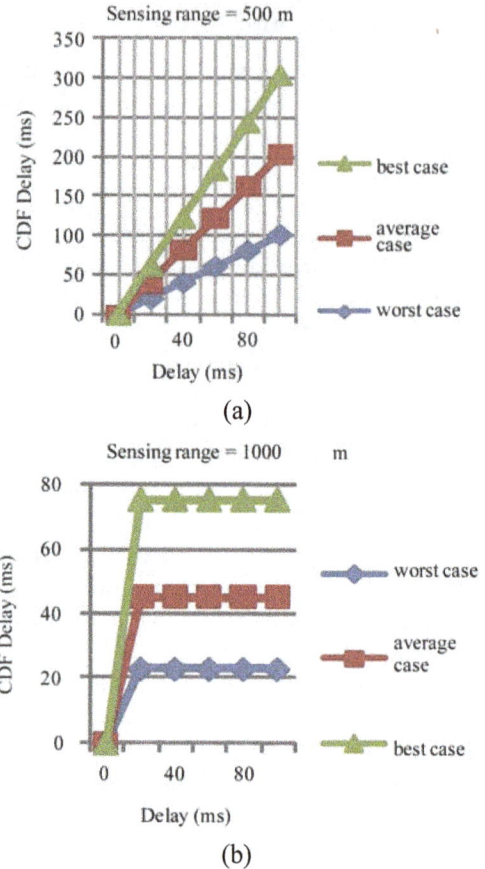

(a)

(b)

Figure 16. (a) Sensing range 500 meters; (b) Sensing range 1000 meters.

Figure 19 the CDF for the minimum distance between two node which utilizing the same slot within the sensing range is depicted for different packet lengths. When smaller packets size more nodes can be handled by the network. When long packets are used, the distance between two nodes intentionally reusing the same slot is reduced. In CSMA/CA, all channel requests did not make it to a channel access and then nodes drop packets. In CSMA/CA there is risk when nodes gets a channel access someone else also sends the packet and collision occurs. This is due to the fact that nodes can experience the channel idle at the same time, or ongoing transmission is not detected.

In **Figure 20**, the CDF for minimum distance between two nodes in CSMA/CA highway scenario sending at the same time for three different packets lengths with different ratio as shown in table above. The minimum distance can be interpreted as the distance between the nodes whose packets will, on the average, interfere the most with each other. 500 bytes, 1 km sensing range scenario, about 47% of the channel requests were granted and hence we conclude that the transmitted packets will be interfered by another transmission within 500 meters in 53% of the cases.

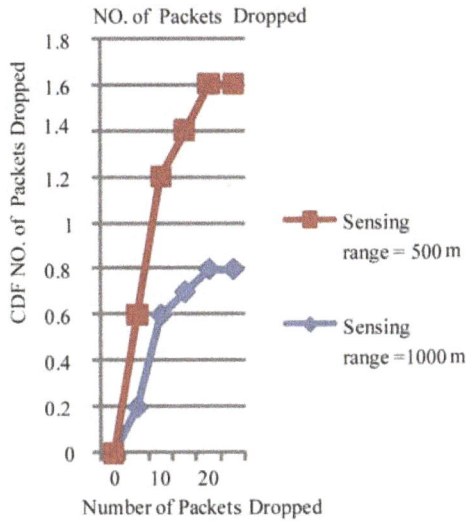

Figure 17. Number of Packets dropped due to no channel access.

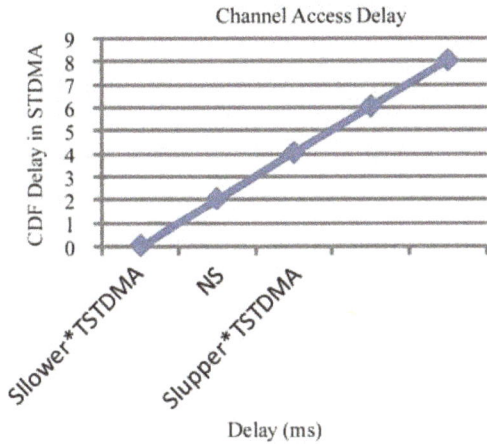

Figure 18. CDF for channel access delay in STDMA.

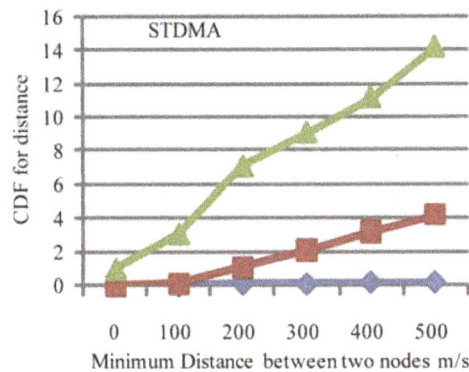

Figure 19. Utilizing the same time slot in STDMA to find minimum distance between two nodes.

5. Conclusion

In this paper we have analyzed how 802.11p and STDMA can be used for vehicle to vehicle performance

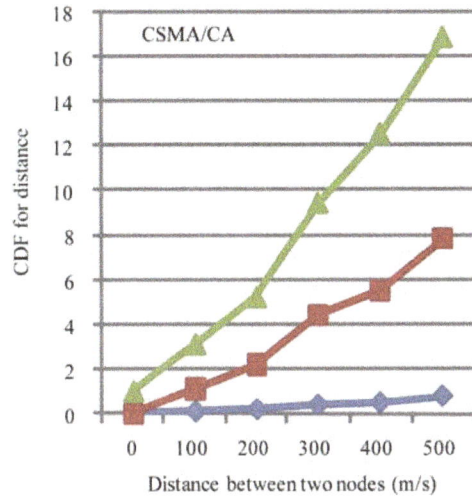

Figure 20. Sending at the same time in CSMa/CA using 500 bytes packets. 10 Hz, sensing range 1 km.

for the safety application on highway. We have considered different scenario and tested their performance and also compared their performance in form of best average and worst case and identified the packets drops are less in percentage in best case when the packet size is small and sensing range can also cover more number or nodes to be broadcasted. It also possible to test the real implementation using 802.11p/DSRC and STDMA and hope we can measure the same performance of DSRC and STDMA as a part of real implementation for future scope.

6. Acknowledgements

I would like to give my sincere thanks to my family members and friends who has helped me and guided me to complete my work and submit the paper on time.

REFERENCES

[1] Intelligent Transportation Systems Joint Program Office, "Reports from the Vehicle Infrastructure Integration Proof of Concept Project," 2012. http://www.its.dot.gov/vii/index.htm

[2] C. M. Krishna and K. G. Shin, "Real-Time Systems," McGraw-Hill, New York, 1997.

[3] IEEE 802.11 Working Group, "IEEE 802.11e Standard for Information Technology—Telecommunications and Information Exchange between Systems—Local and Metropolitan Area Networks—Specific Requirements Part 11: Wireless LAN Medium Access Control (MAC) and Physical Layer (PHY) Specifications Amendment 8: Medium Access Control (MAC) Quality of Service Enhancements," 2005.

[4] J. J. Blum, A. Eskandarian and L. J. Hoffman, "Challenges of Intervehicle Ad Hoc Networks," *IEEE Transactions on Intelligent Transportation Systems*, Vol. 5, No. 4, 2004, pp. 347-351.

[5] K. Bilstrup, "A Survey Regarding Wireless Communication Standards Intended for a High-Speed Vehicle Environment," Technical Report IDE 0712, Halmstad University, Halmstad, 2007.

[6] L. Stibor, Y. Zang and H.-J. Reumermann, "Evaluation of Communication Distance of Broadcast Messages in a Vehicular Ad-Hoc Network Using IEEE 802.11p," *Proceedings of IEEE Wireless Communications and Networking Conference*, Hong Kong, 11-15 March 2007, pp. 254-257.

[7] M. Wellen, B. Westphal and P. Mähönen, "Performance Evaluation of IEEE 802.11-Based WLANs in Vehicular Scenarios," *Proceedings of IEEE 65th Vehicular Technology Conference*, Dublin, 22-25 April 2007, pp. 1167-1171.

[8] W. Xiang, P. Richardson and J. Guo, "Introduction and Preliminary Experimental Results of Wireless Access for Vehicular Environments (WAVE) Systems," *Proceedings of 3rd Annual International Conference on Mobile and Ubiquitous Systems: Network and Services*, San José, 17-21 July 2006, pp. 1-8.

[9] F. Bai and H. Krishnan, "Reliability Analysis of DSRC Wireless Communication for Vehicle Safety Applications," *Proceedings of IEEE Intelligent Transportation Systems Conference*, Toronto, 17-20 September 2006, pp. 355-362.

[10] S. Eichler, "Performance Evaluation of the IEEE 802.11p WAVE Communication Standard," *Proceedings of IEEE 66th Vehicular Technology Conference*, Baltimore, 30 September-3 October 2007, pp. 2199-2203.

[11] N. Choi, *et al.*, "A Solicitation-Based IEEE 802.11p MAC Protocol for Roadside to Vehicular Networks," *Proceedings of Workshops on Mobile Networking for Vehicular Environments*, Anchorage, 11 May 2007, pp. 91-96.

[12] C. Suthaputchakun and A. Ganz, "Priority Based Inter-Vehicle Communication in Vehicular Ad-Hoc Networks Using IEEE 802.11e," *Proceedings of IEEE 65th Vehicular Technology Conference*, Dublin, 22-25 April 2007, pp. 2595-2599.

[13] S. Shankar and A. Yedla, "MAC Layer Extensions for Improved QoS in 802.11 Based Vehicular Ad Hoc Networks," *Proceedings of IEEE International Conference on Vehicular Electronics and Safety*, Beijing, 13-15 December 2007, pp. 1-6.

[14] H. Lans, "Position Indicating System," US Patent 5506587, 1996.

[15] R. Kjellberg, "Capacity and Throughput Using a Self Organized Time Division Multiple Access VHF Data Link in Surveillance Applications," Master Thesis, University of Stockholm, Stockholm, 1998.

[16] D. Jiang and L. Delgrossi, "IEEE 802.11p: Towards an International Standard for Wireless Access in Vehicular Environments," *IEEE Vehicular Technology Conference*, Singapore, 11-14 May 2008, pp. 2036-2040.

[17] K. Sjöberg, "Standardization of Wireless Vehicular Communications within IEEE and ETSI," *IEEE VTS Workshop on Wireless Vehicular Communications*, Halmstad, 9 November 2011, 25p.

[18] A. Alonso, K. Sjöberg, E. Uhlemann, E. G. Ström and C. F. Mecklenbräuker, "Challenging Vehicular Scenarios for Self-Organizing Time Division Multiple Access," European Cooperation in the Field of Scientific and Technical Research, Lund, 2011.

[19] P. Mackenzie, B. Miller, D. D. Coleman and D. A. Westcott, "CWAP Certified Wireless Analysis Professional Official Study Guide," John Wiley & Sons, Hoboken, 2011.

[20] V. Shivaldova, "Implementation of IEEE 802.11p Physical Layer Model in SIMULINK," Master's Thesis, Vienna University of Technology, Vienna, 2010.

[21] M. El Masri, "IEEE 802.11e: The Problem of the Virtual Collision Management within EDCA," *Proceedings of 25th IEEE International Conference on Computer Communications*, Barcelona, 23-29 April 2006, pp. 1-2.

[22] K. Bilstrup, E. Uhlemann, E. G. Ström and U. Bilstrup, "On the Ability of the 802.11p MAC Method and STDMA to Support Real-Time Vehicle-to-Vehicle Communication," *EURASIP Journal on Wireless Communications and Networking*, Vol. 2009, No. 13, 2009, Article No. 5.

[23] K. Bilstrup, E. Uhlemann, E. G. Ström and U. Bilstrup, "On the Ability of the 802.11p and STDMA to Provide Predictable Channel Access," *Proceedings of the 16th World Congress on ITS*, 2009, 10p.

[24] K. S. Bilstrup, E. Uhlemann and E. G. Ström, "Scalability Issues for the MAC Methods STDMA and CSMA/CA of IEEE 802.11p When Used in VANETs," *IEEE International Conference on Communications (ICC2010)*, Cape Town, 23-27 May 2010.

10

Ethernet Implementation of Fault Tolerant Train Network for Entertainment and Mixed Control Traffic

Tarek K. Refaat, Mai Hassan, Ramez M. Daoud, Hassanein H. Amer
Electronics Engineering Department, American University in Cairo, New Cairo, Egypt

ABSTRACT

This paper studies the integration of the control system and entertainment on board of train wagons. Both the control and entertainment loads are implemented on top of Gigabit Ethernet, each with a dedicated controller/server. The control load has mixed sampling periods. It is proven that this system can tolerate the failure of one controller in one wagon. In a two wagon scenario, fault tolerance at the controller level is studied, and simulation results show that the system can tolerate the failure of 3 controllers. The system is successful in meeting the packet end-to-end delay with zero packet loss in all OPNET simulated scenarios. The maximum permissible entertainment load is determined for the fault tolerant scenarios.

Keywords: Ethernet; Networked Control Systems; Fault-Tolerance; Railways; Intelligent Transportation Systems; WiFi

1. Introduction

Networked Control Systems (NCS) is a rapidly expanding field, with many applications, from industrial automation to intelligent transportation systems [1-6]. The origins of NCS can be traced back to CAN, PROFIBUS and PROFINET; however Ethernet has spread rapidly in the past few decades, introducing non-deterministic protocols to the world of NCS [7-9]. Ethernet is a non-deterministic protocol, with several sources of randomness due to the use of Carrier Sense Multiple Access with Collision Detection (CSMA/CD) [10]. Such probabilistic nature is undesirable in real-time NCS implementations until certain modifications were made in order to accommodate real-time applications. Packet formats were modified in order to give certain messages priority over others [11,12]. Other modifications can be found implemented by Rockwell Automation, the ODVA, EtherNet/IP, CIP, TT Ethernet and FTT Ethernet. Some of these solutions are in course of standardization [13-19]. One famous form of NCS exists in terrestrial transportation systems and studies on the subject are numerous and extensive [1,2,20,21]. Recent studies have researched a specific implementation that utilizes the Ethernet protocol (IEEE 802.3) in trains [22-28].

Entertainment is now common in most forms of transportation ranging from terrestrial transportation systems to sea and air travel. The existence of an entertainment load alongside the control network on the same infrastructure may introduce an increased level of traffic congestion. Such a concept was tested in previous studies and it was shown that the control network performed within required deadlines [22,23].

Another growing trend in NCS is the incorporation of Fault Tolerance. Reference [3] studied the use of Fast and Gigabit Ethernet in advanced networked control systems. It was proven that the use of redundant control nodes minimizes down-time. Reference [29] investigated the effect of failures on the productivity of fault-tolerant networked control systems under varying loads. Reference [30] studied the availability of the Pyramid architecture in the context of Networked Control Systems. Reference [31] also studied the Mean Time to Failure (MTTF) of a fault-tolerant two-machine production line in the context of Networked Control Systems (NCS). More details about fault-tolerance in NCS can be found in [32].

This paper presents a study of a train control network utilizing the Ethernet protocol without modifications. It includes all previously mentioned issues namely the use of unmodified protocol communicating both real-time (control) and non-real-time applications (entertainment). It also incorporates fault-tolerance aspects. The study consists of several OPNET Network Modeler [33] simulations which model two train wagons for control and entertainment loads. The main contribution of this paper is that the control load proposed is more realistic, consisting of a mixture of different sampling periods for sensors and actuators (SA) [27]. The entertainment load is comprised of video streaming and bounded WiFi access to the Internet. The proposed design also incorporates a fault-tolerance study at the controller level. It will be shown that the simulated model is successful in both the fault-free and fault-tolerant scenarios.

The rest of this paper is organized as follows. Section 2 gives a recap of previous work on train control networks. Section 3 will introduce the newly proposed model. Section 4 presents the simulated scenarios and shows the results. Section 5 concludes this paper.

2. Previous Work

There have been several research studies modeling train networks such as LonWorks and Train Control Networks (TCN) [25-28,34]. Recent studies have modeled train wagons using Switched Ethernet [22-24,35]. The research revolved around the feasibility of a single network carrying both control and entertainment, fault-tolerance on the controller level and sensor level.

Reference [22] tested the performance of a single Ethernet infrastructure supporting both a control and an entertainment system in train wagons. Using OPNET, several scenarios with different settings were simulated. SAs were set up to use a unified sampling period across the model. Depending on the scenario, some were simulated using a sampling period of 1ms and others using 16ms, each simulated independently. The results of the study guaranteed that the packet end-to-end delay is always within constraints of the maximum allowable delay.

Reference [23] successfully introduced fault-tolerance at the controller level to the system proposed by [22]. A follow-up study introduced fault-tolerance at the sensor level using Triple Modular Redundancy (TMR), a concept described in [36,37], achieving successful results [35].

3. Proposed Model

The proposed model utilizes a 1 Gigabit Ethernet infrastructure without modifications based on the IEEE 802.3 standard for the whole network and follows the regulations described in the IEC 61375 [27]. Several sampling periods were mentioned in the standard, concerning the SAs, however the most common values were 1ms (the smallest sampling period) and 16 ms. References [22,23] studied these values independently. However, the currently proposed model incorporates both sampling periods in a single control network, representing the different possible applications of SAs. A typical number of SAs on a single train wagon is around 250 with sampling periods 1ms (minority) and 16 ms (majority) [27]. These will be broken down into 3 groups. Group 1 (G1) consists of 30 sensors and 30 actuators (1:1 ratio) operating with the most demanding sampling period of 1ms. Group 2 (G2) consists of 100 sensors and 50 actuators (2:1 ratio)

operating at 16ms sampling period. Finally, Group 3 (G3) also operating at 16ms sampling period, consists of 30 sensors and 10 actuators (3:1 ratio). The distribution is illustrated in **Figure 1**. This design minimizes the number of switches, using standard 128 port switches, readily available in the market. Also note that the locations of the SAs have been chosen to increase the distance between the switches and the controller, maximizing the trip distance to simulate a worst case scenario. The main switch (labeled MS1) utilizes a forwarding rate of 6.6 Mbps, which is much lower than rates available in models such as the Cisco Catalyst 3560 Gigabit Ethernet switches [38].

The entertainment load in the train model can be described in terms of the number of streams, the quality of the video screens as well as the number of WiFi nodes and the number of applications per node. The worst case scenario, in a large wagon of 60 seats [25], supports 60 different and simultaneously playing DVD quality video streams (one per seat) [39], as well as one WiFi user per seat. Each WiFi user runs several simultaneous applications: Database access, email sending/receiving, web browsing and file transfer. This gives a total of 60 Video streams and 60 WiFi users, each running 4 simultaneous applications requiring access to the network. These WiFi applications are simulated using the generic built-in heavy load applications of OPNET. This WiFi load requires a maximum bandwidth of 6Mbps and therefore only 6Mbps are allocated for the WiFi traffic. The WiFi access is provided via the Access Point (AP) in the middle of the wagon, maximizing coverage area. Lastly, the model has 2 dedicated controllers per wagon. The existence of two controllers allows the incorporation of fault-tolerance at the controller level as explained next. Wagon 1 has K1 and E1, where K1 is a controller, handling the control load only, and E1 is the entertainment server. The same is true for Wagon 2.

Each controller can readily take over in the case of failure of the other. This is achieved by having sensors send the sampled data at every sampling instant to all available controllers. Should any given controller have to take over for a failed controller, it must have the most updated sampled data in order to achieve zero packet loss. All simulations are modeled using OPNET.

4. Simulated Scenarios and Results

The proposed model is used to form 2 main tests, each broken down into 2 scenarios.

4.1. Simulated Scenarios

The first set of simulations is performed for a single wagon model with 60 video streams and 60 WiFi users. The first scenario simulates the fault-free case, where

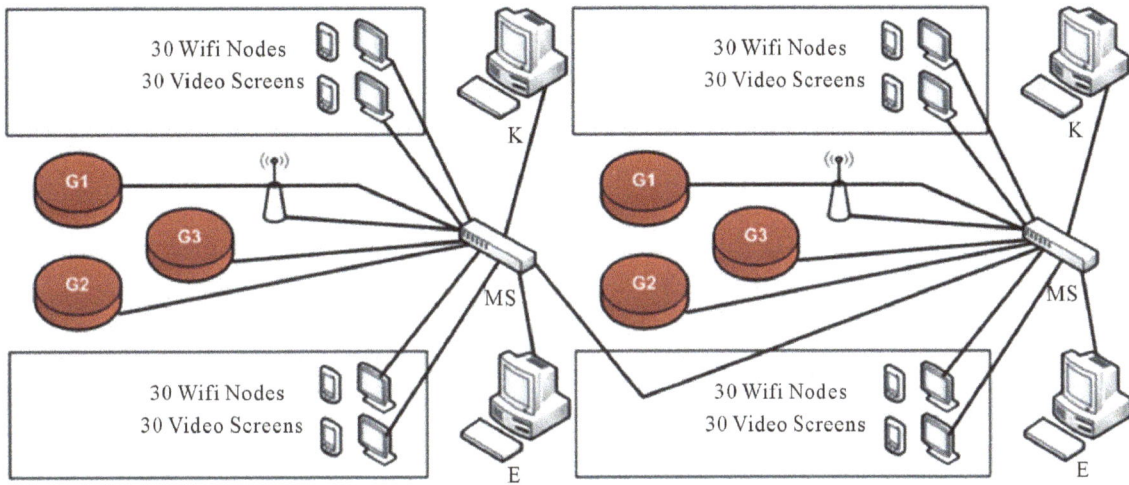

Figure 1. Two-wagon overview.

both controllers are fully functional; K1 handling all control data and E1 handling the entertainment load. In the second scenario (fault-tolerant), one of the two controllers is off, simulating a failure. Both control and entertainment loads are handled by the remaining controller. Then an attempt is made to maximize the number of video streams and WiFi users without jeopardizing the control load.

The second set of simulations takes the model a step further, concatenating two identical wagons and hard-wiring the main switch of each wagon to the other. Modifications are made such that all sensors will send their data to all 4 controllers rather than merely the two on their corresponding wagons. Other communication techniques can be found in [40]. The total number of video streams and WiFi users is now double. Again, the first scenario simulates the fault-free case, where all 4 controllers are fully functional. The next scenario (fault-tolerant) models an extreme case, where 3 of the 4 controllers have failed, and now only 1 controller has to carry the control and entertainment load of both wagons. The goal again is to maximize the number of video streams and WiFi users, without the control load suffering.

4.2. Simulated Results

In order to gauge the performance of the system, end-to-end delay and packet loss must be monitored. In all simulations, zero packet loss (no packets dropped or delayed) is observed and total end-to-end delays across all SAs are within their respective constraints [41]. A 95% confidence analysis is applied to all results. **Figures 2-5** show examples of results from different scenarios. In all figures, the x-axis represents the simulation time in minutes and seconds, while the y-axis shows the delay in seconds. The figures show delay measured at an actuator

from one of the three groups (as indicated in the caption: e.g., G1). These delays represent the time taken for a packet to travel from the K to the A, and in cases like **Figures 3** and **4** the packet delays oscillate between several values, depending on the level of network congestion faced.

With all video streams at DVD quality and all WiFi users running the full load of applications described in Section 3, the maximum total end-to-end delay (with a 95% confidence) for each group of SAs is shown in **Table 1**. The maximum number of streams and WiFi users to be supported in each simulated scenario are shown in **Table 2**. These results guarantee that all SAs operating with a sampling period of 1ms have an end-to-end delay of less than 1ms, and those operating at 16 ms have an end-to-end delay of less than 16 ms (with 95% confidence). It is important to note that the overhead incurred on the network by the WiFi nodes is negligible in comparison to the control and the video streaming loads. As this load is extremely small, no matter what state the network is in, the WiFi network would be unaffected (as can be seen in **Table 2**). This can also be attributed to the fact that the bandwidth allotted to the WiFi users is bounded/restricted to 6 Mbps.

5. Conclusion

Ethernet implementation for Networked Control Systems (NCS) is a growing field. Train networks utilizing Ethernet have been previously studied, carrying both an entertainment load and a control load on a single network. This paper studied a fault-tolerant train NCS utilizing Switched Ethernet without modifications, with an entertainment load of video streams and WiFi access. The model was successfully simulated, showing correct packet transmission and reception, with zero losses in

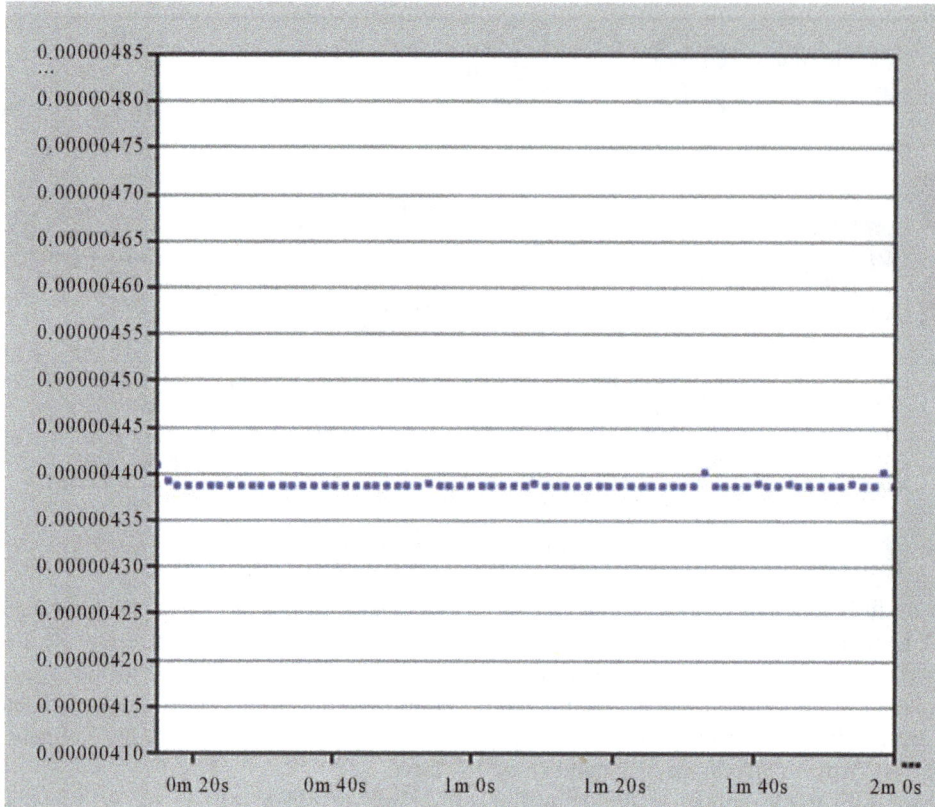

Figure 2. Two wagon fault free scenario (G1).

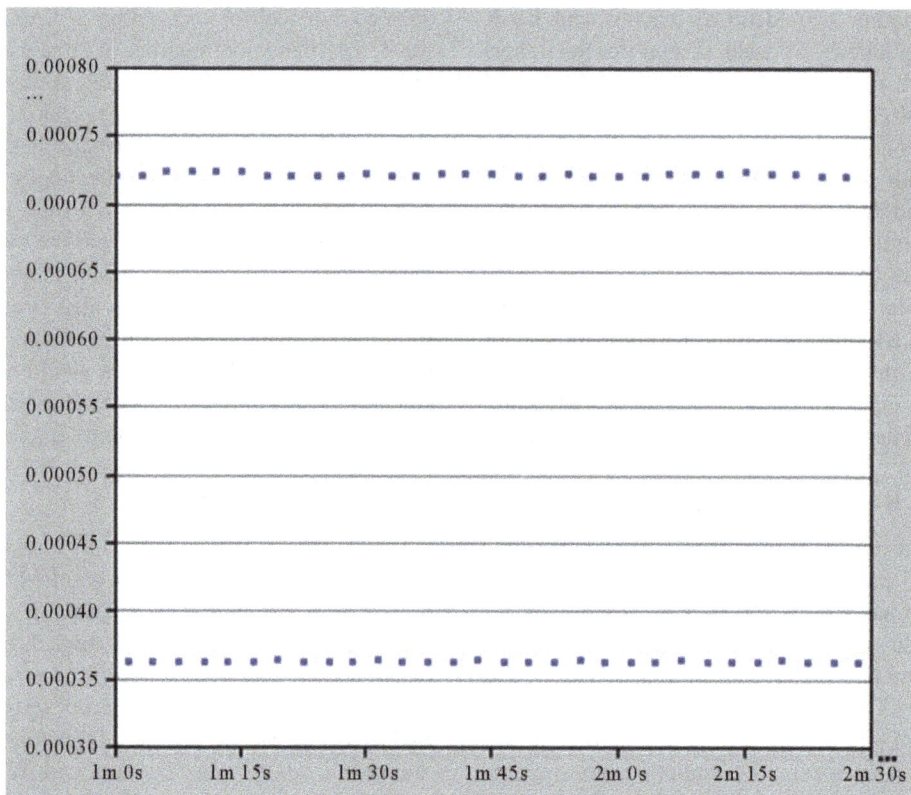

Figure 3. Two wagons fault tolerant scenario (G1).

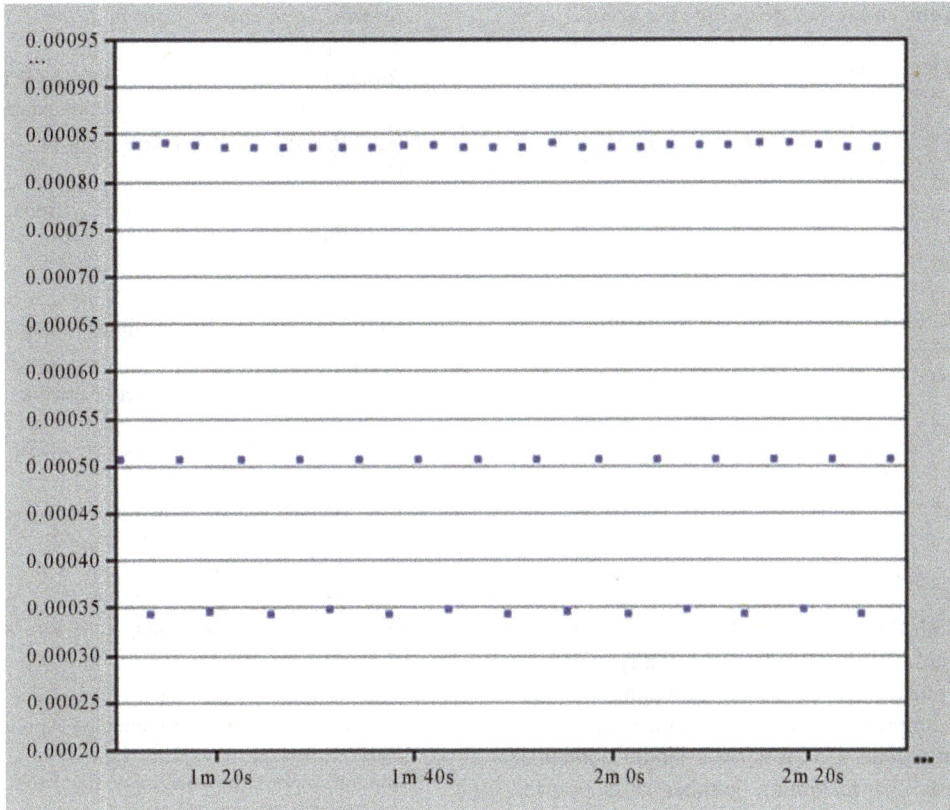

Figure 4. Two wagons fault tolerant scenario (G2).

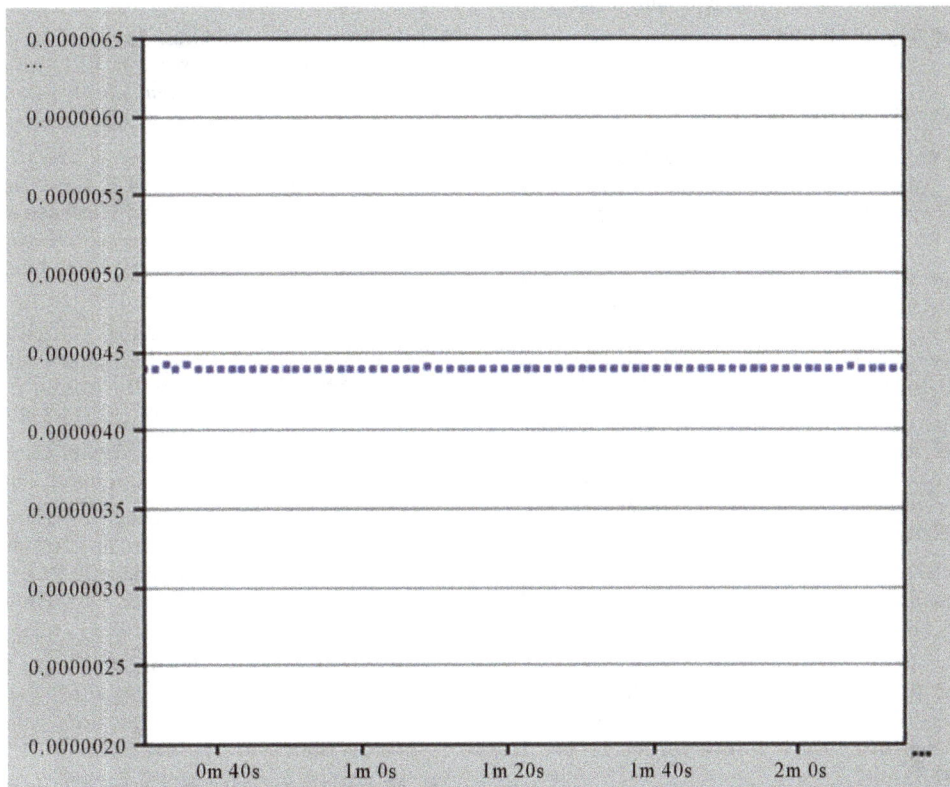

Figure 5. Two wagons fault free scenario (G3).

Table 1. Total end-to-end delay (ms) per group.

Wagons	Operational Ks	G1	G2	G3
1	2	0.010	0.011	0.011
1	1	0.751	0.870	0.876
2	4	0.016	0.013	0.026
2	1	0.772	0.908	0.922

Table 2. Entertainment load per scenario.

Wagons	Operational Ks	Video Streams	WiFi Users
1	2	60	60
1	1	6	60
2	4	120	60
2	1	6	60

control data. Both the single wagon and the two-wagon model were simulated, in fault-free and faulty scenarios showing zero packet loss and delays within the constraints of the sampling periods of the control network. However, when a load increases on a single controller due to a failure, the maximum number of supported streams, is reduced. This does not reduce the number of passengers with a capability of watching a video; it merely limits the number of possible simultaneously playing videos at any given time.

REFERENCES

[1] N. Navet, Y. Song, F. Simonot-Lion and C. Wilwert, "Trends in Automotive Communication Systems," *Proceedings of the IEEE*, Vol. 93, No. 6, 2005, pp. 1204-1223.

[2] R. M. Daoud and H. H. Amer, "Fast Ethernet Implementation of Cost Effective Vehicle On-Board Networks," In: S. Pennacchio, Eds., *Emerging Technologies, Robotics and Control Systems*, Ass. Internationalsar, Bologna, 2007, pp. 81-85.

[3] R. M. Daoud, H. M. Elsayed and H. H. Amer, "Gigabit Ethernet for Redundant Networked Control Systems," *Proceedings of the IEEE International Conference on Industrial Technology ICIT*, Vol. 2, Hammamet, 8-10 December 2004, pp. 869-873.

[4] J. D. Decotignie, "Ethernet-Based Real-Time and Industrial Communications," *Proceedings of the IEEE*, Vol. 93, No. 6, 2005, pp. 1102-1117.

[5] F.-L. Lian, J. R. Moyne and D. M. Tilbury, "Performance Evaluation of Control Networks: Ethernet, ControlNet, and DeviceNet," *IEEE Control Systems*, Vol. 21, No. 1, 2001, pp. 66-83.

[6] M. Tabbara, A. Rantzer and D. Nešić, "On Controller & Capacity Allocation Co-Design for Networked Control

Systems," *Systems & Control Letters*, Vol. 58, No. 9, 2009, pp. 672-676.

[7] Bosch, "CAN in Passenger and Cargo Trains. CAN in Automation," 2011. http://www.can-cia.org/

[8] PI, "PROFIBUS & PROFINET," 2011. http://www.profibus.com/

[9] T. Skeie, S. Johannessen and C. Brunner, "Ethernet in Substation Automation," *IEEE Control Systems*, Vol. 22, No. 3, 2002, pp. 43-51.

[10] IEEE 802.3 Standard. http://standards.ieee.org/about/get/802/802.3.html

[11] S. H. Lee and K. H. Cho, "Congestion Control of High-Speed Gigabit-Ethernet Networks for Industrial Applications," *Proceedings of the IEEE International Symposium on Industrial Electronics ISIE*, Vol. 1, Pusan, 12-16 June 2001, pp. 260-265.

[12] J. S. Meditch and C. T. A. Lea, "Stability and Optimization of the CSMA and CSMA/CD Channels," *IEEE Transactions on Communications*, Vol. 31, No. 6, 1983, pp. 763-774.

[13] ODVA, "EtherNet/IP Adaptation on CIP," CIP Common. http://www.odva.org/

[14] Allen-Bradley, "EtherNet/IP Performance and Application Guide," Rockwell Automation Application Solution, 2003. http://ab.rockwellautomation.com/networks-and-commun ications/ethernet-ip-network

[15] IEC 61784-1. www.iec.ch

[16] IEC 61784-2. www.iec.ch

[17] J. Ferreira, P. Pedreiras, L. Almeida and J. Fonseca, "Achieving Fault-Tolerance in FTT-CAN," *Proceedings of the 4th IEEE International Workshop on Factory Communication Systems WFCS*, Vasteras, August 2002, pp. 125-132.

[18] P. Pedreiras, L. Almeida and P. Gai, "The FTT-Ethernet Protocol: Merging Flexibility, Timeliness and Efficiency," *Proceedings of the IEEE Euromicro Conference on Real-Time Systems ECRTS*, Vienna, 19-21 June 2002, pp. 134-142.

[19] K. Steinhammer and A. Ademaj, "Hardware Implementation of the Time-Triggered Ethernet Controller," *Embedded System Design: Topics, Techniques and Trends*, Vol. 231, Springer, Boston, 2007, pp. 325-338.

[20] Vector CANtech Inc., "Overview of Current Automotive Protocols," 2003. www.vector-cantech.com

[21] B. P. Upender, "Analyzing the Real-Time Characteristics of Class C Communications in CAN through Discrete Event Simulations," SAE Technical Paper 940133, United Technologies Research Center, East Hartford, 1994.

[22] M. Aziz, B. Raouf, N. Riad, R. M. Daoud and H. M. Elsayed, "The Use of Ethernet for Single On-board Train Network," *Proceedings of the IEEE International Conference on Networking, Sensing and Control ICNSC*, Sanya, 6-8 April 2008, pp. 1430-1434.

[23] M. Hassan, S. Gamal, S. Louis, G. F. Za and H. H. Amer, "Fault Tolerant Ethernet Network Model for Control and Entertainment in Railway Transportation Systems," *Pro-*

ceedings of the Canadian Conference on Electrical and Computer Engineering CCECE, Niagara Falls, 4-7 May 2008, pp. 000771-000774.

[24] M. Hassan, R. M. Daoud and H. H. Amer, "Two-Wagon Fault-Tolerant Ethernet Networked Control System," *Proceedings of the Applied Computing Conference*, Istanbul, 2008, pp. 346-351.

[25] H. Kirrmann and P. A. Zuber, "The IEC/IEEE Train Communication Network," *IEEE Micro*, Vol. 21, No. 2, 2001, pp. 81-92.

[26] T. Sullivan, "The IEEE 1473-L Communications Protocol: Experience in Rail Transit," Transportation Systems Design Inc., Oakland, 2002.

[27] International Electrotechnical Committee, "IEC 61375," Train Communication Network, Geneva, 1999. http://www.iec.ch

[28] Siemens, "Siemens AG Transportation Systems Trains," In: *Trains Reference List*, 2006, pp. 41-46. www.siemens.com/trasportation/trains

[29] H. H. Amer, M. S. Moustafa and R. M. Daoud, "Optimum Machine Performance in Fault-Tolerant Networked Control Systems," *Proceedings of the IEEE EUROCON Conference*, Belgrade, 21-24 November 2005, pp. 346-349.

[30] H. H. Amer, M. S. Moustafa and R. M. Daoud, "Availability of Pyramid Industrial Networks," *Proceedings of the Canadian Conference on Electrical and Computer Engineering CCECE*, Ottawa, May 2006, pp. 1862-1865.

[31] H. H. Amer and R. M. Daoud, "Parameter Determination for the Markov Modeling of Two-Machine Production Lines," *Proceedings of the International IEEE Conference on Industrial Informatics INDIN*, Singapore, 16-18 August 2006, pp. 1178-1182.

[32] M. Blanke, M. Kinnaert, J. Lunze and M. Staroswiecki, "Diagnosis and Fault-Tolerant Control," 2nd Edition, Springer-Verlag Berlin Heidelberg, Berlin, 2006.

[33] OPNET Network Modeler, 2011. www.opnet.com

[34] H. Kitabayashi, K. Ishid, K. Bekki and M. Nagasu, "New Train Control and Information Services Utilizing Broadband Networks," 2004. www.hitachi.com

[35] M. Hassan, G. F. Zaki, R. M. Daoud, H. M. ElSayed and H. H. Amer, "Reliable Train Network Using Triple Modular Redundancy at the Sensors Level," In: S. Pennacchio, Ed., *Emerging Technologies, Robotics and Control Systems*, Ass. Internationalsar, Bologna, 2008, pp. 39-44.

[36] D. P. Siewiorek and R. S. Swarz, "Reliable Computer Systems—Design and Evaluation," A. K. Peters, Natick, 1998.

[37] K. S. Trivedi, "Probability and Statistics with Reliability, Queuing, and Computer Science Applications," Wiley, New York, 2002.

[38] Cisco Systems, "Cisco Catalyst 3560 Series Switch," 2011. http://www.cisco.com/en/US/products/hw/switches/ps5528/

[39] Amazon.com, Inc., "V.O.D Streaming Speed," 2011. http://www.amazon.com/gp/help/customer/display.html?nodeId=3748&#speed

[40] G. Marsal, "Evaluation of Time Performances of Ethernet-Based Automation Systems by Simulation of High-Level Petri Nets," Ph.D. Thesis, École Normale Supérieure de Cachan, Cachan, 2006.

[41] R. M. Daoud, "Wireless and Wired Ethernet for Intelligent Transportation Systems," D.Sc. Dissertation, LAMIH-SP, University of Valenciennes and Hainaut-Cambresis, Valenciennes, 2008.

11

Assessing the Impacts of High Speed Rail Development in China's Yangtze River Delta Megaregion

Xueming Chen

L. Douglas Wilder School of Government and Public Affairs, Virginia Commonwealth University, Richmond, USA

ABSTRACT

This paper assesses the impacts of high speed rail (HSR) development in the Yangtze River Delta (YRD) Megaregion, China. After giving an introduction and conducting a literature review, the paper proposes a pole-axis-network system (PANS) model guiding the entire study. On the one hand, the HSR projects in the YRD Megaregion are expected to generate significant efficiency-related transportation and non-transportation benefits. As a result, the spillover effects from Shanghai and other major cities (poles) will greatly promote the urban and regional developments along the major HSR corridors (axes), and the entire megaregion will become more integrated economically, socially, and culturally. But, on the other hand, the HSR projects will also create serious social and geographic inequity issues, which need to be addressed as soon as possible in a proper way. This empirical study confirms the PANS model proposed.

Keywords: High Speed Rail; China; Yangtze River Delta Megaregion; Impacts

1. Introduction

China is building the world's largest high speed rail (HSR) network, which is expected to generate significant though uneven impacts on its mobility, accessibility, socioeconomic development, and others, especially at megaregional levels. According to the "Mid-to-Long Term Railway Network Plan" approved by the State Council in 2004, China will have a total of 100,000 km (revised to 120,000 km in 2008) of railroads by the year 2020, of which 12,000 km (revised to 16,000 km in 2008) are high speed rail lines with an average operating speed of 200 km/h and faster, linking all provincial capital cities and those cities with more than 500,000 population [1]. For reference, as of early 2008, there were only approximately 10,000 km of high speed rail lines in operation in the entire world, including about 2000 km in Japan and about 1900 km in France [2].

Of all Chinese megaregions, the Yangtze River Delta (YRD) Megaregion (1% of national land) occupies the most prominent position in terms of its economic strength (20% of national gross domestic products) and population size (6% of national population). This megaregion is expected to possess about a dozen of existing and proposed rail lines, including trunk passenger dedicated lines (PDLs), intercity HSR lines, and other conventional rail lines. HSR has its relative speed advantage over highway and aviation for the distance between 100 miles and 500 miles, which, to a large extent, matches a typical megaregion's geographic extent in the world. Because of this reason, HSR network directly impacts the megaregion-wide development, and vice versa.

Following this introduction, the paper contains additional six sections. Section 2 reviews the most important and relevant literatures. Section 3 presents a research methodology guiding the entire study. Section 4 describes the HSR lines traversing this megaregion. Section 5 assesses the HSR-generated transportation and non-transportation impacts from the efficiency's standpoints. Section 6 examines the geographic equity issues associated with the HSR developments in this region. Finally, Section 7 summarizes research findings and draws conclusions.

2. Literature Review

Japan and Europe are the leading country and continent to develop HSR technology. Both Japanese and European HSR development experience indicates that HSR has the indisputable transportation efficiency impacts. However, its transportation equity and socioeconomic impacts are more controversial and elusive [3].

In terms of its transportation efficiency, HSR can directly achieve time savings due to its higher operating speeds. For example, the Tokaido Shinkansen began service on October 1, 1964, in time for the Tokyo Olympics. The conventional Limited Express service took six hours and 40 minutes to travel from Tokyo to Osaka, but the Shinkansen made the trip in just four hours in 1964, shortened to three hours and ten minutes in 1965, further down to about two hours at present.

Reference [4] finds that HSR directly contributes to a favorable modal shift towards HSR train, based on his study on the French Train Grande Vitesse (TGV) and Spanish Alta Velocidad Española (AVE) operating performance. From **Table 1**, it can be seen that HSR train and aircraft are direct competitors.

In spite of its obvious transportation efficiency benefits, HSR typically will not benefit different regions equally. Reference [6] evaluates the impact of the future European high speed train network on accessibility by reducing travel time between places and modifying their relative locations. His accessibility indicators consist of calculating a weighted average of the travel times separating each node with respect to the chief economic activity centers, and economic potential model. He found that high speed train will bring the peripheral regions closer to the central ones, but will also increase imbalances between the main cities and their hinterlands. Reference [7] also analyzes the accessibility impact of the Madrid-Barcelona-French border high speed line, which finds that at the national level, the new line will lead to an increase in inequality in the distribution of accessibility, for the cities, which have greatest increases in accessibility, are already highly accessible in the "without the new line scenario".

Reference [8] conducts the entire Chinese HSR accessibility study. Assuming *with HSR* and *without HSR* scenarios, [8] calculates the weighted average travel time by region, HSR access, and city size, which finds that: cities with HSR stations receive higher accessibility benefit than non-HSR cities; and larger cities receive more HSR benefits than smaller cities.

The HSR's socioeconomic impacts are more disputable, though. For example, the Japanese Shinkansen has generated profound impacts on its service sector development. Between 1981 and 1985, the number of employees in information, survey and advertising industries increased by 125% in the areas with both express highway and high speed rail, but only by 63% in the areas with express highway only [9]. In addition, the regions served by the Shinkansen achieved higher population and employment growth rates than those without direct Shinkansen services. However, it should be cautioned that there may be other factors prevailing in these regions as well that could support and affect such an impact, and it is thus unclear if the Shinkansen indeed led to the increase in growth rates or if the Shinkansen was merely constructed in regions where higher growth rates had already existed [10].

Reference [11] acknowledges that spatial dynamics are often associated with the development of physical and social networks in which the nodes (e.g. cities and towns) profit from agglomeration advantages and scale effects. This means that only those cities and places located in the immediate vicinity of HSR stations typically reap benefits. [12] argues that a TGV link is closer to that of an airliner than that of a traditional train when one considers the lengths of journeys from city to city, the seating capacity (and therefore the commercial objectives regarding rate of utilization) and the means of operation (city A-city B). The result is that the structural effects are centered on urban poles and their immediate environs. Because of TGV, Lille, which is the largest industrial city in northern France, has been transformed from an industrial city to a commercial and business-oriented city. Many tertiary sector firms have been using TGV services to travel between Lyon and Paris.

Generally speaking, positive spatial and socio-economic impacts might occur at places connected to the HSR network, yet in those places bypassed by the HSR, negative impacts usually occur because resources are reallocated and gravitated to those HSR-connected places [13]. In another paper, [14] further points out that core metropolitan areas with HSR stations may attract more people and economic activities, while other small cities remain left behind.

With regard to the relationships between rail investment and economic growth, [15] cautions that rail investments do not stimulate economic growth but influence "already-committed" growth and rarely have a significant influence on development patterns. Reference [4]

Table 1. Modal shares before and after the introduction of high speed train services.

Mode	TGV, Paris-Lyon Line			AVE, Madrid-Seville Line		
	Before (1981)	After (1984)	Change	Before (1991)	After (1994)	Change
Aircraft	31%	7%	−24%	40%	13%	−27%
HSR Train	40%	72%	32%	16%	51%	35%
Car and bus	29%	21%	−8%	44%	36%	−8%

Source: [4,5].

warns that cities with bad economies have a difficulty taking advantage of HSR, and may even suffer economic decline. Therefore, HSR is not justified solely based on economic development benefits because the benefits are not great enough. Reference [16] cogently points out that transport infrastructure investment in general acts as a complement to other more important underlying conditions (for example, economic externalities, investment factors, and political factors), which must be met if further economic development is to take place. In the meantime, the potential ability of transport infrastructure investments to produce transport benefits, such as travel time reductions, is not questioned.

Unlike aviation which only impacts a flight's origin and destination metropolises (point effects), HSR generates line effects as well. Reference [17] argues that the HSR not only can be used as a substitute for air traveling, but also can link many cities with an interregional accessibility, thus forming a functional region. In Spain, the HSR network was initially intended to strengthen the relations between distant metropolises (e.g., Madrid, Barcelona, Seville), but it is also showing its usefulness for smaller and closer cities that lie between metropolises.

In summary, there is no universally agreed-on consensus on the socioeconomic impacts of HSR development, which vary from places to places.

3. Research Methodology

This paper utilizes the economic potential model to quantify one of the most important transportation efficiencies: accessibility. According to [7], the economic potential is a gravity-based measure widely used in accessibility studies. According to this model, the level of opportunity (accessibility) between origin node i and destination node j is positively related to the mass of the destination and inversely proportional to some power of the distance or travel time between both nodes. Its mathematical equation is as follows:

$$P_i = \sum_{j=1}^{n} \frac{M_j}{T_{ij}^a}$$

where P_i is the economic potential of origin node i; α is a parameter reflecting the rate of increase of the friction of distance (distance decay. α is assumed to be 1 in most studies); M_j is the mass of destination node j (e.g., gross domestic products, population); T_{ij} is the travel time between origin node i and destination node j.

With respect to quantifying geographic equity issues, this paper uses the Pole-Axis-Network System (PANS) model originally developed by Professor Dadao Lu, who is the renowned Chinese geographer. According to [18-20], the "pole-axis system" is one of the concepts concerning socio-economic spatial structure, and a theoretical model for allocation of productivity, territorial development and regional development. This concept was developed based on the "central place theory" hypothesized by German geographer Walter Christaller [21], the growth pole theory proposed by François Perroux [22], and other regional development theories [23]. In the national and regional development process, most socioeconomic elements are agglomerated at "poles (yielding agglomeration economies)" which are linked up by linear infrastructure to form an "axis". Through axes, socioeconomic factors are diffused along axes from poles to poles, thus turning nodal developments into regional developments. It should be pointed out that both poles and axes have hierarchical orders.

This paper extends this concept into the pole-axis-network system (PANS) concept through an empirical study. As illustrated in **Figure 1**, this schematic diagram assumes the existence of four tiers of cities (its sizes are symbolized by circle sizes) and three tiers of axes (Note: The Chinese province typically has four tiers of cities: provincial capital cities, provincial-ranked cities, regional cities, and county-level cities):

- Tier-1 city: the largest central city, connected to Tier-1, Tier-2 and Tier-3 axes;
- Tier-2 cities: the large-sized cities, connected to Tier-1 and Tier-2 axes;
- Tier-3 cities: the medium-sized cities, connected to Tier-2 and Tier-3 axes; and
- Tier-4 cities: the small cities, connected to Tier-3 axes.

The PANS model proposed above can be applied to the YRD area to quantify the HSR-induced geographic inequity degrees. Since each city is connected to different tiers of axes, this paper defines the following city-level connectivity indices (CI) to compare the HSR-induced accessibilities among different cities (Note: Scale assumptions: Scale for Tier-1 Axis = 5; Scale for Tier-2 Axis = 3; Scale for Tier-3 Axis = 1):

$$CI = 5 * \text{Number of Connected Tier} - 1 \text{ Axes}$$
$$+ 3 * \text{Number of Connected Tier} - 2 \text{ Axes}$$
$$+ 1 * \text{Number of Connected Tier} - 3 \text{ Axes}$$

4. Existing and Future Rail Lines in the YRD Megaregion

4.1. Inventory of Existing and Future Rail Lines

Figure 2 and **Table 2** show the alignments and characteristics of the rail lines in the YRD Megaregion, respectively. In **Figure 2**, the tiers of rail lines are determined based on the economic densities of the cities these rail lines traverse.

The YRD Megaregion has or will have three types of

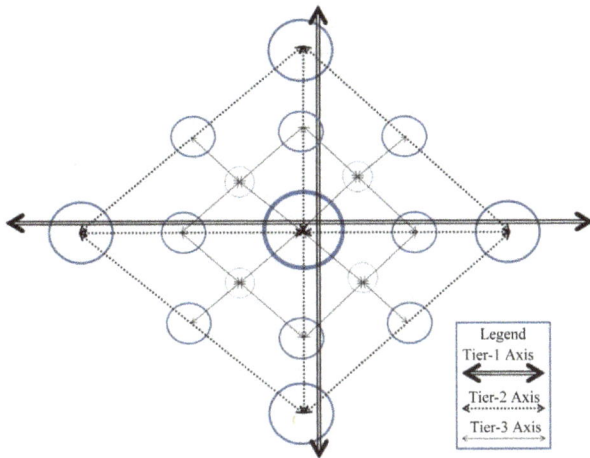

Figure 1. Pole-axis-network schematic diagram.

Figure 2. Alignments of the rail lines in the YRD megaregion.

rail lines, all of which have been approved for construction by the State Council through the National Development and Reform Commission:

- HSR Passenger Dedicated Lines (PDLs): intermediate- and long-distance trunk rail lines, with a design speed around 350 km/h. Examples: Beijing-Shanghai (Jinghu) PDL, and Shanghai-Kunming (Hukun) PDL;
- Intercity HSR Lines: short-distance rail lines serving neighboring cities within the megaregion, with a de-

sign speed of 300 - 350 km/h. Examples: Shanghai-Nanjing (Huning) Line, Shanghai-Hangzhou (Huhang) Line, Nanjing-Hangzhou (Ninghang) Line. This type of HSR line is a new mode lying between trunk rail lines and urban rail transportation, belonging to the category of regional rail system as defined by [24] and primarily serving daily commutes among neighboring cities. Since the travel distance is normally within 400 km, it takes no more than 2 hours to arrive at destinations by using this mode; and

- Conventional rail lines: shared passenger/freight operations, generally with a design speed around or below 200 km/h. Examples: Nanjing-Qidong (Ningqi) Rail Line, Ningbo-Taizhou-Wenzhou (Ningtaiwen) Rail Line, Xuancheng-Hangzhou (Xuanhang) Rail Line.

Since Shanghai-Nanjing and Shanghai-Hangzhou rail lines are the most important rail lines with the highest passenger transportation densities in China and specifically serving the YRD Megaregion, they are described at length below. See **Figure 3** for the rail line alignments and the major cities they link.

4.2. Shanghai-Nanjing Rail Lines

The Shanghai-Nanjing Corridor has three roughly paralleling rail lines: the old conventional rail line in existence since 1908; the new intercity HSR line opening on July 1, 2010 (Huning Line); and the southern Jiangsu segment of the new Beijing-Shanghai PDL opened in 2011 (Jinghu Line). The old conventional rail line will gradually be converted into a freight rail line. The latter two HSR lines will provide exclusive passenger rail services. In this way, the passenger and freight operations will be separated, thus greatly enhancing the corridor's total carrying capacity. **Table 3** lists those HSR stations along the Shanghai-Nanjing Corridor.

Due to their different service types, the Jinghu Line has a much longer average station spacing (41 km/segment) than does the Huning Line (15 km/segment). The former serves large cities and through traffic, whereas the latter serves both large and small cities, and local traffic between Nanjing and Shanghai. For most cities along the corridors, the Huning Line Stations and Jinghu Line Stations typically have different geographic locations, with the former being located much closer to the downtown areas than the latter. The Jinghu Line Stations are typically the brand new stations built in the near suburban areas, with a longer access time.

4.3. Shanghai-Hangzhou Rail Lines

Like the Shanghai-Nanjing Corridor, the Shanghai-Hangzhou Corridor also has three roughly paralleling rail lines: the old conventional rail line in existence since

Table 2. Characteristics of the rail lines in the YRD megaregion.

Rail Line Name	General Orientation	Traversed Municipality and Province(s)	Year Opened or to be Opened	Comments
Changzhou-Suzhou-Jiaxing (Changsujia) Line	West-East-South	Jiangsu, Zhejiang. 202 km.	N/A	In planning stage.
Nanjing-Hangzhou (Ninghang) Line	Northwest-Southeast	Jiangsu, Zhejiang. 251 km long.	2011 (Class I Double Line Railroad)	Design speed: 350 km/h. Total construction cost: ¥23.75 billion yuan. 11 stations.
Nanjing-Qidong (Ningqi) Line	East-West	Jiangsu. 351 km long.	2005 (Class I Single Line Railroad. Phase I from Nanjing to Nantong completed)	Phase II from Nantong to Qidong has not started yet.
Nanjing-Wuhu (Ningwu) Line	North-South	Jiangsu, Anhui. 92 km long.	1958 (Conventional Rail)	Will be connected to Nanjing South Railway Station.
Ningbo-Taizhou-Wenzhou (Yongtaiwen) Line	North-South	Zhejiang. 268 km long.	2009 (Class I Double Line Electrified Railroad.)	Design speed: 250 km/h. Total construction cost: ¥17 billion yuan.
Shanghai-Hangzhou (Huhang) Lines	Northeast-Southwest	Shanghai, Zhejiang. 172 km long.	1909 (Conventional Rail)	Its new HSR line (160 km) opened on October 26, 2010. Design speed: 350 km/h. 9 stations. In addition, the Shanghai-Kunming PDL will also traverse this segment.
Shanghai-Nanjing (Huning) Lines	Northwest-Southeast	Jiangsu. 301 km long.	1908 (Conventional Rail)	This corridor will have two additional paralleling HSR lines. The Shanghai-Nanjing Intercity HSR line opened on July 1, 2010. The Beijing-Shanghai Express Railway traversing this region will open in 2011.
Shanghai-Nantong (Hutong)	North-South	Shanghai, Jiangsu. 249 km long.	N/A	In planning stage.
Xiaoshan-Ningbo (Xiaoyong) Line	East-West	Zhejiang. 147 km long.	1959 (Conventional Rail)	Now it is the Class I Double Line Electrified Railroad. 13 passenger/freight shared stations.
Xinyi-Changxing (Xinchang) Line	North-South	Jiangsu, Zhejiang. 561 km long.	2005 (Conventional Rail)	Design speed: 120 km/h.
Xuancheng-Hangzhou (Xuanhang) Line	East-West	Anhui, Zhejiang. 224 km long.	1992 (Conventional Rail)	Completed.

Source: [25,26].

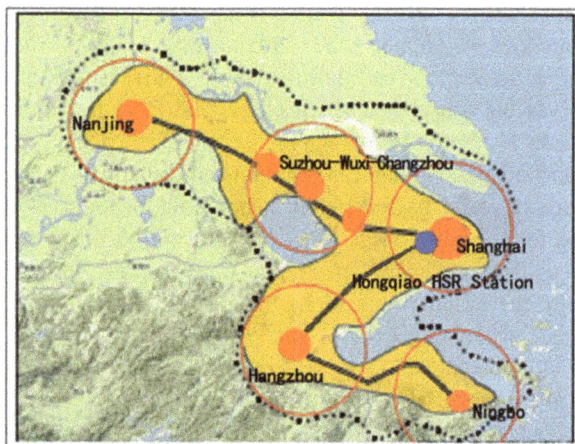

Figure 3. Shanghai-Nanjing and Shanghai-Hangzhou rail lines.

1909; the new intercity HSR line opening on October 26, 2010 (Huhang Line); and the northern Zhejiang segment of the new Shanghai-Kunming PDL scheduled to be open in 2015 (Hukun Line). The old conventional rail line will gradually be converted into a freight rail line. The latter two HSR lines will provide exclusive passenger rail services. **Table 4** lists those stations located along both rail lines.

Same as in the Shanghai-Nanjing Corridor, along the Shanghai-Hangzhou Corridor, the Hukun Line has a much longer station spacing (79 km/segment) than the Huhang Line (20 km/segment) due to their different markets served. The former serves large cities and through traffic, whereas the latter serves both large and medium-sized/small cities, and local traffic between Shanghai and Hangzhou.

Table 3. Stations along the Shanghai-Nanjing corridor.

City/Municipality	Huning Line Stations	Jinghu Line Stations
Shanghai Municipality	• Shanghai • Shanghai West • Nanxiang North • Anting North	• Shanghai Hongqiao
Suzhou City	• Huaqiao • Kunshan South • Yangcheng Lake • Suzhou Industrial Park • Suzhou • Suzhou New Area	• Kunshan South • Suzhou North
Wuxi City	• Wuxi New Area • Wuxi • Huishan	• Wuxi East
Changzhou City	• Qishuyan • Changzhou	• Changzhou North
Zhenjiang City	• Danyang • Dantu • Zhenjiang • Baohuashan	• Danyang North • Zhenjiang West
Nanjing City	• Xianlin • Nanjing	• Nanjing South

Table 4. Stations along the Shanghai-Hangzhou corridor.

City/Municipality	Huhang Line Stations	Hukun Line Stations
Shanghai Municipality	• Shanghai Hongqiao • Songjiang South • Jinshan North	• Shanghai Hongqiao
Jiaxing City	• Jiashan South • Jiaxing South • Tongxiang • Haining West	• Jiaxing South
Hangzhou City	• Yuhang South • Hangzhou East	• Hangzhou East

5. Efficiency Impacts of HSR Lines

This section focuses on assessing the transportation efficiency impacts of HSR lines, especially Huning Line and Huhang Line. Both Huning Line and Huhang Line will generate significant transportation and non-transportation impacts, as summarized below.

5.1. Transportation-Related Impacts

The transportation-related impacts of HSR projects are multifold. Below are some highlights.

5.1.1. Reduced Travel Times
The Huning rail line will bring travel time savings and an improved accessibility to those metropolitan areas and cities affected: Shanghai, Suzhou, Wuxi, Changzhou, Zhenjiang, and Nanjing metropolitan areas. Compared to conventional train, the HSR saves about 70 minutes of one-way travel time. Even compared to multiple unit train (MUT) (D-series HSR train), G-series HSR train

still saves about 40 minutes of one-way travel time.

The Huhang Line will also achieve one-way travel time savings of 30 minutes (compared to multiple Unit train) and 50 minutes (compared to conventional train).

5.1.2. Improved Accessibility
This paper calculates the HSR-related economic potentials of the key YRD cities using the latest train schedules. See **Tables 5** and **6** for details. Two scenarios are assumed: Fast travel time scenario with the high speed rail technology assumed; and the slowest travel time scenario with the conventional rail technology assumed. This calculation confirms the findings of [27] that the core areas of the YRD Megaregion (Suzhou, Wuxi, Changzhou, and Shanghai) on the Jiangsu Province side reap more benefits from the HSR development than the relatively

Table 5. HSR-related economic potentials of the key YRD cities (fastest travel time assumed).

City	GDP (Billion Yuans)	Economic Potential (EP)	Ranks of EP in Descending Order
Suzhou	774	307.95	1
Wuxi	499.2	276.08	2
Shanghai	1490.1	214.36	3
Changzhou	251.9	171.50	4
Jiaxing	191.8	129.46	5
Hangzhou	509.9	128.08	6
Zhenjiang	167.2	126.59	7
Nanjing	423	122.03	8
Shaoxing	237.5	85.49	9
Ningbo	421.5	49.89	10
Taizhou	202.5	36.57	11

Source: [28].

Table 6. HSR-related economic potentials of the key YRD cities (slowest travel time assumed).

City	GDP (Billion Yuans)	Economic Potential (EP)	Ranks of EP
Suzhou	774	87.67	1
Wuxi	499.2	85.70	2
Changzhou	251.9	60.15	3
Shanghai	1490.1	55.90	4
Jiaxing	191.8	35.27	5
Zhenjiang	167.2	32.91	6
Ningbo	421.5	31.25	7
Taizhou	202.5	29.99	8
Nanjing	423	29.73	9
Shaoxing	237.5	29.01	10
Hangzhou	509.9	27.28	11

Source: [28].

outlying areas on the Zhejiang Province side.

5.1.3. Enhanced Corridor Transportation Capacity

In addition to their significant travel time savings, both Huning and Huhang lines will have separate passenger operations (using new lines) from freight operations (using existing lines), thus greatly enhancing the transportation capacity of both corridors, relieving overcrowding and delays caused by sharing rail rights-of-ways.

5.1.4. Improved Intermodal Connection

The HSR lines will help improve intermodal connection in many hub cities, especially Shanghai, Nanjing, Hangzhou, Suzhou, and Wuxi. Take Shanghai for example. As one of the eastern termini of the Huning Line, Shanghai Hongqiao Airport Terminal 2 has become a multimodal interchange station among four high speed rail lines (Beijing-Shanghai PDL, Shanghai-Hangzhou Intercity HSR Line, Shanghai-Wuhan-Chengdu PDL, Hukun PDL), two Shanghai Metro subway lines (Line 2 and Line 10), and Hongqiao International Airport.

5.2. Non-Transportation-Related Impacts

The non-transportation-related impacts are described in this section.

5.2.1. Strengthened Nodal Transit-Oriented Development (TOD)

The Huning Line traverses the entire Southern Jiangsu Province, which is at the core of the Shanghai Megaregion (the largest and most developed one in China). Those cities and areas located in the immediate vicinity of HSR train stations are primary beneficiaries of this new transportation alternative.

In Hangzhou, the City East New Town is the HSR terminal area. At present, this area primarily serves as a transportation terminal. Its other functions in tourism, logistics, business and entertainment have not been fully played yet. Due to the opening of HSR, Hangzhou and Shanghai have formed a one-hour transportation circle. Many tourists may tour Hangzhou during daytime and stay in Shanghai at night. Therefore, Hangzhou's hotel business may be negatively impacted.

5.2.2. Reinforced Corridor Spillover Effects

Table 7 shows the economic strength and density of the cities traversed by HSR lines. Economic strength and economic density (preferred indicator) are measured by total 2009 GDP, and 2009 GDP/km, respectively. From this table, it can clearly be seen that: 1) Nanjing-Hangzhou (Ninghang) Line, Shanghai-Hangzhou (Huhang) Line, and Shanghai-Nanjing (Huning) Line encompass the triangular-shaped area with the strongest economy. The cities traversed by these three lines had the highest

total GDPs in 2009; 2) Shanghai-Hangzhou (Huhang) Line and Shanghai-Nanjing (Huning) Line are associated with the highest economic densities. Therefore, these two lines traverse the most important transportation corridors and can be regarded as Tier-1 rail corridors. Due to its much shorter length, the Huhang Line had a slightly higher economic density that the Huning Line. Along the Huning Line, the eastern cities (Suzhou, Wuxi, and Changzhou) are relatively more developed than the western cities (Nanjing and Zhenjiang); 3) Based on their economic densities, Changzhou-Suzhou-Jiaxing (Changsujia) Line, Nanjing-Hangzhou (Ninghang) Line, Nanjing-Wuhu (Ningwu) Line, and Xiaoshan-Ningbo (Xiaoyong) Line can be regarded as Tier-2 rail corridors. The rest lines are Tier-3 corridors.

According to [29], the industries in Shanghai are capital-intensive, and the others in Zhejiang and Jiangsu provinces are labor-intensive. In expanding their production capacity, many of Shanghai's plants contract out the labor-intensive production processes to plants in other cities, while retaining its capital-intensive industries. Therefore, those low-end labor-intensive industries are more likely to be dispersed, whereas high-end capital-intensive industries will continue to be centralized at growth pole cities like Shanghai. Economic polarization and dispersion coexists in the YRD Megaregion.

In summary, due to the better intercity linkage provided by the HSR, the nodal TOD will eventually be turned into the linear Transit-Oriented Corridor (TOC) along the HSR corridors. TOC is the "pearl necklace-like" linear land development chaining all nodal TODs together, reflecting corridor spillover effects. The polycentric, linear TOCs along the rail corridors are destined to become the ideal urban spatial structures in China due to the pole-axis dispersion development pattern [30].

5.2.3. Resource Optimal Distribution

By decreasing the travel times between cities in the region, the new trains make it practical for companies to spread out their facilities into locations that are most suitable for each type of operation. This means that while a company may need to keep its sales offices in Shanghai for proximity to its customers, it can easily set up production in a lower wage area such as Kunshan which will be less than 30 minutes away, or establish research and development facilities in industrial parks in Suzhou or Wuxi where it can enjoy favorable tax conditions and lower land costs. Also, by placing the cities of the Yangtze River Delta within an easy reach of each other, the new rail system may make it easier to relocate existing management into emerging cities, thus relieving some of the cost pressures on companies operating in Shanghai and other urban center. People working in Shanghai can also buy houses and live in other neighboring cities.

Table 7. Economic strength and density of the cities traversed by HSR lines.

Rail Line Name	Traversed Cities	Economic Strength (2009 GDP in Billions)	Line Length (km)	Economic Density (2009 GDP in Billions/km)	Corridor Economic Tiers
Shanghai-Hangzhou (Huhang) Line	Shanghai, Jiaxing, Hangzhou	¥2191.80	172 km	¥12.74/km	Tier-1
Shanghai-Nanjing (Huning) Line	Shanghai, Suzhou, Wuxi, Changzhou, Zhenjiang, Nanjing	¥3605.40	301 km	¥11.98/km	Tier-1
Shanghai-Nantong (Hutong) Line	Shanghai, Suzhou, Nantong	¥2551.40	249 km	¥10.25/km	Tier-1
Xiaoshan-Ningbo (Xiaoyong) Line	Hangzhou, Shaoxing, Ningbo	¥1168.9	147 km	¥7.95/km	Tier-2
Nanjing-Hangzhou (Ninghang) Line	Nanjing, Changzhou, Wuxi, Huzhou, Hangzhou	¥1795.2	251 km	¥7.15/km	Tier-2
Nanjing-Wuhu (Ningwu) Line	Nanjing, Maanshan, Wuhu	¥579.8	92 km	¥6.30/km	Tier-2
Changzhou-Suzhou-Jiaxing (Changjiasu) Line	Changzhou, Suzhou, Jiaxing	¥1217.70	202 km	¥6.03/km	Tier-2
Ningbo-Taizhou-Wenzhou (Yongtaiwen) Line	Ningbo, Taizhou, Wenzhou	¥876.8	268 km	¥3.27/km	Tier-3
Nanjing-Qidong (Ningqi) Line	Nanjing, Yangzhou, Taizhou, Nantong	¥1061	351 km	¥3.02/km	Tier-3
Xinyi-Changxing (Xinchang) Line	Xuzhou, Suqian, Huaian, Yanzheng, Nantong, Taizhou, Wuxi, Huzhou	¥1688.4	561 km	¥3.01/km	Tier-3
Xuancheng-Hangzhou (Xuanhang) Line	Xuancheng, Huzhou, Hangzhou	¥664.4	224 km	¥2.97/km	Tier-3

Source: From each city's published 2009 statistical data.

6. Equity Issues of HSR Lines

As indicated above, the efficiency impacts of HSR lines are so obvious in the YRD Megaregion. However, its equity issues are often raised.

Following the modeling approach proposed in Section 3, the city-level connectivity indices in the YRD Region can be calculated and shown in **Table 8**.

Table 8 clearly indicates that:

First, the more connected cities are Shanghai and those cities along the Huning Corridor and Huhang Corridor.

Second, the less connected cities are those cities in the north of Yangtze River in Jiangsu Province, and the east/south of Zhejiang Province.

Third, the future HSR plans in the YRD region will aggravate the regional inequalities while lifting the overall economic development levels. The HSR-induced economic and non-economic benefits are apparently unevenly distributed across the YRD cities.

7. Conclusions

Through this empirical analysis, it can be concluded that HSR lines (PDL trunk lines and intercity lines) will generate significant transportation and non-transportation impacts.

With respect to transportation impacts, HSR lines will dramatically reduce travel times, improve accessibility, enhance corridor transportation capacity, strengthen intermodal connection, and realize other benefits.

In terms of its non-transportation impacts, HSR lines can effectively strengthen nodal transit-oriented development (TOD), reinforce corridor spillover effects, optimally distribute regional resources, and others.

In the meantime, it should also be recognized that HSR lines will also generate inequity issues. Geographically, the existing HSR plans, if fully implemented, will only worsen geographic equity, which makes the rich cities richer and the poor cities poorer, even though the overall regional economic development levels will be lifted. The polarization trends seem evident amid the HSR plan implementation.

The proposed PANS model helps conduct this study. The pole-tiers and axis-tiers are highly correlated. Some low-end economic activities and resources may be dispersed and trickled down from higher-tier cities to lower-tier cities. But, the reverse concentration and agglomeration processes may even be stronger, making Shanghai the more dominant central city in the YRD Region.

8. Acknowledgements

This author greatly appreciates the data support provided by Professor Haixiao Pan, Department of Urban Planning, Tongji University, China. The potential remaining errors

Table 8. City-level connectivity indices in the YRD region.

City Tier	City Name	Connected Axis Tier	CI
Tier-1 City	Shanghai	5 Tier-1 Axis(es): Jinghu PDL, Hukun PDL, Huning Intercity HSR Line, Huhang Intercity HSR Line, Hutong Rail Line; 0 Tier-2 Axis(es): N/A; 0 Tier-3 Axis(es): N/A;	25
Tier-3 City	Nanjing	2 Tier-1 Axis(es): Jinghu PDL, Huning Intercity HSR Line; 2 Tier-2 Axis(es): Ninghang Intercity HSR Line, Ningwu Rail Line; 1 Tier-3 Axis(es): Ningqi Rail Line.	17
Tier-3 City	Hangzhou	2 Tier-1 Axis(es): Hukun PDL, Huhang Intercity HSR Line; 1 Tier-2 Axis(es): Xiaoyong Intercity HSR Line; 1 Tier-3 Axis(es): Xuanhang Rail Line	14
Tier-2 City	Suzhou	2 Tier-1 Axis(es): Jinghu PDL, Huning Intercity HSR Line; 1 Tier-2 Axis(es): Changsujia Rail Line; 0 Tier-3 Axis(es): N/A.	13
Tier-4 City	Jiaxing	2 Tier-1 Axis(es): Hukun PDL, Huhang Intercity HSR Line; 1 Tier-2 Axis(es): Changsujia Rail Line; 0 Tier-3 Axis(es): N/A.	13
Tier-3 City	Changzhou	2 Tier-1 Axis(es): Jinghu PDL, Huning Intercity HSR Line; 1 Tier-2 Axis(es): Changsujia Rail Line; 0 Tier-3 Axis(es): N/A.	13
Tier-2 City	Wuxi	2 Tier-1 Axis(es): Jinghu PDL, Huning Intercity HSR Line; 0 Tier-2 Axis(es): N/A; 1 Tier-3 Axis(es): Xinchang Line.	11
Tier-3 City	Zhenjiang	2 Tier-1 Axis(es): Jinghu PDL, Huning Intercity HSR Line; 0 Tier-2 Axis(es): N/A 0 Tier-3 Axis(es): N/A	10
Tier-4 City	Nantong	1 Tier-1 Axis(es): Hutong Rail Line; 0 Tier-2 Axis(es): N/A; 1 Tier-3 Axis(es): Ningqi Rail Line.	6
Tier-3 City	Ningbo	0 Tier-1 Axis(es): N/A; 1 Tier-2 Axis(es): Xiaoyong Intercity HSR Line; 1 Tier-3 Axis(es): Yongtaiwen Rail Line	4
Tier-4 City	Huzhou	0 Tier-1 Axis(es): N/A; 1 Tier-2 Axis(es): Ninghang Intercity HSR Line; 1 Tier-3 Axis(es): Xuanhang Rail Line.	4
Tier-4 City	Shaoxing	0 Tier-1 Axis(es): N/A; 1 Tier-2 Axis(es): Xiaoyong Intercity HSR Line; 0 Tier-3 Axis(es): N/A	3
Tier-4 City	Yangzhou	0 Tier-1 Axis(es): N/A; 0 Tier-2 Axis(es): N/A; 1 Tier-3 Axis(es): Ningqi Rail Line	1
Tier-4 City	Taizhou in Jiangsu	0 Tier-1 Axis(es): N/A; 0 Tier-2 Axis(es): N/A; 1 Tier-3 Axis(es): Ningqi Rail Line	1
Tier-4 City	Taizhou in Zhejiang	0 Tier-1 Axis(es): N/A; 0 Tier-2 Axis(es): N/A; 1 Tier-3 Axis(es): Ningtaiwen Rail Line.	1
Tier-4 City	Zhoushan	0 Tier-1 Axis(es): N/A; 0 Tier-2 Axis(es): N/A; 0 Tier-3 Axis(es): N/A.	0

Notes: CI = 5 * Number of Connected Tier-1 Axis(es) + 3 * Number of Connected Tier-2 Axis(es) + 1 * Number of Connected Tier-3 Axis(es).

of this paper are mine.

REFERENCES

[1] http://news.21cn.com/domestic/yaowen/2010/03/13/7402 813.shtml

[2] J. Campos and G. de Rus, "Some Stylized Facts about High-Speed Rail: A Review of HSR Experiences around the World," *Transport Policy*, Vol. 16, No. 1, 2009, pp. 19-28.

[3] R. Knowles, J. Shaw and I. Docherty, "Transport Geographies: Mobilities, Flows and Spaces," Wiley-Blackwell, Malden, 2008.

[4] M. Givoni, "Development and Impact of the Modern High-

Speed Train: A Review," *Transport Reviews*, Vol. 26, No. 5, 2006, pp. 593-611.

[5] COST 318, "Interaction between High Speed Rail and Air Passenger Transport," Directorate General of Transport, European Commission, 1998.

[6] J. Gutierrez, R. Gonzalez and G. Gomez, "The European High-Speed Train Network: Predicted Effects on Accessibility Patterns," *Journal of Transport Geography*, Vol. 4, No. 4, 1996, pp. 227-238.

[7] J. Gutierrez, "Location, Economic Potential and Daily Accessibility: An Analysis of the Accessibility Impact of the High-Speed Line Madrid-Barcelona-French Border," *Journal of Transport Geography*, Vol. 9, No. 4, 2001, pp. 229-242.

[8] M. Diao and Y. Zhu, "High Speed Rail Network, Accessibility and Regional Development in China," *6th Annual Conference of the International Association for China Planning*, Wuhan, 18 June 2012.

[9] R.-Q. Yao, "Researching the Influence toward Hangzhou of the High Speed Railway of Shanghai to Hangzhou," *Modern Urban Research*, No. 6, 2010, pp. 16-24.

[10] B. Sands, "The Development Effects of High-Speed Rail Stations and Implications for California," *Built Environment*, Vol. 19, No. 3-4, 1993, pp. 257-284.

[11] H. Priemus, P. Nijkamp and D. Banister, "Mobility and Spatial Dynamics: An Uneasy Relationship," *Journal of Transport Geography*, Vol. 9, No. 3, 2001, pp. 167-171.

[12] A. Bonnafous, "The Regional Impact of the TGV," *Transportation*, Vol. 14, No. 2, 1987, pp. 127-137.

[13] K. Spiekermann and M. Wegener, "The Shrinking Continent: New Time-Space Maps of Europe," *Environment and Planning B*, Vol. 21, No. 6, 1994, pp. 653-673.

[14] R. Vickerman, "High-Speed Rail in Europe: Experience and Issues for Future Development," *The Annals of Regional Science*, Vol. 31, No. 1, 1997, pp. 21-38.

[15] R. Cervero and M. Bernick, "High-Speed Rail and Development of California's Central Valley: Comparative Lessons and Public Policy Considerations," Institute of Urban and Regional Development, University of California, Berkeley, 1996. http://iurd.berkeley.edu/sites/default/files/ wp/675.pdf

[16] D. Banister and Y. Berechman, "Transport Investment and the Promotion of Economic Growth," *Journal of Transport Geography*, Vol. 9, No. 3, 2001, pp. 209-218.

[17] U. Blum, K. E. Haynes and C. Karlsson, "The Regional and Urban Effects of High-Speed Trains," *The Annals of Regional Science*, Vol. 31, No. 1, 1997, pp. 1-20.

[18] D. Lu, "An Analysis of Spatial Structure and Optimal Regional Development," *Scientia Geographica Sinica*, Vol. 56, No. 2, 2001, pp. 127-135.

[19] D. Lu, "Formation and Dynamics of the 'Pole-Axis' Spatial System," *Scientia Geographica Sinica*, Vol. 22, No. 1, 2002, pp. 1-6.

[20] D. Lu, "Objective and Framework for Territorial Development in China," *Chinese Geographical Science*, Vol. 19, No. 3, 2009, pp. 195-202.

[21] W. Christaller, "Die Zentralen Orte in Suddeutschland," Gustav Fischer, Jena, 1933. Translated (in part), by C. W. Baskin, as Central Places in Southern Germany, Prentice Hall, Upper Saddle River, 1966.

[22] F. Perroux, "Economic Space: Theory and Applications," *Quarterly Journal of Economics*, Vol. 64, No. 1, 1950, pp. 89-104.

[23] A. O. Hirschman, "The Strategy of Economic Development," Yale University Press, New Haven, 1958.

[24] V. R. Vuchic, "Urban Transit Systems and Technology," John Wiley & Sons, Inc., Hoboken, 2007.

[25] http://www.china-mor.gov.cn/dzdt/tlyyxl_small.html

[26] http://www.hudong.com/categorypage/show/%E4%B8% AD%E5%9B%BD%E9%93%81%E8%B7%AF%E7%B A%BF/

[27] Y.-L. Huan and C.-L. Xu, "Study on the Influence of Traffic Integration of Yangtze River Delta on the Potential of Local Economic Development (in Chinese)," *East China Economic Management*, Vol. 25, No. 8, 2011, pp. 10-11.

[28] http://chaxun.shike.org.cn/FindResult.asp

[29] J. Yang, "Spatial Planning in Asia," In: C. L. Ross, Ed., *Megaregions: Planning for Global Competitiveness*, Island Press, Washington DC, 2009, pp. 35-52.

[30] Z. Sun and Y. Yang, "Inter-City Rail Transit and Urban Development (in Chinese)," *Modern Urban Research*, No. 12, 2005, pp. 38-42.

12

Review of the Effectiveness of Vehicle Activated Signs

Diala Jomaa, Siril Yella, Mark Dougherty
Department of Computer Engineering, Dalarna University, Borlänge, Sweden

ABSTRACT

This paper reviews the effectiveness of vehicle activated signs. Vehicle activated signs are being reportedly used in recent years to display dynamic information to road users on an individual basis in order to give a warning or inform about a specific event. Vehicle activated signs are triggered individually by vehicles when a certain criteria is met. An example of such criteria is to trigger a speed limit sign when the driver exceeds a pre-set threshold speed. The preset threshold is usually set to a constant value which is often equal, or relative, to the speed limit on a particular road segment. This review examines in detail the basis for the configuration of the existing sign types in previous studies and explores the relation between the configuration of the sign and their impact on driver behavior and sign efficiency. Most of previous studies show that these signs have significant impact on driver behavior, traffic safety and traffic efficiency. In most cases the signs deployed have yielded reductions in mean speeds, in speed variation and in longer head-ways. However most experiments reported within the area were performed with the signs set to a certain static configuration within applicable conditions. Since some of the aforementioned factors are dynamic in nature, it is felt that the configurations of these signs were thus not carefully considered by previous researchers and there is no clear statement in the previous studies describing the relationship between the trigger value and its consequences under different conditions. Bearing in mind that different designs of vehicle activated signs can give a different impact under certain conditions of road, traffic and weather conditions the current work suggests that variable speed thresholds should be considered instead.

Keywords: Vehicle Activated Signs; Variable Message Signs; Threshold Trigger Value; Speed Variation; Traffic Volume; Mean Speed

1. Introduction

Excessive or inappropriate speeds are often a reason for traffic fatalities. Therefore the primary consideration for relevant traffic authorities is to reduce speeding either by setting up additional signage or by improving roadways infrastructure. Improving road infrastructure however is quite expensive and hence may not be viable all the time. Therefore, vehicle activated signs (VASs) are recently being used in roadways to increase traffic safety and road efficiency. A VAS is a digital road sign that is mainly used for speed enforcement or to provide the road users with a warning about a hazard. Typically, VAS consists of radar that is mounted inside the sign to detect vehicle or driver speed. The sign displays a message when vehicle speed exceeds a pre-set threshold. It should be noted that VASs belong to a much bigger class of signs known as variable message signs (VMSs). Hence in the current work the terms VAS and VMS have been used interchangeably for the sake of simplicity.

A VMS is a digital board that displays one of a number of messages that may be changed as required in order to inform or warn travelers for a specific event. The message either indicates road and traffic conditions, alternates routes, construction activities or states the appropriate road speed limit. The signs are in general linked to a control center through one to one communication, a local network or radio link to provide real time information on the oncoming road. In the control center, a variety of traffic monitoring and surveillance systems shall extensively be done to provide the right information in order to be displayed to motorists. According to other researcher's definitions, VMS was also known by various other names such as:
- Dynamic Message Sign, DMS;
- Changeable Message Sign, CMS;
- Electronic Message Sign;
- Variable Speed Limit Sign, VSL.

It should be noted that information displayed by a VAS is triggered on an individual basis and provides information targeted to a specific individual of the driver population as opposed to VMS which mainly provide information to drivers in general.

Researches reporting the usage of VAS and the more general VMS have both been investigated in this review for the sake of completeness. This review examines in detail the basis for the configuration of the existing sign types in previous studies and explores the relation between the configuration of the sign and their impact on driver behavior and sign efficiency.

Studies reviewing the effect of variable message signs and or vehicle activated signs have been reported by Nygårds and Helmers in 2007 [1]. Such studies have reviewed relevant work published between 2000 and 2005 and have mainly investigated the influence of VMS over human behavior. The authors concluded that VMS have more effect on driver behavior as opposed to static road signs. This is because VMS have higher luminance and better contrast and are hence better at attracting driver's attention. Further the aforementioned study has illustrated that the perceived credibility of the message displayed in the sign plays an essential role in influencing driver behavior. Hence it has been suggested that VMs only display the most relevant information.

More recently another study has reported the happenings within the area during the years 2006-2009 [2]. However, the influence of weather has not been greatly emphasised in the above reported studies *i.e.* taking into account local weathering conditions prior to displaying information on the VMS. Another major issue that has not been carefully studied concerns the configuration of the sign. Configuration of the signage is deemed extremely important because relevant information collected and displayed by the sign is largely affected as a result of incorrect configuration which in turn greatly affects drivers' compliance. Therefore, once VMS or a VAS is installed, the sign shall not be used as a simple communication device to display various safety messages or unnecessary warning for promoting road safety. The sign must have a solely purpose of providing essential travel advices to drivers. In order to decide whether the traffic advices shall be displayed on or not, most of VMS systems use algorithms based on some threshold values such as traffic flow, occupancy or mean speed or combination of them. These threshold values shall be obviously dynamic and require fine-tuning because different locations in different time can yield different threshold values [3]. Finding a reasonable pre-set threshold value to trigger either VMS or VAS is challenging because an early or late activation of these signs can reduce their eligibility. Besides the reduction of sign eligibility provides to poor compliance between the driver and the established sign.

To apply right configuration on these signs is not a simple task and becomes a key target for transport agencies. The threshold should be set depending on road, traffic conditions otherwise the sign face the same problem as in the case of the static signs. Traffic agencies established VMS and VAS in several test sites to evaluate their effect on driver behavior and traffic efficiency but they did not carefully focused on the configuration of the sign in different road conditions. The main research question is how the threshold values were set on these signs and how the signs were configured? At which trigger threshold value should be activated to give positive effect on driver's behavior? Can different threshold values give different impact under adverse conditions? This literature review aims to figure out the following statements:

- The effects of Variable message signs and vehicle activated sigs on driver behavior proving the signs necessity to be related to various weather, road and traffic conditions;
- The parameters used to configure the threshold values that were set on the established signs;
- The parameters used to assess the effectiveness of variable message signs and vehicle activated signs.

The rest of the paper is organised as follows. First, an overview about the assessment of the effectiveness of VMS and VAS has been provided. Next, the effects of variable message sign on driver behavior and headways are provided in Section 3. The effect and evaluation of variable speed limit is presented and discussed in Section 4. Section 5 further discusses the effect of vehicle activated sign. The paper finally presents concluding remarks and provides future research directions.

2. Assessment of the Effectiveness of VMS and VAS

There are no universal criteria for measuring the effectiveness of variable message signs. One of the measures is the relationship of the sign to crashes. The collection of crash data correlated to the presence of specific signage is valuable but crashes to a selected test site can be relatively rare. Sometimes may be required several years to determine whether the introduction of the sign had a beneficial effect or not [4]. Therefore the relation between vehicle speed and the frequency of injury collisions had been discussed and interested many researchers. Taylor *et al.* deducted that for mean speed between 25 mph to 35 mph, the reduction in collisions per 1 mph is 4% [5]. However, Nilson proposed another relation to estimate the collision reduction based on speed reduction. At 30 mph, he concluded that the reduction in speed of 1 mph lead to reduce by 6.6% in injury accidents, by 0.7% in serious injury accidents and by 12.7% in fatal accidents [6]. Another measure that had been used in the effectiveness is changes in drivers' speed. This measure is not straightforward in the case for warning signs such as bend curves warning where the speed reduction is not required. The most often measure used in determining warning signs effectiveness is the degree of how much the drivers notice them according to their visibility and comprehensibility. This measurement can be obtained by recall or recognition question to drivers that had been passed recently the signs [4].

3. Effect of Variable Message Sign on Driver Behaviour and Headways

A large number of studies had been established to test the effectiveness of VMSs in reducing vehicle speed and in improving safety on a work zone environment. Richard and Alex established a quasi-experiment to examine the effects of different safety messages displayed on VMS on driver's attitudes and on-road traffic speed. The quasi-experiment was chosen on an inter-city Highway 2 between the cities of Edmonton and Calagary. Besides, a questionnaire survey was developed and managed to a sample of 97 drivers to report their responses to the messages displayed on VMS [7]. The results of the survey showed that a small proportion of the respondents reported that the messages increased their likelihood of obeying the speed limit. In addition, there was a small beneficial effect on driver's speed. The safety messages in Richard and Alex study were not reflecting any traffic related information. Their messages were simply a reminder about traffic safety. The messages shown in VMS should only be activated when essential conditions deteriorate otherwise they led to a slightly noticeable effect on driver's speed.

However, Jihzen et al. investigated VMS traffic guide information that was based on real-time detection of traffic flow and the actual conditions in Beijing Olympics. In their study, Jihzen et al. introduced a logical structure of the traffic guidance VMS information system. The system was put into service and performed well during Beijing Olympic Games. The system collected the actual data from the road network, processed it and released timely traffic information to VMS display. The authors claimed to have achieved good practical effects but the assessment of the effectiveness of the system was not shown in this study. The information shown in VMS display was obtained after a complicated storage, processing and computational model. The time taken from the proposed model could lead to late release of the information where it could lead to decreases in VMS reliability.

Firman et al. determined the effectiveness of portable changeable message signs PCMS in rural highway work zones. An experiment was conducted in Seneca, Kansas in US to evaluate PCMS under two different conditions: 1) PCMS was switched on, and 2) PCMS was switched

off. The data was collected by two Smart Sensor HD radar sensor systems having the capability of data storage and wireless data downloading. Standard deviation, mean and standard deviation error mean were calculated and analyzed by using statistical software called the Statistical Package for Social Science (SPSS). The results for this experiment showed that when the PCMS was turned on, it reduced vehicle speeds by 4.7 mph over 500 feet distance on average. Besides, when PCMS was turned off, the average speed was reduced to 3.3 mph [8]. Firman et al. did not concerned that there is a reduction in the average speed even if the sign is off. In this experiment, it was not mentioned when and how PCMS was activated. The activation of the sign could be a challenging task. Furthermore, the study did not measure what effect the speed reduction had in relation to the volume and time of day. Despite the fact that the time periods, especially peak hours or weekday, can have clearly consequences in the result.

There are a number of messages that may be posted in dynamic message signs, DMS but which type can be the most useful and effective one. Some researchers suggested that an active DMS that display warning messages could directly affect driver performance because warning messages attracted the attention of the drivers. Borello and Ornitz grouped different types of messages given on a DMS in three different classes. Based on this classification, the authors measured what effect the message of sign had on driver performance with the measure of speed. In order to test the immediate influence of the DMS on driver performance, Borello and Ortniz defined a cone of influence of the DMS. The cone of influence was defined as a distance between the minimum and maximum line of sight for the sign to determine the best location of the speed detector for ideal readings. The cone was based on an angle of inclination of 7.5 degree in order that the driver may read the message in 918.0 ft [9]. By carefully setting the cone parameters, it could be easily excluded drivers who could not see the message in the DMS from the analysis.

Rämä and Kulmala investigated the effectiveness of variable message signs warning of slippery road conditions and a minimum headway sign on driver behavior. Driver behavior was measured in terms of speed and headways variations under weather conditions, road surface and traffic conditions. The data was collected by using loop detectors at three test sites in Finland. To examine the effects of driver adaption and any possible novelty, an after study was established within two winters. Driver behavior was monitored at three measurement points, 536 - 1800 m upstream before the sign, 360 - 1100 m downstream after the sign and 7670 - 13,000 m downstream. The operators at the Traffic Management Centre TMC classified the road surface conditions in three categories, good, possibly slippery and verified slippery. The TMC operators switched on or off the sign to flashing mode depending on the presence of slipperiness. Results showed that the VMS decreased mean speed by 1 - 2 km/h and increased the following distances. The positive effects were not found at all test sites. The effects were more significant before the sign than after the sign. However, at a distance 3 - 14 km after the signs, the mean speed increased slightly [10]. Rämä and Kulmala concluded that the variable message signs had other effects on driver behavior besides speed and headway.

Luoma, *et al.* designed a complementary study to investigate the other possible effect on reported driver behavior. In Luoma, *et al.* study, a combination of roadside and telephone interviews were done to drivers who encountered either of the signs that was used Rämä and Kulmala test sites. 2% of drivers who encountered the minimum headway sign declared that the sign had no effect on their behavior. Drivers reported other frequently effect of the slippery road conditions signs with a change in focus of attention, more concentration on their own driving, testing the road slipperiness and careful overtaking behavior. Many drivers informed that variable message signs improved their driving comfort. Luoma *et al.* found that the slippery road condition sign had more effects during black ice than during snowfall conditions [11]. Therefore, the authors proposed dynamic weather-controlled systems that can perform successfully in further study.

Rämä carried on another experiment on the Finish E18 test site to study the effect of weather-controlled variable speed limits and warning signs on driver behavior. The speed limits were automatically controlled by two road weather station in order to estimate the effects of road and weather conditions on speed. The road and weather conditions were classified as good, moderate or poor. These classifications were based on different parameters such as rain or snowfall, rain intensity, road surface, visibility and wind velocity. The main results from her study showed that the weather-controlled system decreased the mean speed and the standard deviation of speeds. The system had more effect on mean speed during summer season when the higher speed limits were allowed but weather conditions were not obvious, similar as winter conditions. Rämä reported that lowering the speed limit decreased the mean speed where the weather-controlled system increased the homogeneity of driver behavior [12]. The experiments done by the author were carefully established and analyzed. She mentioned that there is a need for a more advanced system to recognize adverse road and weather conditions and low frictions.

The effect of VMS signs on drivers' behavior can vary with the ages of drivers. Older driver are more risky for crashes in work zones claimed Heaslip *et al.* in their research study. The authors studied the effectiveness of the work zones features implemented in Greenfield, Manssachusetts by collecting speed and video data over a four-month period. Video data was used to be able to approximate the age of the drivers and their maneuvers. The results showed that VMS had a positive effect for all drivers regardless that older drivers' speeds were obviously different than other drivers [13]. Heaslip *et al.* advised to develop advanced traffic analysis tools to get better understanding of driver behavior in work zone. They suggested the use of micro simulation and visualization techniques that should concern with new theory and structure by simulating expected driver behavior changes with other characteristics.

4. Effect and Evaluation of Variable Speed Limits

Variable speed limits VSL are variable messages signs that display speed limits. The displayed speed limits are determined by VSL algorithms that are based on real time traffic data. The VSL algorithms can have different design and provide different impact depending on the type and the control objective of the VSL system. Jiang *et al.* designed three types of algorithms for activating VSL on. These algorithms were based on high flow, queuing and weather conditions. For high flow conditions, the logic for determining new speed limit was retrieved from the speed-flow relationship curve, but for queuing conditions, the logic is done by reducing the upstream speed to a medium level speed level, between normal speed and the activated low speed limit. Jiang *et al.* did not present the algorithm that is based on weather conditions. However, the authors demonstrated the effectiveness of VSL in high flow and queuing conditions by using a micro simulation model. The results from their model showed that VSL was able to achieve speed harmonization for motorway sections with low ramp and could contribute to the reduction of secondary crashes [14].

Nissan and Koutsopoulosb focused on the evaluation of the impact of advisory variable speed limit VSL on motorway capacity. A motorway control system was implemented E4 in Stockholm. The system was equipped with an Automatic Incident Detection to detect serious disturbances in the traffic streams. When the system noticed the disturbance, it generated automatically a suitable set of advisory speed limits for the approaching traffic [15]. The new displayed speed limit was based on a comparison to two speed threshold with low and high values. The values of these two speed threshold are not clearly identified in this study while these values are the main elements in triggering the VMS sign. However, the results from this study showed that advisory VSL did not give any important effect after its implementation. Nissan and Koutsopoulosb concluded that the obtained result was related to the advisory system used in Stockholm. The system should be obvious to motorists that the recommended speed limit was based on the serious disturbance detected where the system did not seem to be focus in this point. Besides, motorists ignored the advisory speed limit if it was not motivated by the traffic situation.

A two stage variable advisory speed limit system was developed and implemented by Kwon et al. The system was established for only three week period in February to March at one of the I-494 work zones in Twin Cities. The proposed system used real-time measurements at both downstream and upstream speed levels. It tried to reduce the speed of the upstream flow to reach the same level as that of the downstream traffic. Therefore, the advisory speed limit was determined from a function based on both upstream and downstream speed levels. Kwon et al. reached a promising result in reducing the speed differences by 25% to 35% along the work zone area during the 6:00 to 8:00 a.m. periods on weekdays. The estimation of driver compliance level by correlating the speed differences in both upstream and downstream was 20% to 60% [16]. In this research, the effects of the system were only checked for a short-time period. The effects of the signs should be measured on longer time period to study driver's compliance. Weather conditions play an essential role in the long term effectiveness of the advisory speed limit where the authors excluded this factor from their study.

Sandberg et al. investigated long term effectiveness of Dynamic Speed Monitoring Display, DSMD signs that were permanently installed in Minnesota, Washington and Dakota County. The study was conducted at speed reduction transition zones from a rural highway to urbanized area and was kept in one year duration. The overall results across all the test sites were clearly consistent in reductions in the 50th, 85th and 95th speeds averaging 6.3, 6.9 and 7.0 mph, respectively [17]. The authors mentioned that in order to draw conclusions for any long-term study the potential external influences other than the DSMS sign should be reviewed. However the mentioned review was not shown in their work. Simply they examined traffic speed and traffic volume with no check for the speed variation in relation with seasonal conditions. Sandberg et al. analysed the 24-hour average traffic speed and volume but vehicle speed could be extremely varied in rush hours than other time of the day. The variability of speeds during night time and daytime should also be analysed in their study.

McMurry et al. studied the effects of Variable Speed Limit VSL signs at work zones in Utah. VSL signs were tested against the existing static sign for about 3 months.

Night time standard deviation was analysed to examine the speed variation without construction interruption compared to daytime when work time was present. The results for this study showed that there were wide variations of average speed during the night time with regular static signs as compared to VSL signs [18]. Nevertheless the authors showed that night drivers and day drivers were more compliant to VSL signs than the regular static sign. However, the variability of speeds was only examined at short term but these variations might be very different at longer term. In this study, the speed limit posted in the sign at night time was 10 mph higher than daytime meanwhile 10 mph lower than the regular speed limit that usually posted on the road. The authors did not showed why the speed should be reduced by this amount and why the night posted speed in VSL could not be stepped back to the original speed limit. The effects of VSL on driver behaviour are summarized in **Table 1**.

5. Effect of Vehicle Activated Sign

Automatic speed warning signs, vehicle activated speed limit sign, or Active speed warning sign were considered as Vehicle Activated Sign, VAS. In fact, VAS has a system that is activated by driver's speed that exceeds a pre-set threshold. The sign had been developed and established in typical locations like dangerous entrances and curves, work zones and school zones. Different studies had been done in this area by presenting the effects of the use of VAS. Kathmann presented different approaches on how to assess the effectiveness of VAS. He pointed out that the approaches should be taken under two essential considerations that are important when assessing the effectiveness of VAS. The first consideration was that the observations must be done in a way that cannot be discovering by the driver. The second one was that the measurement for collecting and analysing the data should be cost-effective. The approaches could be either an inductive loop measurement or other empirical measurements. As empirical measurements, Kathmann proposed three methods such as car following method, video camera surveillance and voice recording. Car following method is used by a test vehicle with Datron sensor to get speed profile along the way instead of local speed data. Video camera surveillance and voice recording are used for analysing respective drivers braking behaviour and his familiarity with the road. Kathmann concluded that speed measurements should be collected before and after the installations of a VAS otherwise the assessment become difficult and time and work consuming. The best result reached from the methods was the car-following method [19]. Kathmann showed that speed reduction is not always based on the presence of VAS. Therefore there was a need to check the speed distribu-

Table 1. Effect of VSL on driver behaviour.

Authors	Year	Effect	Parameters used in the study
Richard and Alex	2010	Small beneficial reduction in mean speed and standard deviation	Traffic volume and speed for some days
Firman et al.	2009	Reduction in mean speed by 4.7 mph and when VMS is turned on and in 3.3 mph when VMS is off	Traffic volume and speed
Borello and Ornitz	2010	Type of messages have different effect on mean speed, standard deviation	Weekly average speed for each 20s
Rämä and Kulmala	2000	1 - 2 km/h reduction in mean speed and increasing the following distance in headways	Speed, headways, road surface conditions in three categories, good, possibly slippery and verified slippery
Rämä	1999	More effect on mean speed and standard deviation in summer season, increase the homogeneity of driver behavior	Speed, road and weather conditions in three categories good, moderate and poor
Kwon et al.	2007	25% to 35% speed reduction between 0600 and 0800 hrs	Speed, volume for every 30 s
Sandberg et al.	2008	Average speed reduction by 7 mph	Vehicle speed and volume
Nissan and Koutsopoulosb	2011	No significant impact	Vehicle speed and flow density
McMurry et al.	2008	Variations in speed was reduced in general	Vehicle speed at day and night time
Jiang et al.	2011	Speed harmonization and better environmental impacts	High flow, queuing and weather

tion instead of average speed. He suggested integrating data from different VAS sources. That would give better knowledge for assessing the effectiveness of VAS.

Winnett and Wheeler established a full scale study of the effectiveness of over 60 signs installations of different types of vehicle-activated signs on rural single carriage way roads. This study had been conducted by TRL for the Department of Transport. They aimed to assess the effectiveness of the signs on both driver's speed and injury accident. Besides, they aimed to evaluate driver's understanding of the signs [20]. Vehicle Activated signs was depended on the trigger speed or pre-set threshold that switches on the sign. The speed threshold was set at the 50th percentile speed detected before the sign was installed. That was supposed to target half of drivers. In other previous studies, the speed thresholds were set at between the 75th and 81st percentile speeds. In that case, the signs targeted a very small population. However, when the threshold were set at between the 20th to 30th percentile speeds, thus targeted more drivers. The effect of the sign is clearly related to the trigger speed value in order that activating VAS. An early or late switch on could provide a negative or positive impact on the sign. So, what is the most suitable threshold trigger speed for a specific site and how can be determined? The activation threshold speeds should be dependent on different time periods under various conditions. To find out the right trigger speed that activates the sign, there is a need to analyse various trigger speed distribution in relation to average speed under diverse conditions.

In other study, Winnett et al. evaluated the effectiveness of an interactive fibre optic sign at a rural cross road. The main objective of this evaluation was in general to reduce speeds consistently and in particular target mo-

torists at the top end of the speed distribution. The sign used was a warning sign with SLOW DOWN message, were switched on when driver's speed exceeded 46 mph. In previous studies in TRL, Winnett et al. suggested that it is possible to control the traffic speed by varying the threshold. The threshold established in their study was fixed to 46 mph. That was chosen to warn driver travelling above 50 mph with 10% error margin on the detector used [21]. There was no basis why the threshold was fixed to 46 mph. VAS should only be activated when it should be activated, otherwise it could decrease its impact on the driver behaviour. In order to define the threshold of VAS, diverse potential factors should be checked.

In 2007, Mattox et al. conducted research study in South California DOT for development and evaluation of speed-activated sign by lowering speed in work zones. This study was based on a depth literature review of several speed reductions measures in work zones such as speed monitoring displays, changeable message signs with and without radar, vehicle activated signs. As a result of the assessment of these signs, it had been proven to provide remarkable effect in speed reduction, but due to their high cost, many transportation agencies did not established in all their work zones. The sign used in Mattox et al. research was vehicle activated sign, VAS which was the most cost effective sign among all other mentioned signs [22]. The predetermined speed threshold was set to the post speed limit with a 3 mph buffer. Results from this research study showed a reduction in mean speed, 85 percentile speeds and percentage of vehicles exceeding the speed limit. This study was only evaluated short term effectiveness. However VAS might lose its effectiveness over time as drivers become habituated to seeing them regularly.

The evaluation of effectiveness of the sign was established by studying before-after studies and concluded that the sign have an effect on reducing speeds, standard deviations or headways. The sign has its largest effect on drivers exceeding the speed limit excessively. All these evaluations were concerned when the sign is in operation but a few studies investigated the evaluation after removing the sign from the test site. Walter and Broughton observed the effect of speed indicator devices, SIDs after the sign is removed. SID is vehicle activated signs that display the real-time speeds of vehicle passing the device. In the study, Walter and Broughton installed ten SID at ten sites in South London in 2008 across all weekdays with different level of traffic flow and with periods of installation that are randomly assigned to weekdays. They designed the experiment to be balanced as far as possible by reducing the possible effect of external factors on the results. Across all sites, an overall speed reduction was 1.4 mph [23]. The effect varied over time and differed across sites. SID showed a significantly effect on speeding drivers at all sites. However, there was no effect in speed after the sign was removed. Only a small reduction was lasting at those sites where the SID was in operation and had the most effect. **Table 2** summarise the effect of VAS on driver behaviour and present the trigger speed value applied in the configuration of the sign.

6. Summary and Conclusions

In previous studies, authors pointed out that the effectiveness of VMS is related to the credibility of the message displayed on the sign. The displayed message should reflect road, traffic or weather conditions and VMS should only be triggered when such conditions deteriorate. The majority of drivers travel at a speed they consider reasonable, and safe for road, weather and environmental conditions. Therefore, unnecessary activation of the sign can provide negative effect of the sign on driver's behaviour. Additionally, the information shown in VMS were obtained after a long process in order to gather the traffic data, process it and compute an advisory message or a new speed limit. The time taken from this process could lead to late release of the information where it could lead to decreases in VMS reliability.

Weather conditions play an essential role in the effec-

tiveness of VMS where most of the researchers excluded this factor from their study. There is a need for a more advanced system to recognize adverse road and weather conditions and low frictions. Therefore, dynamic weather-controlled systems that can perform successfully in further study were proposed by some authors.

To assess the effectiveness of VMS or VAS, vehicle speed and traffic flow are the most used data in the previous studies. These data were basically analysed by averaging the speed or checking the standard deviation in relation to a specific time per days or days of week However, a reduction on the average speed is not always based on the presence of the sign. The sign should be placed at a site where it is not influenced by other factors such as junctions, roundabout or any traffic calming. That can lead to a misinterpretation of the results. The evaluation of the effectiveness of the VMS or VAS systems were conducted by using either empirical data from sensors such as loop detectors, camera or radar or by using micro simulation techniques. The empirical analyses were based on before-and-after VMS implementation on a specific test site. The results from this analysis were shown that VMS and VAS had an impact on driver behaviour in order on the effectiveness of both signs.

The relation between the design of VMS algorithm and the appropriate conditions are not clearin previous research studies [14]. Besides, for VAS, the trigger speed value was set with no basis. Authors did not consider the importance of the value of trigger speed on the impact of the sign on driver behaviour. The threshold value was set either 3 mph or 5 mph above the posted limit or 4 mph under the posted speed limit. Which consequence could lead if the sign was trigger by all vehicle speed or vehicle speed that is above the posted speed limit or under the posted speed limit? All the experiments were established by providing a constant threshold value under different conditions with no consideration for a dynamic threshold value.

7. Future Work

Developing dynamic activation threshold value has not been thoroughly investigated in previous studies. The idea is to establish a model that attempt to predict the appropriate activation threshold value to the corresponding traffic and weather situations. Therefore, a systematic data collection will be focused in first place where traffic

Table 2. Effect of VAS on driver behavior.

Authors	Year	Effect	Trigger speed
Mattox *et al.*	2007	Average speed reduction range 2.0 - 6.0 mph	3 mph over the posted speed limit
Winnett *et al.*	1999	Nearly 70% speed reductions for high end speeders.	46 mph for 50 posted speed limit
Winnett and Wheeler	2002	Mean speed reductions between range 1.2 - 13.8 mph	5 mph above the posted speed limit

data will be collected but by varying the value of threshold activation value. The data is archived by various threshold values. Four databases shall be available for each threshold value: the database for daytime traffic; the database for night time traffic; the database for weekday's traffic and the database for weekends. Meanwhile, the traffic data shall further be integrated with weather data sources. Besides, data is analysed partly with the help of descriptive methods which may describe, organize and present the raw data and partly with the help of association rules which find common relationships between attributes that may be difficult to discover. Another aspect is to extract the speed distribution pattern of the road speed traffic data finding the similarities between vehicle speed and time. The similarities between different attributes can be obtained by using k-means clustering algorithms.

REFERENCES

[1] S. Nygårds and G. Helmers, "VMS: Variable Message Signs. A Literature Review," *Swedish National Road and Transport Research Institute VTI Rapport 570A*, Linköping, 2007.

[2] S. Nygårds, "Literature Review on Variable Message Sign 2006-2009," *Swedish National Road and Transport Research Institute VTI Rapport 15A*, Linköping, 2011.

[3] A. Nissan, "Evaluation of Variable Speed Limits: Empirical Evidence and Simulation Analysis of Stockholm's Motorway Control System," Ph.D. Dissertation, Royal Institute of Technology, Stockholm, 2010.

[4] S. G. Charlton and P. Baas, "Assessment of Hazard Warning Signs Used on New Zealand Roads," *Land Transport New Zealand Research Report 288*, New Zealand, 2006.

[5] M. C. Taylor, A. Baruya and J. V. Kennedy, "The Relationship between Speed and Accidents on Rural Single-Carriageway Roads," *TRL Report*, Wokingham, 2002.

[6] G. Nilsson, "Traffic Safety Dimensions and the Power Model to Describe the Effect of Speed on Safety," *Lund Bulletin, 221*, Lund Institute of Technology, Lund, 2004.

[7] T. Richard and D. B. Alex, "Effectiveness of Road Safety Messages on Variable Messages Signs," *Journal of Transportation Systems Engineering and Information Technology*, Vol. 10, No. 3, 2010, pp. 18-23.

[8] U. Firman, Y. Bai and Y. Li, "Determining the Effectiveness of Portable Changeable Message Signs in Work Zones," *Proceedings of the 2009 Mid-Continent Transportation Research Symposium*, Ames, 2009.

[9] C. Borello and S. E. Ortniz, "Dynamic Message Sign Influence on Driver Performance," *ITE 2010 Annual Meeting and Exhibit*, Vancouver, 2010.

[10] P. Rämä and R. Kulmala, "Effects of Variable Message Signs for Slippery Road Conditions on Driving Speed and Headways," *Transportation Research, Part F*, Vol. 3, No. 2, 2000, pp. 85-94.

[11] J. Luoma, P. Rämä, M. Penttinen and V. Anttila, "Effects of Variable Message Signs for Slippery Road Conditions on Reported Driver Behavior," *Transportation Research Part F: Traffic Psychology and Behaviour*, Vol. 3, No. 2, 2000, pp. 75-84.

[12] P. Rämä, "Effects of Weather-Controlled Variable Speed Limits and Warning Signs on Driver Behavior," *Transportation Research Record*, Vol. 1689, 1999, pp. 53-59.

[13] K. Heaslip, J. Collura and M. Knodler, "Evaluation of Work Zone Design Features to Aid Older Drivers," *88th Annual Meeting of the Transportation Research Board*, Washington DC, 11-15 January 2009.

[14] R. Jiang, E. Chung and J. Lee, "Variable Speed Limits: Conceptual Design for Queensland Practice," *Proceedings of the Australasian Transport Research Forum*, Adelaide, 28-30 September 2011.

[15] A. Nissan and H. N. Koutsopoulosb, "Evaluation of the Impact of Advisory Variable Speed Limits on Motorway Capacity and Level of Service," *Procedia Social and Behavioral Sciences*, Vol. 16, 2011, pp. 100-109.

[16] E. Kwon, D. Brannan, K. Shouman, C.Isackson and B. Arseneau, "Development and Field Evaluation of Variable Speed Limit System for Work Zones," *Journal of the Transportation Research Board*, No.2015, Transportation Research Board of the Natioanl Academies, Washington DC, 2007, pp. 12-18.

[17] W. Sandberg, T. Schoenecker, K. Sebastian and D. Soler, "Long-Term Effectiveness of Dynamic Speed Monitoring Displays (DSMD) for Speed Management at Speed Limit Transitions," *15th World Congress on Intelligent Transport Systems and ITS America's 2008 Annual Meeting*, New York, 16-20 November 2008.

[18] T. McMurtry, M. Saito, M. Riffkin and S. Heath, "Variable Speed Limits Signs: Effects on Speed and Speed Variation in Work Zones," *Transportation Research Circular: Maintenance Management*, E-C135, 19-23 July 2009, pp. 159-174.

[19] T. Kathmann, "Assessment of the Effectiveness of Active Speed Warning Signs-Use of Inductive Loop Data or Empirical Data," *12th International Conference on Traffic Safety on Three Continents*, Moscow, 19-21 September 2001.

[20] M. A. Winnett and A. H. Wheeler, "Vehicle-Activated Signs: A Large Scale Evaluation," *TRL Report TRL 548*, TRL Limited, Crowthorne, 2002.

[21] M. A. Winnett, E. Woodgat and N. Mayhew, "Interactive Fibre Optic Signing at a Rural Crossroad (B1149 Felthorpe, Norfolk)," *Transport Research Laboratory*, TRL Report TRL 401, Crowthorne, 1999.

[22] J. Mattox, W. Sarasua, J. Ogle, R. Eckenrode and A. Dunning, "Development and Evaluation of a Speed-Activated Sign to Reduce Speeds in Work Zones," *Transportation Research Record of the National Academies*, Washington DC, 2007, pp. 3-11.

[23] L. Walter and J. Broughton, "Effectiveness of Speed Indicator Devices: An Observational Study in South London," *Accident Analysis and Prevention*, Vol. 43, No. 4, 2011, pp. 1355-1358.

13

Cloud-Based Information Technology Framework for Data Driven Intelligent Transportation Systems

Arshdeep Bahga, Vijay K. Madisetti
Electrical and Computer Engineering, Georgia Institute of Technology, Atlanta, USA

ABSTRACT

We present a novel cloud based IT framework, CloudTrack, for data driven intelligent transportation systems. We describe how the proposed framework can be leveraged for real-time fresh food supply tracking and monitoring. Cloud-Track allows efficient storage, processing and analysis of real-time location and sensor data collected from fresh food supply vehicles. This paper describes the architecture, design, and implementation of CloudTrack, and how the proposed cloud-based IT framework leverages the parallel computing capability of a computing cloud based on a large-scale distributed batch processing infrastructure. A dynamic vehicle routing approach is adopted where the alerts trigger the generation of new routes. CloudTrack provides the global information of the entire fleet of food supply vehicles and can be used to track and monitor a large number of vehicles in real-time. Our approach leverages the advantages of the IT capabilities of a computing cloud into the operations and supply chain.

Keywords: Cloud Computing; Vehicle Routing; Supply Chain; Tracking; Hadoop

1. Introduction

Intelligent transportation systems (ITS) have evolved significantly in recent years. Modern ITS are driven by data collected from multiple sources which is processed to provide new services to the stakeholders. In a recent survey paper, Zhang *et al.* [1] describe how conventional intelligent transportation systems (ITS) have transformed into data-driven ITS. By collecting large amount of data from various sources and processing the data into useful information, data-driven ITS can provide new services such as advanced route guidance [2,3], dynamic vehicle routing [4], etc.

Cloud computing has been implemented in various domains such as healthcare [5], education [6], smart grids [7], etc. Recent publications have demonstrated the benefits of cloud computing for intelligent transportation systems [8,9]. In our previous work [10], we demonstrated how cloud computing technologies can be used for massive scale sensor data collection and analysis for predicting faults in industrial machines. The successful adoption of cloud computing paradigm in various domains provides the motivation to implement a cloud-based framework for data driven intelligent transportation systems.

Collection and organization of data from multiple sources in real-time and using the massive amounts of data for providing intelligent decisions for operations and supply chains, is a major challenge, primarily because the size of the databases involved is very large, and real-time analysis tools have not been available. However, recent advances in massive scale data processing systems, utilized for driving business operations of corporations provide a promising approach to massive ITS data storage and analysis.

In this paper we propose a cloud-based IT framework, CloudTrack, for data driven intelligent transportation systems. CloudTrack is built using proven open source cloud-based technologies that are already deployed in other domains. The proposed framework allows efficient storage, processing and analysis of real-time data collected from various sources. The global information available when utilizing a cloud-based IT environment allows a scalable, efficient, and optimized integration of the IT environment into the operations environment.

Fresh food can be damaged during transit due to unrefrigerated conditions and changes in environmental conditions such as temperature and humidity, which can lead to microbial infections and biochemical reactions or mechanical damage due to rough handling. In emerging countries, such as in India that is second largest producer of fruits and vegetables in the world, as much as 30% - 35% of fruits and vegetables perish during harvest, storage, grading, transport, packaging and distributions [11]. Since fresh foods have short durability, tracking the supply of fresh foods and monitoring the transit conditions can help identification of potential food safety hazards. The analysis and interpretation of data on the environmental conditions in the container and food truck positioning can enable more effective routing decisions in real time. Therefore, it is possible to take remedial measures such as: 1) the food that has a limited time budget before it gets rotten can be re-routed to a closer destinations; 2) alerts can be raised to the driver and the distributor about the transit conditions, such as container temperature exceeding the allowed limit and corrective actions can be taken before the food gets damaged. **Table 1** provides a comparison of the published approaches for fresh food tracking.

2. Current Challenges & Contributions

Collecting and organizing location and sensor data from vehicles in transit and using the data for raising alerts about violation of certain conditions is a major challenge for the following reasons: 1) wide coverage is needed for collection of location and sensor data from vehicles carrying fresh food supply; 2) data needs to be collected from a large number of vehicles in real-time to raise timely alerts; 3) the collected data is massive scale, since the real-time data from a large number of vehicles is collected simultaneously; 4) the massive scale data needs to be organized and processed in real-time; 5) the

infrastructure used for data collection should be low cost and easily deployable to ensure wide popularity. The major contributions of this paper are: 1) We propose, CloudTrack, a framework for organization and analysis massive scale data generated by data-driven ITS (such as vehicle location and sensor data) in a computing cloud, that allows efficient collection of data on vehicle locations and container conditions and creation of alerts based on the global information; 2) An efficient cloud-based deployment architecture for data driven ITS that leverages a distributed batch processing infrastructure; 3) A dynamic vehicle routing approach that is triggered by the alerts which are generated by CloudTrack. Cloud-Track can support a wide variety of dynamic vehicle routing algorithms; 4) A cloud-based vehicle location and container conditions tracking Software as a Service (SaaS). Vehicles can register with the CloudTrack service on-demand. CloudTrack is flexible to scale up or scale down resources based on the number of vehicles registered with the service; 5) A global approach to collect the data from a large number of vehicles at a centralized location, which can be analyzed for detecting bottlenecks in the supply chain such as traffic congestions on routes, reorganization of assignments and generation of alternative routes, and supply chain optimization.

3. Deployment Architectures for Data Driven ITS

Figure 1 shows a typical three tier web-based deployment architecture used by conventional data driven intelligent transportation systems [12-14]. Tier-1 or the front end servers consists of the web servers, tier-2 consists of

Table 1. Comparison of related work.

Reference	Data Collection	Data Storage	Data Analysis	Control Mechanism	Coverage & Scalability
Pang et al. [12]	Multiple sensor nodes in container, master sensor node in each vehicle.	Relational sensor database in an operation center.	Operation center. The user can monitor and track visualized data on webpage through the Data Visualization Engine.	No automated control mechanism described.	Global knowledge of the entire fleet captured in a relational database. Scalability is limited due to the use of the relational databases (management database and sensor database).
Xi et al. [13]	Multiple slave nodes in container, mother node, GPS receiver in each vehicle.	Data logger and a local PC in the vehicle. Option for transmitting the data to a remote server.	Local PC in the vehicle. Option for transmitting the data to a remote server monitoring center.	No automated control mechanism described.	Real-time miniature wireless monitoring system. No mechanism for capturing global knowledge of the entire fleet and scalability described.
CloudTrack	Multiple sensor nodes in the container, one master node and one Android device in each vehicle.	In the cloud. CloudTrack uses the Hadoop Distributed File System (HDFS) for storing the massive sensor and location data.	CloudTrack (MapReduce jobs in the cloud). The real-time location and sensor data is continuously analyzed by CloudTrack and alerts are generated when any abnormal conditions are observed.	CloudTrack dynamic vehicle routing module and controller. Automated and passive control mechanism. Corrective measures are taken by generation of new routes based on the current locations of the vehicles. Alerts trigger the generation of new routes.	CloudTrack provides the global information of the entire fleet of food supply vehicles and can be used to track and monitor a large number of vehicles in real-time. CloudTrack has been designed to scale up on demand with very little effort.

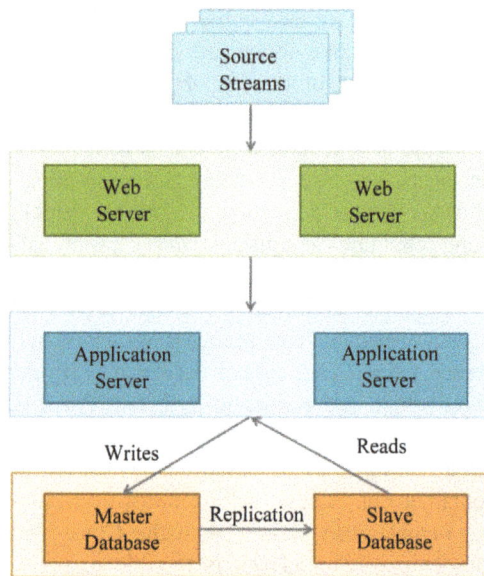

Figure 1. Three-tier web-based architecture used by conventional data driven intelligent transportation systems.

application servers and tier-3 consists of data-base servers. **Figure 2** shows a cloud deployment architecture used in our proposed framework. In this deployment architecture, tier-1 consists of the web servers and load balancers, tier-2 consists of application servers and tier-3 consists of a cloud based distributed batch processing infrastructure such as Hadoop. Compute intensive tasks such as data processing are formulated as MapReduce jobs which are executed on Hadoop. This deployment is suitable for massive scale data analytics. Data is stored in a cloud based distributed storage such as Hadoop Distributed File System (HDFS). The advantages of cloud-based architecture shown in **Figure 2** as compared to the traditional web-based architecture shown in **Figure 1**, are as follows.

3.1. Rapid Elasticity

Cloud-based deployment architecture leverages the dynamic scaling capabilities of computing clouds. Two types of scaling options are available for the cloud-based deployment, described as follows:

1) *Horizontal Scaling* (*scaling-out*): Horizontal scaling or scaling-out involves launching and provisioning additional server resources for various tiers of the deployment.

2) *Vertical Scaling* (*scaling-up*): Vertical scaling or scaling-up involves changing the computing capacity assigned to the server resources while keeping the number of server resources constant.

3.2. Massive Data Analysis

Cloud-based distributed batch processing infrastructure

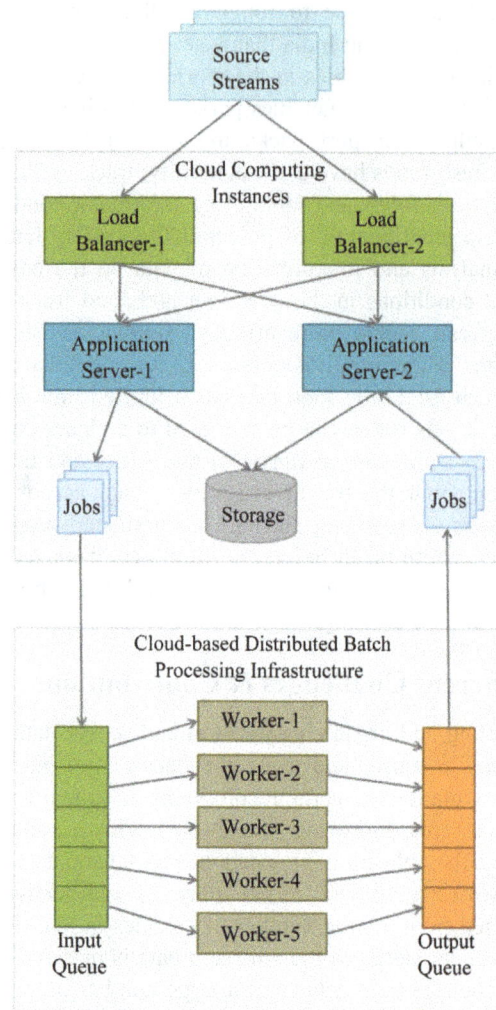

Figure 2. Cloud-based architecture used by our proposed framework for data driven intelligent transportation systems that leverages a distributed batch processing infrastructure.

such as Hadoop allows processing large scale data. Thus Hadoop is well suited for location and sensor data analysis. The Hadoop Distributed File System (HDFS) allows storing large files as multiple blocks which are replicated on multiple nodes to provide reliability. The scale of location and sensor data is so large that it is not possible to fit the data on a single machine's disk. HDFS not only provides reliable storage for large amount of data but also allows parallel processing of data on machines in a cluster.

3.3. Ease of Programming

Programming models used by cloud-based distributed batch processing infrastructures such as Hadoop allow parallel processing of data. For example, with Hadoop, the location and sensor data analysis algorithms can be implemented as MapReduce jobs. Scaling out the computation on a large number of machines in a cluster is simple with Hadoop. The same computation that runs on a single machine can be scaled to a cluster of machines with few configuration changes in the program.

3.4. Flexibility in Data Analysis

Cloud-based distributed batch processing infrastructure such as Hadoop allows scaling the data analysis jobs up or down very easily which makes analysis flexible. With this flexibility in data analysis jobs, the frequency of analysis jobs can be varied.

4. Proposed Cloud-Based IT Framework

Figure 3 shows the proposed system architecture for real-time fresh food supply tracking and monitoring. The major hardware and software components of the proposed system architecture are described as follows:

1) *Sensor Node*: Sensor nodes are deployed in the container carrying food for monitoring temperature, humidity, etc.

2) *Master Node*: Master Node collects sensor data from the sensor nodes in the container and transmits the data to the Android Device using a USB or Bluetooth interface.

3) *Android Device*: An Android operating system based mobile device is used for capturing the sensor data collected by the Master Node, capturing GPS location data using an in-built GPS sensor and transmitting the data over a Wireless Wide Area Network (WWAN) to the data center.

4) *Communication Infrastructure*: The Android devices use cellular network technologies such as WIMAX, GPRS, EDGE, 3G, etc provided by a wireless service provider which have nationwide or even global coverage.

5) *Cloud Based Data Organization and Analysis Infrastructure*: The data transmitted by the Android devices deployed in vehicles is collected and organized in a computing cloud. The proposed CloudTrack framework is used for data organization and analysis.

Figure 4 shows the architecture of the CloudTrack framework for real-time fresh food supply tracking and monitoring. CloudTrack is based on Hadoop [15] which is a framework for running applications on large clusters built of commodity hardware. Hadoop comprises of two major components:

1) *Hadoop Distributed File System* (*HDFS*): HDFS stores files across a collection of nodes in a cluster. Large files are split into blocks (64 MB by default) and each block is written to multiple nodes (default is three) for fault-tolerance [15].

2) *MapReduce*: MapReduce is a parallel data processing model which has two phases: Map and Reduce. In the Map phase, data is read from a distributed file system (such as HDFS), partitioned among a set of computing nodes in the cluster, and sent to the nodes as a set of key-value pairs. The Map tasks process the input records independent of each other and produce intermediate results as key-value pairs. The intermediate results are stored on the local disk of the node running the Map task. When all the Map tasks are completed, the Reduce phase begins in which the intermediate data with the same key

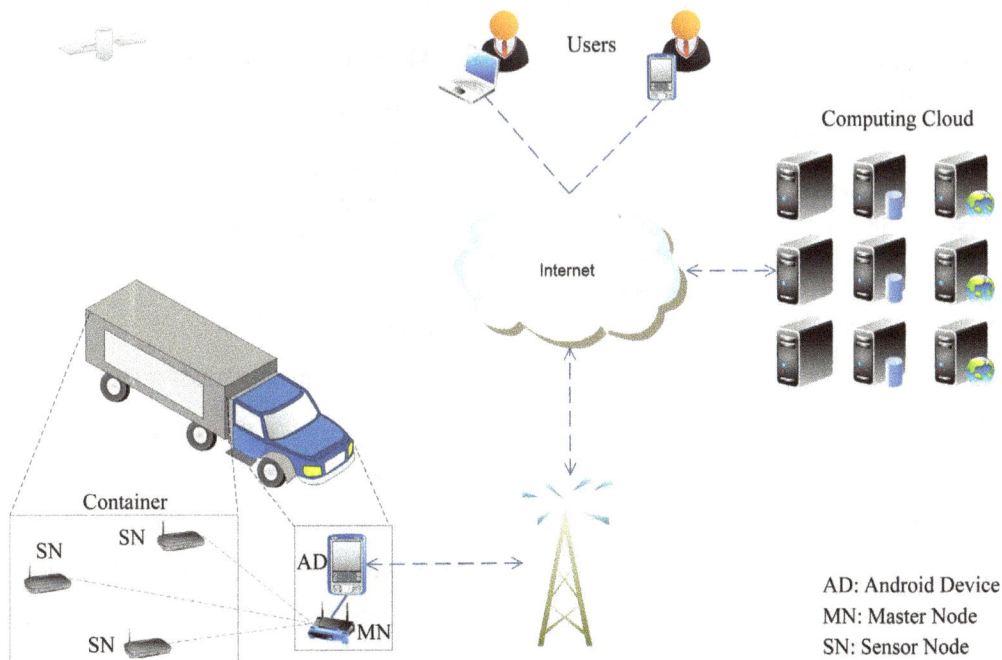

Figure 3. Proposed system architecture for real-time fresh food supply tracking and monitoring.

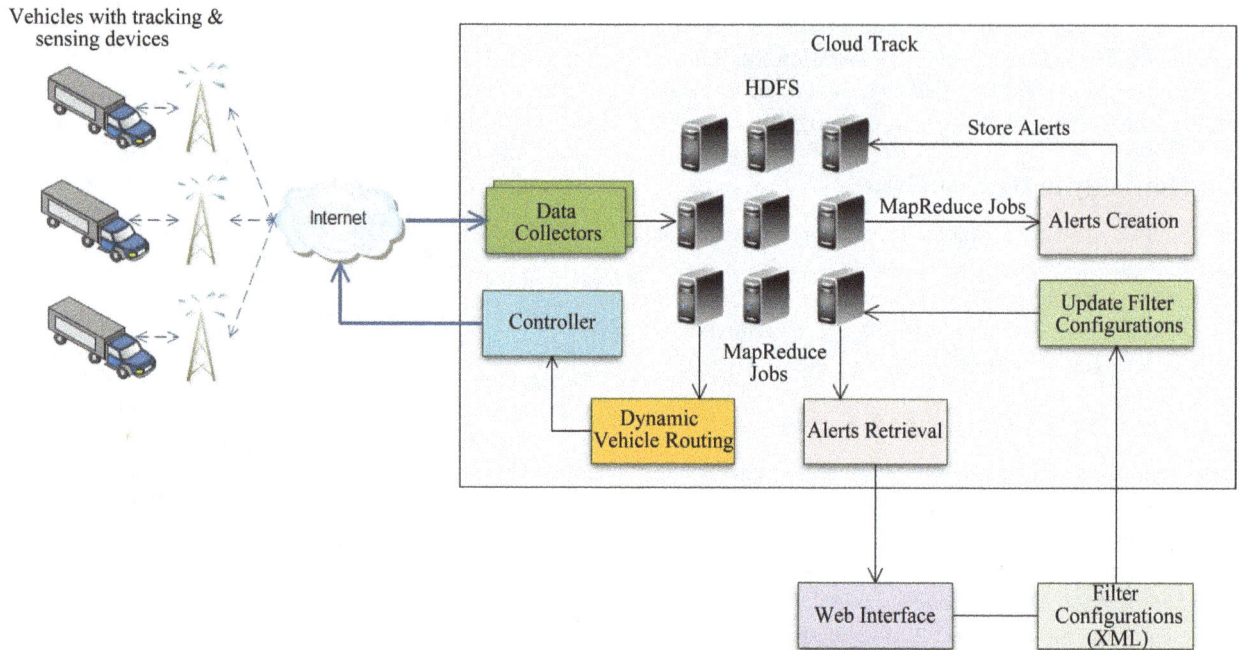

Figure 4. Proposed framework for real-time data organization and analysis in a computing cloud-CloudTrack.

is aggregated.

4.1. Data Collectors

The Data Collectors in the CloudTrack framework collect the streaming time-series data from the master nodes in the vehicles. Each incoming data stream is mapped to one of the Data Collector nodes as shown in **Figure 4**. Each Data collector node has a Data Aggregator, Data Filter and Data Archiver module. Since the raw location and sensor data comes from a large number of vehicles in the form of data streams, the data has to be preprocessed to make the data analysis using Hadoop more efficient. The Hadoop MapReduce data processing model works more efficiently with a small number of large files rather than a large number of small files. The Data Collectors buffer, preprocess and filter the streaming data into larger chunks (called Sequence Files) and store it in HDFS. Data Collectors use Hadoop's Sequence File class which provides a persistent data structure and serves as a container for multiple records. Since HDFS and MapReduce are optimized for processing large files, packing records into a Sequence File makes processing of data more efficient. The Data Aggregator aggregates streams of location and sensor data into Unstructured-Sequence Files on the local disk of the Data Collector node. The Data Filter converts the Unstructured-Sequence Files into structured records by parsing the records (lines) in Unstructured-Sequence Files and extracting the sensor readings. The Data Filter also filters out bad records in which some sensor readings are missing. The Data Archiver moves the Structured Records to HDFS.

4.2. Alerts Creation Module

The data collected is processed to generate alerts based on the user specified filters for alerts creation. This Alerts Creation Module collects the alerts into an alerts-base (alerts database) which is organized into a manageable structure in HDFS. Real-time alerts are created using the real-time location and sensor data collected in a small time window. Offline alerts can also be created from the past location and sensor data.

4.3. Alerts Retrieval Module

The Alerts Retrieval Module retrieves the alerts for displaying them in the CloudTrack dashboard. The user can search for a particular vehicle by vehicle number, vehicle type, arrival or departure locations. The Alerts Retrieval Module then retrieves the alerts for that particular vehiclefrom the alerts-base.

4.4. Controller

The Controller module sends the new routes generated by the Dynamic Scheduler to all the vehicles. Alternatively, the vehicles can also pull the new routes and additional information on the nearby vehicles from the controller.

4.5. Dynamic Vehicle Routing Module

The Dynamic Vehicle Routing Module generates routes for the vehicles based on the real-time data collected in order to minimize the spoilage of fresh food. Deviations in the planned schedule occur due to changing traffic conditions. Moreover, there may be changes in the container conditions such as an increase in temperature due to a fault in the cooling system, etc. With CloudTrack it is possible to have a global view of all the vehicles in transit. The Dynamic Vehicle Routing Module creates new routes for the vehicles when alerts occur. For example, a vehicle that is bound to miss the deadline for a scheduled destination and has a limited time window left before the spoilage of food starts, can be re-routed to a closer destination. The knowledge about the state variables of the vehicle (such as truck capacity, location, speed, container temperature, etc.) and the vehicles nearby is important for creation of new routes. The routes are generated to minimize the food spoilage and the costs involved in transportation. Savings come due to sharing of transportation costs on common routes by better utilization of vehicles and better re-routing of vehicles in the event of delays.

Instead of proposing new algorithms for vehicle routing (which is an established area of research within transportation systems), our effort behind CloudTrack is to provide a cloud-based framework that supports a wide variety of vehicle routing algorithms within a cloud architecture. We now describe a typical use case of dynamic vehicle routing. We have used Tabu Search [16] algorithm for the use case. CloudTrack is flexible to support other dynamic routing algorithm as well. For the use case we have chosen Tabu Search as it has been applied widely for various types of optimization problems, with very good results.

Figure 5 shows an example of food supply vehicle routing. The routing problem involves a set of food supply pickup points (shown as sources) and the delivery points (shown as destinations). The sources can be either collection centers for fresh food produce or warehouses where food is temporarily stored before distribution. The destinations can be either retail stores where fresh food is sold or warehouses where the food is kept refrigerated before it is transported to other locations. The problem described in this section involves a number of vehicles which can start and end their routes at different locations. The number of routes in the problem is equal to the number of vehicles. Each vehicle has a limited capacity and can serve a limited number of delivery points. An initial schedule is obtained such that each delivery point is visited only once by one of the vehicles. The Cloud-Track framework is used to monitor the real-time data captured from all the vehicles and the vehicle routes are updated if there are changes in conditions that can lead to spoilage of food supply during transit.

Figure 6 shows an example of a dynamic schedule which is generated after analysis of real-time data from the vehicles. The example shows a scenario where the vehicle on route-1 generates an alert for a delayed deliv-

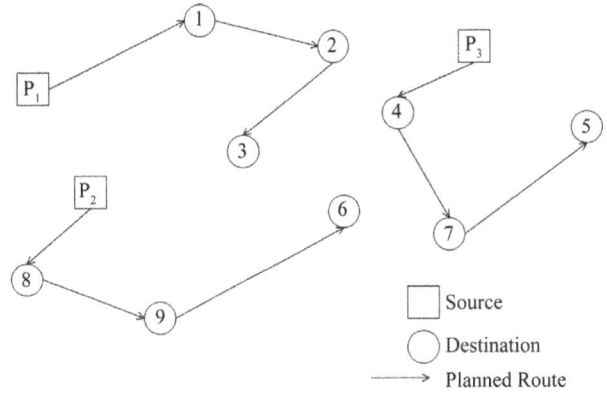

Figure 5. Example of an initial schedule.

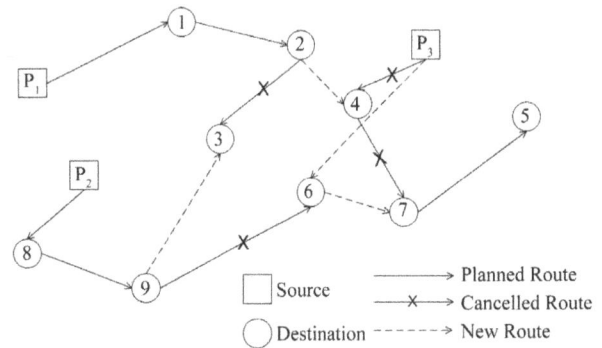

Figure 6. Modified schedule based on real-time alerts.

ery to destination-3. This alert triggers the creation of a new routes in which the vehicle-1 is re-routed to destination-4, vehicle-2 is re-routed to destination-3 and vehicle-3 is re-routed to destination-6.

We now describe the mathematical model for the vehicle routing problem shown in **Figure 5**. **Figure 7** shows the time sequence of a route. The parameters involved in the model are as follows:

N is the total number of destinations, K is the number of vehicles, M is the maximum number of destinations that can be served by a vehicle on one route, D_i is the time taken to travel from source to destination-i, T_{mk0} is the start time of the mth trip of vehicle k, T_{mk1} is the reaching time at the destination for the mth trip of vehicle k, T_{mk2} is the end time of the mth trip of vehicle k, S_i is the handling time at destination-i, R_i is the time window for delivery at the destination-i, W_{i0} is the start time of the time window at destination-i and W_{i1} is the end time of the time window at destination-i. $X_{ikm} \in \{0,1\}$ is the decision variable where $i \in \{1, \cdots, N\}$, $k \in \{1, \cdots, K\}$, $m \in \{1, \cdots, M\}$. $X_{ikm} = 1$ if destination-i is served by vehicle k on its mth trip and 0 otherwise. The objective function that represents the total cost of transporting the food supply is defined as follows:

$$F = \sum_{i=1}^{N} D_i \sum_{k=1}^{K} \sum_{m=1}^{M} X_{ikm} + \sum_{i=1}^{N} \alpha_i \left(T_{mk1} - W_{i1} \right) \quad (1)$$

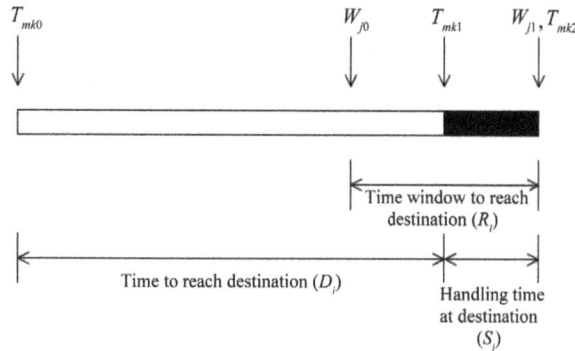

Figure 7. Time sequence of a route.

where α_i is lateness penalty coefficient associated with destination-i. The objective function involves the transportation cost from the source to the destination and the penalty due to late delivery at the destination. For the use case, to minimize the objective function given the constraints as described in the mathematical model, we adopt a meta-heuristic approach called Tabu Search [16].

The solution obtained by minimizing the objective function is a schedule specifying the routes for each vehicle and the sequence of the destinations to be served. Tabu search is a memory based meta-heuristic method that uses a memory structure called tabu list to store the recent moves or solutions. During the search process, attempts that produce the moves or solutions in the tabu list are denied. The moves or solutions in the tabu list can be overridden sometimes, when an *aspiration criteria* is satisfied which produces a globally best solution. To obtain an initial solution we adopt the Push Forward Insertion Heuristic (PFIH) described by Solomon [17]. The algorithm for generating initial solution based on Push Forward Insertion Heuristic (PFIH) is shown in **Table 2**.

The Tabu Search process proceeds by a sequence of intensifications and diversifications. Intensification is a strategy that aims at a detailed exploration of the neighborhood of the solutions that are historically found to be good. Whereas, diversification is a strategy that aims at driving the search into new regions. After obtaining an initial solution, it is intensified using a λ-interchange local search method introduced by Thangiah *et al.* [18]. The algorithm for vehicle routing based on Tabu Search [16] is described in **Table 3**.

The algorithm for dynamic vehicle routing which is used in the use case, is described in **Table 4**. A vehicle routing algorithm such as tabu search is used to obtain an initial static schedule. Whenever a new alert occurs, a new schedule is generated based on the current locations and the existing filled capacities of the vehicles. If a feasible solution is found, the new schedule is pushed to all the vehicles by the Controller module. If no feasible solution is found, a local fix is obtained for the vehicle that generated the alert. For example, a local fix can be re-

Table 2. Algorithm for generating initial solution based on PFIH [17].

1. Begin with an empty route starting from a depot and set R = 1.
2. If all delivery points have been routed goto step-10.
3. Else for all delivery points which haven't been routed compute of the cost of inserting them as the first node and sort them in ascending order.
4. Select first delivery point d1 from the sorted list which has the least cost and is feasible in terms of time and capacity constraints.
5. Append d1 to the current route R and update the total cost of the route.
6. For all delivery points which haven't been routed, for all edges {m, n} in the current route, compute the cost of inserting each of the unrouted delivery points between m and n.
7. Select the delivery point d and edge {m, n} that has the least cost.
8. If the insertion of delivery point d between m and n is feasible in terms of time and capacity constraints, insert the d between m and n and update the cost of the current route. Goto step-6.
8. Else goto step 9.
9. Begin a new route from a depot and set R = R + 1. Goto step-2.
10. Stop.

Table 3. Algorithm for vehicle routing based on Tabu Search [16].

1. Obtain an initial solution using Push Forward Insertion Heuristic (PFIH) and set the global best solution equal to the current solution, *i.e.* $S_b = S$.
2. Initialize the tabu list and candidate list and add the current solution to the tabu list.
3. Do intensification, *i.e.* exploration of the neighborhood of the current solution S using 2-interchange local search and update the candidate list (with the best solution at the top of the list).
4. Set S_0 equal to the first solution in the candidate list.
5. If S_0 is in the tabu list select the next best solution from the candidate list and set it as S_0.
6. If Cost(S_0) < Cost(S_b) set the best solution $S_b = S_0$ and update the tabu list.
7. Do diversification and update the candidate list with the random solutions produced by the random hops
8. If the total number of iterations is less than maximum allowed iterations, go to step-3.
9. Else terminate the search.

Table 4. Algorithm for dynamic vehicle routing.

1. Obtain an initial static schedule using a vehicle routing algorithm (such as tabu search).
2. If an alert occurs, obtain the current locations and filled capacities of the vehicles.
3. Generate a new solution using the vehicle routing algorithm with the current locations and filled capacities of the vehicles as input.
4. If a feasible solution is found send the new schedule to all the vehicles.
5. If no feasible solution is found, mark the alert for a local fix and obtain a local solution for the vehicle that raised the alert.

routing the vehicle to the nearest delivery point such as a retail store or warehouse which is not on the planned route of other vehicles.

4.6. CloudTrack Dashboard

Figure 8 shows a screenshot of the CloudTrack Dashboard that is used to visualize the tracking and monitoring data for a particular vehicle. The dashboard has wid-

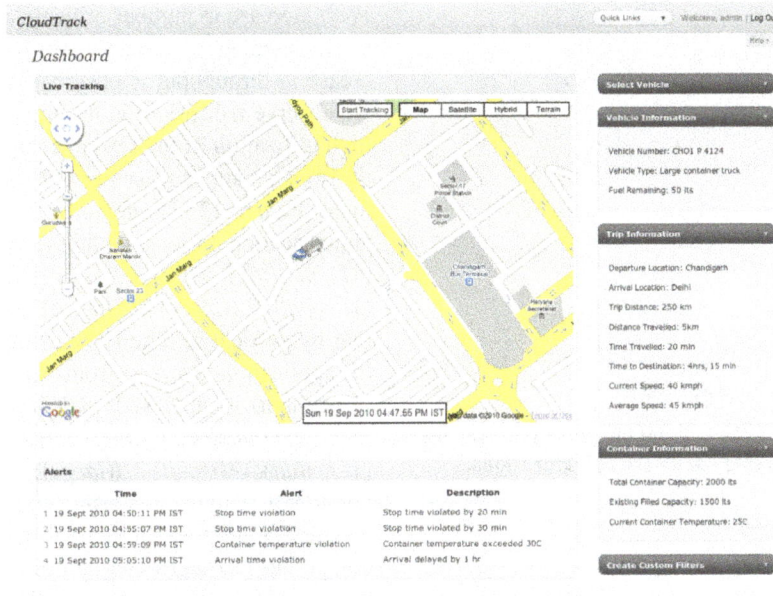

Figure 8. Screenshot of CloudTrack dashboard.

gets for displaying the following information: 1) vehicle information, such as vehicle number, vehicle type, etc.; 2) trip information such as departure and arrival locations, trip distance, distance and time traveled, time to destination, current and average speeds, etc.; 3) container information such as total container capacity, existing filled capacity, container temperature and humidity, etc. The dashboard also allows creation of custom filters for alerts.

5. Evaluation

To model the data collection architecture of CloudTrack, we developed a prototype system that used an Arduino Uno development board, Sensirion SHT21 temperature and humidity sensor, and an Android OS based mobile device. On the Android device an application was written to read in the sensor readings. A service within the application captures the sensor data from the Arduino board and pushes the sensor data along with the GPS data obtained from the Android device to CloudTrack. Using the empirical data collected from the prototype system a model for a food truck was developed. This model was then used to generate a large number of virtual trucks using a data generator from which synthetic data was generated. A large data set (for upto 1000 delivery points and 100 vehicles) was generated synthetically, which allowed us to validate the scalability, flexibility and control mechanism of CloudTrack.

In order to evaluate the scalability of the proposed CloudTrack framework, we performed a series of experiments with varying number of delivery points and vehicles, using the Amazon Elastic Compute Cloud (EC2) infrastructure [19]. For simplicity in describing Cloud Track's multi-tier deployment configuration we use the naming convention-(#L (*size*)/#A (*size*)/#H (*size*)), where #L is the number of instances running load balancers and web servers, #A is the number of instances running application servers, #H is the number of instances running the Hadoop cluster and (*size*) is the size of an instance. **Figure 9** shows comparisons of alerts creation times for varying number of records on varying hardware configurations (Amazon EC2 compute units). Each compute unit provides an equivalent CPU capacity of 1.0 - 1.2 GHz 2007 Opteron processor or 2007 Xeon processor. We observe that even with a large number of vehicles and records, the alerts can be created in a time-scale of few seconds to minutes.

With Hadoop it is possible to analyze such massive scale data efficiently. We also observed that when the amount of data to be analyzed is small, the Hadoop's non-significant startup costs dominate the execution time.

However as the amount of data to be processed increases the startup costs are dwarfed by the execution time. For analysis of 1,000,000 records with CloudTrack, experimental measurements show a speed up of upto 4 times using a computing cluster (with 33.5 EC2 compute units) as compared to a single node (with 1EC2 compute unit). From the results in **Figure 9** it is observed that CloudTrack is well suited for massive scale vehicle location and sensor data processing. For example, if the location and sensor readings are collected from 100 vehicles every 5 seconds then 1 million records correspond to approximately 14 hours of data. **Figure 10** shows the comparisons of vehicle route generation times for varying number of delivery points and vehicles on varying

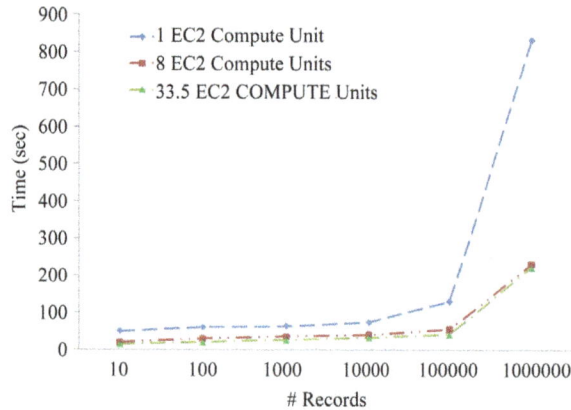

Figure 9. Comparisons of alerts creation times for varying number of records on varying hardware configurations.

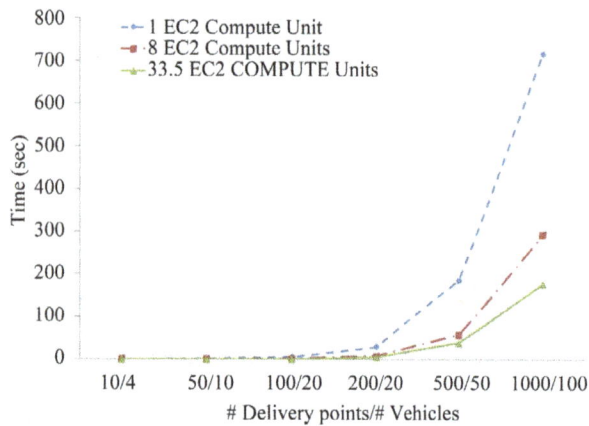

Figure 10. Comparisons of vehicle route generation times for varying number of delivery points and vehicles on varying hardware configurations.

hardware configurations. From the results in **Figure 10** we observe that CloudTrack can generate vehicle routes on timescale of few seconds to minutes. Furthermore, from the results in **Figures 9** and **10**, it is observed that the CloudTrack framework can be scaled up by adding additional computing resources. **Figure 11** shows the relative costs of the vehicle routing solutions generated by the CloudTrack's vehicle routing module at different iterations. The costs shown in the plot have been normalized with the cost of the final solution. These costs are calculated using Equation (1). For simplicity, we choose D_i in Equation (1) to be the distance from source to destination-i. Normalization of costs is done by dividing the cost of a solution with the cost of the final solution. From the results in **Figure 11** it is observed that the relative costs of the solutions decrease in each successive iteration as the vehicle routing algorithm progresses.

Figure 12 shows the average throughput for the CloudTrack dashboard for three different deployment configurations. We observe that throughput continuously increases as the number of tracked vehicles increase.

With increase in number of vehicles CloudTrack services higher number of requests per second, therefore an increase in throughput is observed. Beyond 1600 vehicles, we observe a decrease in throughput, which is due to the high utilization of the resources (CPU, Disk I/O, etc.) for the web or application tiers of the CloudTrack deployment. **Figure 12** also demonstrates the vertical and horizontal scaling options of CloudTrack. Comparing (1*L*(*small*)/2*A*(*small*)/2*H*(*large*)) and (1*L*(*small*)/2*A*(*large*)/2*H*(*large*)) deployments, we observe that by vertical scaling (increasing the compute capacity of application servers from small to large) a higher throughput is achieved. Similarly comparing deployments (1*L*(*small*)/2*A*(*small*)/2*H*(*large*)) and (1*L*(*small*)/3*A*(*small*)/2*H*(*large*)), we observe that by horizontal scaling (increasing the number of application servers), a higher throughput is achieved. Additional computing resources can be provisioned for larger number of vehicles, depending on the scale of the problem. **Figure 13** shows the average response time for the CloudTrack dashboard for three different configurations. With increase in number of vehicles the mean request arrival rate increases since CloudTrack services higher number of requests per second, therefore an increase in response time is observed. **Figure 13** also demonstrates the vertical and horizontal scaling options of CloudTrack. Comparing the three different deployment configurations of CloudTrack we observe that lower response times are achieved by vertical and horizontal scaling.

To sum up, the experiments done using 1) the prototype system for sensor and location data collection; 2) virtual vehicle model developed using empirical data collected from the prototype system; 3) synthetic data generated using the virtual vehicle model and a data generator, were able to test the complete CloudTrack architecture including data collection, data analysis for alerts creation and dynamic route generation.

6. Conclusion

In this paper, we propose a cloud-based framework, CloudTrack, and provide a case study of how the IT infrastructure may be efficiently integrated into the supply chain and operational systems. We demonstrated the feasibility of CloudTrack as a scalable platform for data driven intelligent transportation systems, based on new cloud-based programming models and data structures, Hadoop and MapReduce. The fresh food supply vehicles are equipped with Sensor Nodes, a Master Node and an Android Device to collect and transmit real-time location and sensor data. A distributed batch processing infrastructure, Hadoop, is used for running data analysis jobs on clusters of machines. Data analysis jobs are formulated using Hadoop MapReduce programming model,

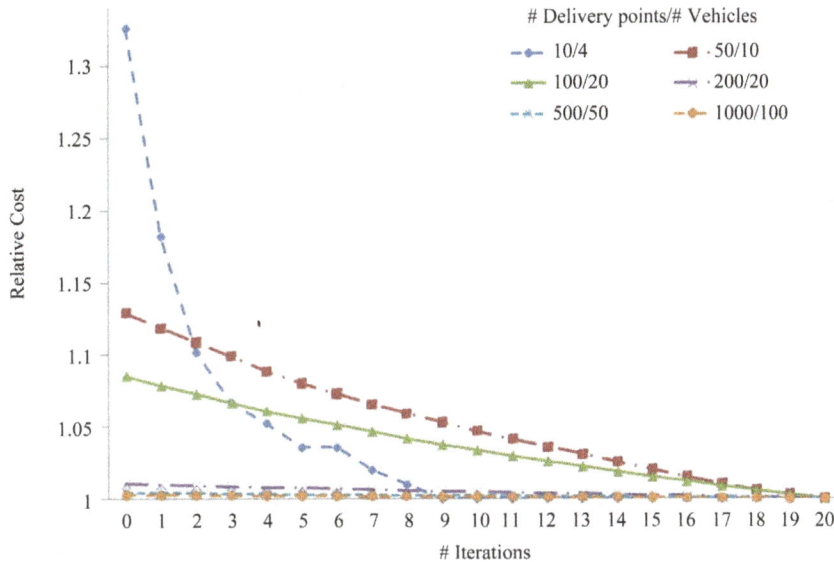

Figure 11. Relative total cost of generated routes. Tabu search algorithm was used for experimental evaluation.

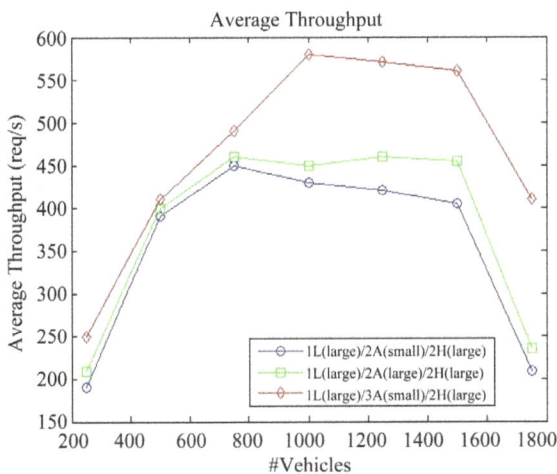

Figure 12. Average throughput for CloudTrack dashboard for different deployment configurations.

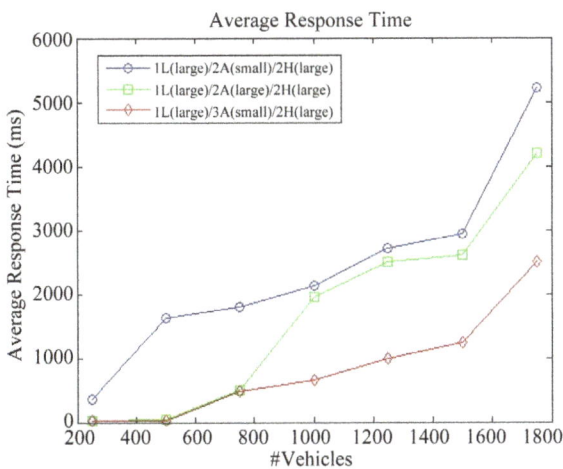

Figure 13. Average response time for CloudTrack dashboard for different deployment configurations.

which allows jobs to be run in parallel. Experimental measurements showed that the MapReduce implementation can create alerts on a timescale of few seconds to minutes on the cloud. Moreover, upon the creation of alerts, new vehicle routes can be generated in a timescale of few seconds to minutes. With the proposed approach the spoilage of fresh food during transit can be reduced due to better re-routing of vehicles based on the real-time information.

REFERENCES

[1]   J. Zhang, F. Wang, K. Wang, W. Lin, X. Xu and C. Chen, "Data-Driven Intelligent Transportation Systems: A Survey," *IEEE Transactions on Intelligent Transportation Systems*, Vol. 12, No. 4, 2011, pp. 1624-1639.

[2]   R. Claes, T. Holvoet and D. Weyns, "A Decentralized Approach for Anticipatory Vehicle Routing Using Delegate Multiagent Systems," *IEEE Transactions on Intelligent Transportation Systems*, Vol. 12 No. 2, 2011, pp. 364-373.

[3]   D. A. Steil, J. R. Pate, N. A. Kraft, R. K. Smith, B. Dixon, L. Ding and A. Parrish, "Patrol Routing Expression, Execution, Evaluation, and Engagement," *IEEE Transactions on Intelligent Transportation Systems*, Vol. 12 No. 1, 2011, pp. 58-72.

[4]   E. Schmitt and H. Jula, "Vehicle Route Guidance Systems: Classification and Comparison," *Proceedings of IEEE ITSC*, Toronto, 2006, p. 242247.

[5]   M. T. Nkosi, "Cloud Computing for Enhanced Mobile Health Applications," *IEEE Second International Conference on Cloud Computing Technology and Science (CloudCom)*, Indianapolis, 30 November-3 December 2010.

[6]   M. A. H. Masud, "Cloud Computing for Higher Education: A Roadmap," *IEEE 16th International Conference*

on Computer Supported Cooperative Work in Design (CSCWD), Wuhan, 23-25 May 2012.

[7] X. Fang, S. Misra, G. L. Xue and D. J. Yang, "Managing Smart Grid Information in the Cloud: Opportunities, Model, and Applications," IEEE Network, Vol. 26, No. 4, 2012, pp. 32-38.

[8] Z. J. Li, "Cloud Computing for Agent-Based Urban Transportation Systems," IEEE Intelligent Systems, Vol. 26, No. 1, 2011, pp. 73-79.

[9] P. Jaworski, "Cloud Computing Concept for Intelligent Transportation Systems," 14th International IEEE Conference on Intelligent Transportation Systems (ITSC), Washington DC, 5-7 October 2011.

[10] A. Bahga and V. K. Madisetti, "Analyzing Massive Machine Maintenance Data in a Computing Cloud," IEEE Transactions on Parallel & Distributed Systems, Vol. 23, No. 10, 2012, pp. 1831-1843.

[11] Department of Scientific & Industrial Research, "Fruits & Vegetables Sector: An Overview," Department of Scientific & Industrial Research Report, India, 2011.

[12] Z. B. Pang, J. Chen, Z. Zhang, Q. Chen and L. R. Zheng, "Global Fresh Food Tracking Service Enabled by Wide Area Wireless Sensor Network," IEEE Sensors Applications Symposium (SAS), Limerick, 23-25 February 2010.

[13] Y. Xi, W. Yang, N. Yamauchi, Y. Miyazaki, N. Baba and H. Ikeda, "Real-Time Data Acquisition and Processing in a Miniature Wireless Monitoring System for Strawberry during Transportation," TENCON, Hong Kong, 2006, pp. 1-4.

[14] Y. L. Bu and L. Wang, "Leveraging Cloud Computing to Enhance Supply Chain Management in Automobile Industry," International Conference on Business Computing and Global Informatization, Shanghai, 29-31 July 2011.

[15] Apache Hadoop. http://hadoop.apache.org

[16] F. Glover, "Tabu Search Part I," ORSA Journal on Computing, 1989.

[17] M. M. Solomon, "Algorithms for the Vehicle Routing and Scheduling Problems with Time Window Constraints," Operations Research, Vol. 35, No. 2, 1987, pp. 254-265.

[18] S. R. Thangiah, I. H. Osman, R. Vinayagamoorthy and T. Sun, "Algorithms for the Vehicle Routing Problems with Time Deadlines," American Journal of Mathematical and Management Sciences, Vol. 13, No. 3-4, 1993, pp. 323-355.

[19] http://aws.amazon.com/ec2/instance-types

14

Wi-Fi Traffic Enforcement System (WiTE)

Fady W. Gendi[1], Tarek K. Refaat[1], Amir H. Sadek[1], Ramez M. Daoud[1], Hassanein H. Amer[1], Chahir S. Fahmy[1], Omar M. Kassem[1], Hany M. ElSayed[2]

[1]Electronics Engineering Department, American University in Cairo, New Cairo, Egypt
[2]Electronics and Communication Department, Cairo University, Giza, Egypt

ABSTRACT

This paper proposes a single integrated traffic enforcement system that is able to recognize and report various traffic violations. It consists of a Wi-Fi infrastructure that enables communication between moving vehicles and a central node. Unlike existing solutions, which address single violations, the proposed model encompasses several issues like exceeding speed limits, entering a no entry street, car theft, congestion and tolling. OPNET simulations were run to test the Wi-Fi model and define its different characteristics and limitations. A proof-of-concept case was modeled, and the proposed architecture succeeded in meeting all design requirements.

Keywords: Wi-Fi; Traffic Enforcement; Vehicular Networking; Vehicle to Infrastructure (V2I); Infrastructure to Vehicle (I2V)

1. Introduction

Traffic enforcement systems are very important implementations of different technologies used worldwide. However, there is no single solution that involves a unified integrated system that is able to enforce all traffic laws. Existing solutions include Radio Frequency Identification (RFID) used for tolling [1], Camera-based used for red light crossing violations [2]. Also, Radars [3], Wireless Magnetic Sensors [4] and Induction Loops are used for speeding violations [5]. There are also Global Positioning System (GPS)-based traffic monitoring technologies; however, such solutions are mainly used for congestion reporting rather than traffic enforcement [6, 7].

Each of the currently available solutions addresses mainly a single violation, requiring a combination of several solutions to address them all. The lack of a single comprehensive system is the motivation for this study.

This paper proposes an integrated traffic enforcement system that is able to recognize and report various violations using Wi-Fi. The system consists of Wi-Fi access points that connect to Wi-Fi-enabled vehicles in urban environments. Using this system, the access points are able to communicate with the vehicles and using novel algorithms, and the vehicles are able to identify the various violations and report them to the access point, which

then connects to the local server. The servers and access points are distributed across the city grid. Some of the violations will require real time communication while others will be able to cope with delay. The issues that are addressed by this system are exceeding speed limits, entering a no entry street, car theft, congestion and tolling. This study focuses on the feasibility of a system that addresses these issues, presenting several solutions for each problem. It is important to note that while not addressed in this paper, the effects of implementing such a system in a non-free space environment, considering for example fading, are currently being studied.

This paper will introduce a literature survey of the existing traffic enforcement systems in Section 2. Section 3 describes the proposed system. The specific architecture shall be discussed in Section 4 and the results of the simulations will be presented in Section 5. Section 6 will conclude the study.

2. Existing Solutions

Traffic enforcement systems are important systems used throughout the world. There are several possible solutions that would allow such traffic enforcement. RFID systems are a common example of such systems [1]. This technology utilizes radio waves to relay data from a tag attached to an object to a reader for identification. The range of RFID can be several millimeters up to a few meters. It is mostly used for tolling; however, it was recently used for traffic management and enforcement [1]. This involves tracking and surveillance. Active RFID tags can also act as readers and can detect multiple tags simultaneously [8].

Another technology used for traffic enforcement is a camera-based traffic enforcement system. This is the most basic and common system, which relies on image processing for traffic violation detection. Camera-based systems take pictures upon violation, such as red-light-crossing, and then send the image to a processing unit, which recognizes the violation [2]. These systems can be expensive to implement [8].

One speed detection systems is the radar system. Traffic radar units send out a wide radar beam, which widens as it travels and can reach widths of hundreds of meters. The radar unit detects the vehicles when the beam bounces back. The speed is calculated using the Doppler shift [3]. Due to the use of low power, radars suffer from detection problems with out-of-range vehicles [3]. Also, they mistakenly identify violations due to antenna positioning errors [3].

Another system that is used for vehicle detection uses wireless magnetic sensors. It is able to sense road vehicles due to the fact that they have significant amounts of ferrous metals in the chassis. The wireless magnetic sensors are sensitive, small and immune to environmental factors. When a vehicle passes by the detector, it affects the flux lines of the magnetic fields of the earth, which are then detected [4]. These systems are used in applications such as parking lot space detection. However, a problem occurs when the vehicle does not emit sufficient magnetic fields to be detected by the sensors. Another similar system is the inductive loop system, which consists of a loop, its extension and a detector. When the detector is powered up, electricity will flow through the loop creating a magnetic field that resonates at a constant frequency. When a vehicle passes above the loop it increases the resonating frequency. The difference in resonation can also distinguish between a large vehicle and a compact car [5].

GPS-based systems used for congestion monitoring are also available. Such systems can utilize on-board GPS units or in more recent times, mobile device-based GPS software [6,7]. While the accuracy or resolution of such devices is sufficient for congestion monitoring, it may not be accurate enough for other applications. There is also the problem of available bandwidth, as the currently implemented cellular networks may not be able to handle more demanding real-time loads. The lack of suitability of a GPS-based system will also be addressed for each of the proposed violations.

Seeing as none of the previously mentioned solutions can efficiently address *all the issues* mentioned in the

introduction, a Wi-Fi-based system is proposed next that integrates several of the major traffic enforcement techniques using standard protocols [9,10].

3. Proposed System

The main concept behind the system proposed in this paper, Wi-Fi Traffic Enforcement (WiTE), is placing a Wi-Fi card on-board of all vehicles. This card is able to communicate with various Wi-Fi Access Points (APs) in an infrastructure along the roads of a city, where APs will be available at all intersections. These APs can either be an existing infrastructure or a specifically built infrastructure for the WiTE system. When simulating the system on OPNET Network Modeler [11], a Free Space Wi-Fi environment is used (path loss exponent of 2).

It is critical to note, that a Free Space environment is not an accurate representation of the practical scenario. For that reason, the following section details the countermeasures and calculations used in order to build a network on OPNET which more realistically represents the practical scenario. Equation (1) is the Free Space propagation model (with path loss exponent n = 2). In this case, n will be left as a variable rather than a constant to enable modeling of different environments.

$$P_r = P_t G_t G_r \left(\frac{\lambda}{4\pi R} \right)^n \qquad (1)$$

where:

P_r : R_x power (W);
P_t : T_x power (W);
G_r : R_x gain;
G_t : T_x gain;
λ : wavelength (m);
R: distance between T_x and R_x (m);
n: path-loss exponent.

A vehicle speed of 80 Km/h is higher than any vehicle speed allowed in most downtown areas. In most countries the top speed is 60 Km/h and has a 10% acceptable margin of increase (about 48 Km/h in urban environments in Minnesota and 50 Km/h France [12,13]).

Modeling a moving Wi-Fi node at 80 Km/h on OPNET with an assumed transmit power of 1 mW, it was found that the required antenna sensitivity at the receiver was −80 dBm. Plugging these values into Equation (1) [14], the distance needed for a node to stay within the coverage of an AP in order to achieve successful communication was calculated to be 99.47 m. This distance, as a coverage radius, with the speed of 80 Km/h indicated that the time of connecting/disconnecting and sending a single packet was 4.48 s.

In order to verify the previously calculated values, an experiment was conducted on OPNET Network Modeler. The scenario involved a mobile node sending one packet to an AP while moving at 80 Km/h along the diameter of the AP coverage area (with a 95% confidence level). The outcome of the experiment showed an AP coverage radius of 98.19 m and connection duration of 4.42 s. These results confirm the previously calculated values.

Whether on OPNET simulations or in an actual implementation, the main requirement for the system is that an AP must have a coverage radius of at least 98.19 m, accommodating a maximum vehicle speed of 80 Km/h. This radius can be controlled by two parameters: transmit power and antenna sensitivity. To achieve 98.19 m on OPNET, a transmit power of 1 mW was used with a R_x sensitivity of −80 dBm. These values shall be recalculated using Equation (1) for different values of n, to determine the required transmit power and R_x sensitivity in an urban environment, guaranteeing a 98.19m coverage radius. The values of n used must lie between 3 and 3.5 [15] and the resulting transmit power and R_x sensitivity must conform to the Wi-Fi standard.

The distance (radius of 98.19 m), and P_r (receiver sensitivity) set to −80 dBm and the wavelength at 2.4 GHz is 0.125 m. From Equation (1), it can be concluded that the harshest path loss exponent representing an urban environment that could use the WiTE system with the maximum allowable Wi-Fi transmit power of 100 mW (in EU states) is 2.5 [10]. On the other hand, in the United States [16], the maximum allowed power is 1000 mW and this can be achieved in an environment of path loss exponent 2.75.

The previously stated path loss exponents are quite optimistic and the WiTE system, if proposed for urban environments, must at least be able to perform within a path loss exponent of 3.0.

A new technology presented in [17] can deliver sensitivity as high as −100 dBm and this value is reused for the calculations. The final set of results, for the highest sensitivity available, is shown in **Table 1**. It can be concluded that using a sensitivity of −100 dBm, a path loss exponent of 3.0 for the EU and 3.254 for the US can be reached.

The next section presents the actual architecture to be used in the WiTE system. The presented architecture is a proof-of-concept with preliminary results, to guarantee feasibility of the system. Effects of fading and interference are not considered yet but are currently being investigated. The presented system also addresses a scenario where the only Wi-Fi nodes needed are located at the intersections. In many cities, there is widespread Wi-Fi coverage, and it would be even more optimistic.

4. System Architecture

The proposed system architecture requires the placement of Wi-Fi APs at traffic junctions and roundabouts. These are linked together using wired Switched Ethernet to "ZONES" representing a small geographical area. These ZONES in turn are connected to larger centers "AREAS"

Table 1. Path loss exponent vs. power needed for −80 dbm and −100 dbm antenna sensitivity.

Path Loss Exponent (n)	Power (W) for −80 dBm	Power (W) for −100 dBm
2.4	0.0385	0.0004
2.5	0.0967	0.0010
2.6	0.2425	0.0024
2.7	0.6084	0.0061
2.8	1.5262	0.0153
2.9	3.8285	0.0383
3.0	9.6037	0.0960
3.1	24.0909	0.2409
3.2	60.4321	0.6043
3.3	151.5942	1.5159
3.4	380.2745	3.8027
3.5	953.9198	9.5392

representing larger geographical areas. In turn, all of these AREAS are connected to a central administrative traffic unit that contains the central database with all the vehicles' records. A simple representation can be seen in **Figure 1**. These wireless access points transmit a beacon to the vehicles in the intersection or roundabout repeatedly at fixed intervals. On the receiving end, all vehicles are equipped with a processor, connected to the Wi-Fi card. Each beacon sent from an AP contains traffic regulations regarding that area that are received by the vehicle processor. This onboard processor uses the beacon packet to monitor the vehicle and report any violations. If a violation takes place, the vehicle immediately sends a packet to the nearest AP, which can then be relayed to the central station and added to the vehicle violations log in the central database.

The WiTE system's main strength is the integration of several different traffic violations. The violations are monitors and reported by the same system. These are: speeding, no entry, and theft. In addition, solutions for tolling and congestion reporting are also presented. In the AP Beacon, the current AP ID, adjacent 4 AP IDs, vehicles allowed to pass and their respective speed, and stolen car IDs are all included. This information takes up less than the smallest Ethernet data payload, 46 bytes. The violation packet is even smaller containing the violation ID, time and date as well as the vehicle and AP ID. The algorithms proposed for each of the different traffic violations supported by the WiTE system are presented below.

Speeding Violation

1) A default maximum speed of 60 km/hr is preprogrammed into all vehicles.

2) As a vehicle enters an intersection, it receives a beacon from the Zone AP, containing the speed limit data for

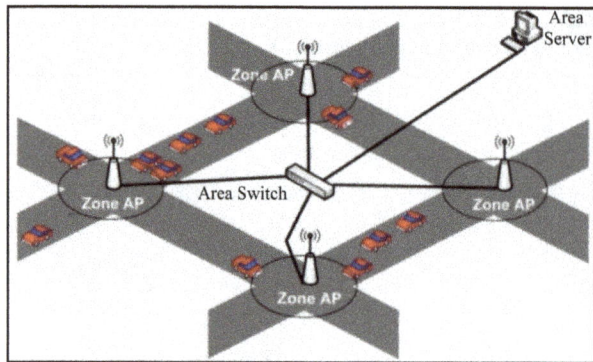

Figure 1. Proposed system architecture.

different vehicle types.

3) This speed limit is updated within the vehicle if it is different from the default 60 km/hr.

4) Vehicle onboard speedometer monitors speed and processing unit continuously monitors speed and compares to limit.

5) When the vehicle speed exceeds the limit for a certain duration of time, a violation is logged.

6) The logged violation saves the zone where the vehicle was at pre-violation and post-violation, the speed, time, violation ID and vehicle ID.

7) The next zone the vehicle joins, will receive the violation, over TCP/IP. The acknowledgment of receipt will initiate clearing of onboard logged violation. If no acknowledgement is received, the vehicle will continue to report the violation.

8) Once successfully received by Zone AP, the violation is forwarded to the central office.

9) Processing is finalized at the central office, logging of data, mapping location, time, date, fine and ID.

To enable the detection of speeding violations, all wireless access point beacons for a certain junction are programmed with a preset maximum speed limit depending on the zone.

No Entry Violation

1) Vehicle receives beacon upon entering intersection (*i.e.*, beginning of no entry zone).

2) For this vehicle type, the beacon speed limit data will indicate a speed of zero.

3) Automatically, upon entering the zone, the vehicle will record a violation (if in motion).

4) The violation is reported to the nearest Zone AP to which it connects after violation.

5) Violation contains location, time, date, vehicle type and ID.

6) Again over TCP/IP protocol.

Stolen Vehicle Violation

1) A stolen vehicle is reported by its owner.

2) Be it reported in a certain area or over the entire city, the authorities will utilize the zones that form this area/ city.

3) A section of the beacons sent out, will be reserved for searching for stolen vehicles.

4) If more than 1 vehicle is being searched through the system, Time Division Multiplexing (TDM) will be used. At any given time interval, only one specific vehicle will be searched for.

5) These beacons will be continuously sent until the stolen vehicle replies to an AP, therefore giving its location.

6) When the vehicle's internal system reads the packet from the AP and recognizes itself as the stolen vehicle, a certain internal flag is set, which will cause the vehicle to continuously send beacons. This will make the tracking of the vehicle possible. This will only stop when the vehicle is reset (by the authorities).

7) The beacon sent by the vehicle contains a unique ID for tracking purposes. This gives these packets priority over other packets.

8) The AP knows to directly forward these received packets to the central office of the area to notify the authorities in real time.

Congestion Monitoring

1) A vehicle in a congested area (e.g., where the speed of the vehicle is 10 Km/h or less for more than 2 minutes) will log its location based on the nearest Zone AP.

2) A congestion packet is sent to the Zone APs as if it was a violation and will await an acknowledgement. At the area level, when the system receives a large number of congestion packets, either authorities will be notified of an issue in the area, or merely a state of traffic congestion will be declared.

3) If an AP receives a packet indicating congestion outside its area, then that AP will forward that information to the station connected with that certain area.

Tolling Usage

1) Entering a toll station, a vehicle receives a beacon from the corresponding station AP, containing location, time and date.

2) At the exit toll station, the car receives the beacon from the exit AP and sends the AP ID and time of entry into the tolled road to the exit AP.

3) Once the exit AP receives the exit message, it sends the AP ID and time of the exit and entry points to the central station.

4) The central station calculates the fare according to the system implemented in the country and bills the vehicle ID accordingly.

The flowchart presented in **Figure 2** can help visualize the proposed algorithm sequence. The system first checks if there is a saved packet or if the packet is new. Throughout the process, if the system is shutdown (for example after parking), the most recent packet is saved for use upon restart. A GPS-based system would fall short in several ways in comparison to the system proposed in this

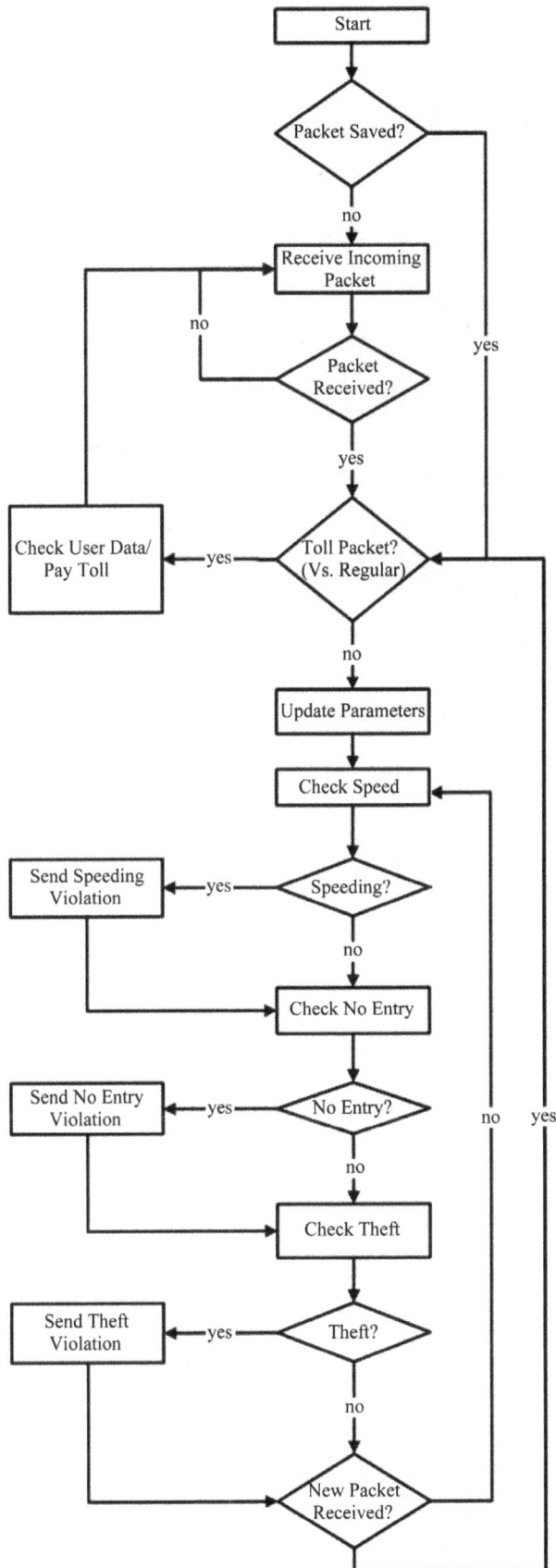

Figure 2. Proposed system algorithm flowchart.

paper. For example, in the case of no entry violations, the accuracy of current GPS systems is not high enough to determine the location (and specifically the lane) of the vehicle. For high path-loss exponents (as previously used in Equation (1)), the GPS signal accuracy deteriorates. Moreover, in the case of theft, due to the lack of a cellular downlink (infrastructure to vehicle communication), a very large overhead will be added for cars to constantly notify the infrastructure of their IDs and locations. The use of the uplink (vehicle to infrastructure communication) for reporting will also congest the cellular network, which is not designed to accommodate this load, affecting network subscribers and hence, degrading Quality of Service (QoS).

5. Simulations and Results

To verify the effectiveness and proof-of-concept for the system, several simulations were conducted using OPNET to model the worst-case real life scenarios for the system. The goal was to ensure that the system could withstand the worst conditions, with 95% confidence, and be able to detect all possible violations.

The experiments revolved around two testing criteria: the maximum number of vehicles the AP can withstand (Experiment I) and the maximum speed the data packets (beacons and violations) can be sent and received successfully (Experiment II).

In Experiment I, the maximum number of vehicles that can simultaneously communicate with one AP at a traffic junction with no delayed or dropped packets according to the predefined data was determined to be 62 vehicles. Based on the delay/drop criteria mentioned, a number of vehicles greater than 62 would not be satisfied, and the AP would not guarantee that all vehicles are able to communicate correctly.

This value was obtained by modeling a basic intersection layout. Vehicles were equally distributed along the four lanes of the intersection. All vehicles are evenly spaced and are all static, simulating a traffic jam.

An AP (modeling the Zone AP) is positioned in the middle of the intersection, wired to a switch, and finally to the server (modeling the central office). This layout can be seen in **Figure 3**. Each of the vehicles is programmed to

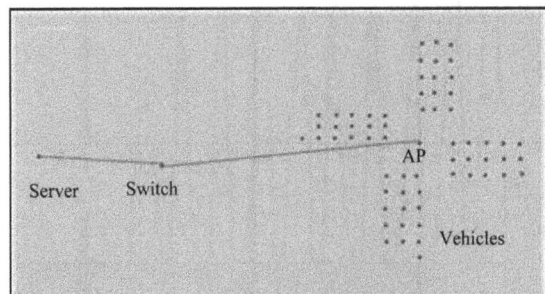

Figure 3. Experiment I layout.

report a range of the different violations, to maximize congestion and the number of vehicles is gradually increased to determine the maximum number of vehicles that the AP can serve, without violating system requirements.

In Experiment II, a case study is presented as a proof-of-concept. The experiment is an arbitrarily deployed model of a street, spanning ten intersections with a number of vehicles of varying speeds moving along the street. The main purpose of this experiment is to demonstrate that within a generic example, with miscellaneous violations occurring, all vehicles, including those traveling at high speeds, will achieve two-way communication.

The experiment consisted of 66 vehicles in 6 parallel paths (3 paths heading East and 3 West) with each path having 11 vehicles arranged serially. Each group of 6 vehicles was initially placed at one of the ten Access Points. Each path is assigned a different speed (40 Km/h, 60 Km/h, 80 Km/h). The six parallel paths pass beside ten APs that are equally dispersed and then return to the origin using the same paths. The layout of the experiment can be seen in **Figure 4**.

Figures 5-7 show the communication undergone by vehicles moving at different speeds in Experiment II. In **Figures 5-7**, the x-axis is simulation time in minutes and seconds. The y-axis is traffic received in bytes/sec. These

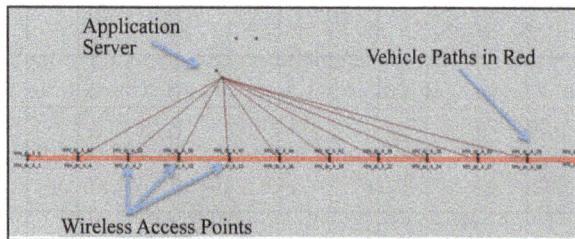

Figure 4. Experiment II layout.

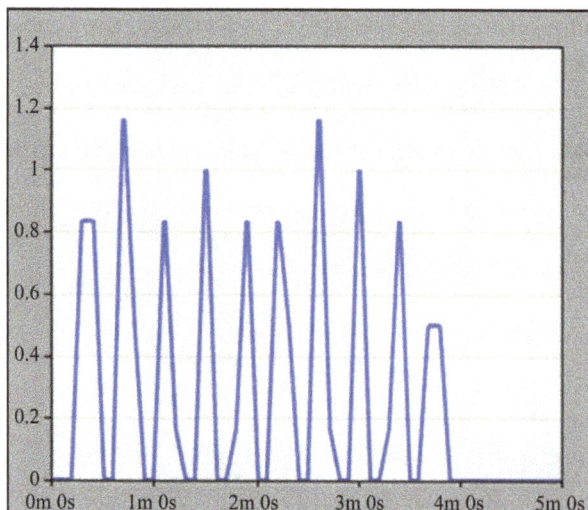

Figure 5. Traffic sent by one of the vehicles (80 Km/h).

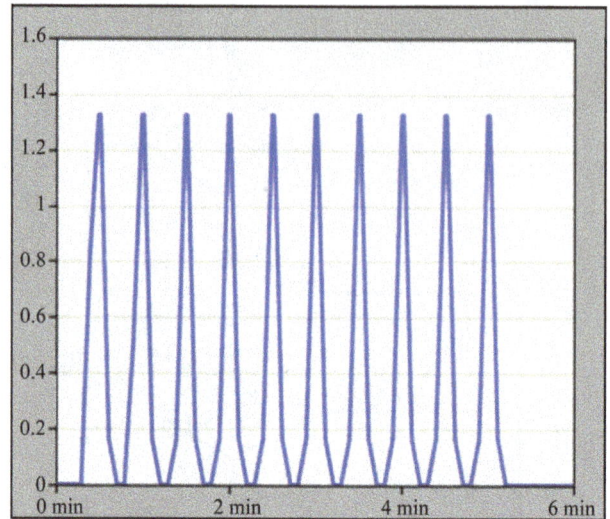

Figure 6. Traffic sent by one of the vehicles (60 Km/h).

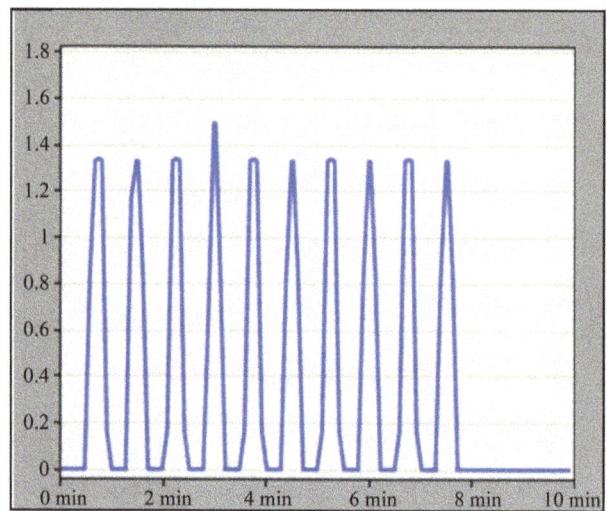

Figure 7. Traffic sent by one of the vehicles (40 Km/h).

graphs indicate that even at the unlikely (and illegal) speed of 80 Km/h in the proposed environment, the vehicle is still able to communicate with each zone AP.

6. Conclusions and Future Work

There are several existing solutions that address different types of traffic violations. These systems incorporate different technologies to tackle violations such as speeding, theft, no-entry and other issues. None of the existing systems contain a comprehensive solution that provides lower costs and reduced technical segmentation.

In this paper, a Wi-Fi-based traffic enforcement system that provides comprehensive enforcement is proposed, addressing speeding, theft, no-entry, tolling and congestion issues within the same system. The system simulations, tested on OPNET Network Modeler, show that all violations are reported in a timely fashion and a single access point (located at intersections) can support up to 62 vehicles in a worst-case scenario. The effects of interference on the system are currently being investigated.

REFERENCES

[1] Y. A. Kathawala and B. Tueck, "The Use of RFID for Traffic Management," *International Journal of Technology, Policy and Management*, Vol. 8, No. 2, 2008.

[2] Road Flow, "Fixed/Static Traffic Enforcement System," 2011. http://www.roadflow.co.uk/Traffic_Enforcement_Solutions/ROADflow_Fixed.aspx

[3] TIPMRA, "How Speed Trap Radar Works," 2011. http://tipmra.com/new_tipmra/how_Speed_Trap_Radar_works.htm

[4] E. Sifuentes, "Wireless Magnetic Sensor Node for Vehicle Detection with Optical Wake-Up," *IEEE Sensors Journal*, Vol. 11, No. 8, 2011, pp. 1669-1676.

[5] MARSH Products Inc, "The Basics of Loop Vehicle Detection," 2011. http://www.marshproducts.com/pdf/Inductive%20Loop%20Write%20up.pdf

[6] Y. Byon, A. Shalaby and B. Abdulhai, "Travel Time Collection and Traffic Monitoring via GPS Technologies," *Proceedings of the IEEE Intelligent Transportation Systems Conference ITSC*, Toronto, September 2006.

[7] J. C. Herrera, D. B. Work, R. Herring, X. Ban and A. M. Bayen, "Evaluation of Traffic Data Obtained via GPS-Enabled Mobile Phones: The Mobile Century Field Experiment," *Journal of Transportation Research Part C: Emerging Technologies*, Vol. 18, No. 4, 2010.

[8] S. Bandyopadhyay, "Traffic Congestion Management Using RFID & Wireless Technologies," Indian Institute of Management, Calcutta, 2011. http://www.intranse.in/its1/sites/default/files/7-Congestion%20Management-Somprakash.pdf

[9] IEEE 802.3 Std.

[10] IEEE 802.11 Std.

[11] Official Site for OPNET Network Modeler. http://www.opnet.com

[12] Minnesota Department of Transportation. http://www.chico.ca.us/building_development_services/traffic/speed_limits.asp

[13] "Driving in France, French Monitoring Laws." http://driving.drive-alive.co.uk/driving-in-france.htm

[14] T. S. Rappaport, "Wireless Communications: Principles and Practice," 2nd Edition, Prentice Hall, Upper Saddle River, 2002.

[15] T. Chrysikos and S. Kotsopoulos, "Impact of Channel-Dependent Variation of Path Loss Exponent on Wireless Information-Theoretic Security," *Proceedings of the IEEE Wireless Telecommunications Symposium WTS*, Prague, April 2009.

[16] Official Site for FCC Regulations. www.fcc.gov

[17] Tropos, "Receive Sensitivity: A Practical Explanation," 2007. http://www.tropos.com/pdf/technology_briefs/tropos_techbrief_rx_sensitivity.pdf

15

Life Cycle Assessment of Creosote-Treated Wooden Railroad Crossties in the US with Comparisons to Concrete and Plastic Composite Railroad Crossties

Christopher A. Bolin[1], Stephen T. Smith[2]
[1]AquAeTer, Inc., Division of Sustainability, Centennial, USA
[2]AquAeTer, Inc., Division of Sustainability, Helena, USA

ABSTRACT

Creosote-treated wooden railroad crossties have been used for more than a century to support steel rails and to transfer load from the rails to the underlying ballast while keeping the rails at the correct gauge. As transportation engineers look for improved service life and environmental performance in railway systems, alternatives to the creosote-treated wooden crosstie are being considered. This paper compares the cradle-to-grave environmental life cycle assessment (LCA) results of creosote-treated wooden railroad crossties with the primary alternative products: concrete and plastic composite (P/C) crossties. This LCA includes a life cycle inventory (LCI) to catalogue the input and output data from crosstie manufacture, service life, and disposition, and a life cycle impact assessment (LCIA) to evaluate greenhouse gas (GHG) emissions, fossil fuel and water use, and emissions with the potential to cause acidification, smog, ecotoxicity, and eutrophication. Comparisons of the products are made at a functional unit of 1.61 kilometers (1.0 mile) of railroad track per year. This LCA finds that the manufacture, use, and disposition of creosote-treated wooden railroad crossties offers lower fossil fuel and water use and lesser environmental impacts than competing products manufactured of concrete and P/C.

Keywords: Creosote; Environmental Impact; Railroad Crossties; Life Cycle Assessment (LCA); Concrete; Plastic Composite

1. Introduction

Railroads are a critical transportation element of the US economy, distributing large quantities of material goods and oftentimes in a more efficient manner than road-based transportation [1]. This transportation efficiency is often measured by the equipment moving goods [2-4], but to understand the burdens associated with various modes of transportation, one must consider the system as a whole, including not only the equipment moving the goods, but the surface the equipment moves upon. The structural components that make up the railway line include the rail, rail tie-plate, crossties, supportive ballast, and subgrade [5]. Railroad crossties are the base members, to which steel rails are attached to transfer load from the rails to the underlying ballast. The ties also provide the critical function of keeping the rails at the correct gauge and alignment. Wooden crossties have been

the backbone of this system for more than 150 years, a system that, in the US, has an estimated 273,700 track kilometers (170,000 miles) [6].

While non-durable wood products are susceptible to degradation when left untreated [7], wood preservative treatments can extend the useful life of a wood product by 20 to 40 times that of untreated wood [8] when used in weather-exposed or wet environments subject to microbial or insect attack. Wood preservation with coal-tar creosote became commercially viable when a patent was taken out by John Bethell in 1838 [9]. Creosote "empty cell treatment" was introduced by Rueping in 1902 and refined in 1907 to a process in which a large cylinder is filled with compressed air, creosote is pumped in while maintaining air pressure, injection occurs under pressure, preservative is pumped out, and then a vacuum is applied at the end of the process so that air contained in wood cells will expel excess preservative. Lowry

Life Cycle Assessment of Creosote-Treated Wooden Railroad Crossties in the US with Comparisons to Concrete
and Plastic Composite Railroad Crossties

149

introduced in 1906 a quick vacuum at the end of the pressure process [9]. Today, most ties are treated with creosote using the empty cell Rueping process. Coal tar creosote treated wood products have a long history of proven performance in transportation systems [10].

Consumer and regulatory agency concern about environmental impacts resulting from the manufacture, use, and disposal of infrastructure products, such as coal-tar creosote treated crossties, understandably has resulted in increased scrutiny during selection of transportation construction products. Products such as creosote-treated wooden crossties are, in some cases, being replaced with concrete and plastic composite (P/C) crossties for various reasons, but at least partially based on perception rather than scientific or quantitative consideration of these concerns.

2. Goal and Scope

The goal of this study is to provide a comprehensive; scientifically-based; and fair, accurate, and quantifiable understanding of environmental burdens associated with the manufacture, use, and disposition of creosote-treated wooden crossties using primary data collected at US treating plants and secondary data from other sources.

The scope of this study includes investigation of cradle-to-grave life cycle environmental impacts for creosote-treated wooden railroad crossties in US Class 1 railroads using life cycle assessment (LCA) methodologies. The results of the creosote-treated crosstie LCA are compared to LCA findings for alternative products: concrete and P/C crossties. LCA is the preferred method for evaluating the environmental impacts of a product from cradle to grave, and determining the environmental benefits one product might offer over its alternative [11].

The LCA methodologies used in this study are consistent with the principles and guidance provided by the International Organization for Standardization (ISO) in standards ISO 14040 [12] and 14044 [13]. The study includes the four phases of an LCA: 1) Goal and scope definition; 2) Inventory analysis; 3) Impact assessment; and 4) Interpretation. The environmental impacts of creosote-treated, concrete, and P/C railroad ties are assessed throughout their life cycles, from the extraction of the raw materials through processing, transport, primary service life, reuse, and recycling or disposal of the product.

Crosstie alternatives are produced by many different manufacturers using differing materials and manufacturing processes. Therefore, a "typical product" has been estimated for both concrete and P/C crossties. The concrete and P/C typical products have approximately the same dimensions as, and generally are used as direct alternatives to, creosote-treated railroad ties. However, concrete ties have a different spacing requirement and

cannot be interspersed with other types of ties. The LCAs for concrete and P/C ties do not include independent manufacturing inventory data (primary data). Consequently, a general comparison of LCIA impact indicators is done to understand how the creosote-treated crosstie and alternative product life cycles compare. Additional alternative product data collection and analysis are needed to fully detail the comparability of specific alternative products.

3. Life Cycle Inventory Analysis

The Railway Tie Association [14] estimates that North American railroads purchased 20,394,000 new wood ties in 2007. The creosote-treating industry reports that approximately 314 million liters (82.9 million gallons) or 345 million kilograms (760 million pounds) of creosote were used in the US in 2007 to treat 2.86 million cubic meters (101 million cubic feet) of wood, of which approximately 71% was produced for railroad applications, most of which was for creosote-treated crossties [15].

Primary data and information for the life cycle inventory (LCI) are obtained from US treaters of wooden railroad ties using creosote preservative. Secondary data are obtained from the scientific literature and from the US Life Cycle Inventory Database maintained by the National Renewable Energy Laboratory (NREL). LCI inputs and outputs for the creosote-treated wood tie are quantified per 28.3 cubic meters (1000 cubic feet (Mcf)). The cubic foot (cf) unit is a standard unit of measure for the US tie industry and is equivalent to 0.028 cubic meters (m^3). Inventory data are converted to a functional unit of per 1.61 kilometers (1.0 mile) of Class 1 railroad per year of use, allowing assessment of the impacts of tie spacing and service life. The cradle-to-grave life cycle stages considered in this LCI are illustrated in **Figure 1**.

This life cycle assessment allocates manufacturing inputs on both volumetric and mass basis and outputs on a mass basis. In most cases life cycle process modules were downloaded from NREL. The NREL modules include allocations needed to determine applicable inputs and outputs associated with material acquisition and manufacturing processes. At disposition, some of the product leaves the system as thermal energy and is allocated as a credit to the use of fossil fuel.

3.1. Creosote-Treated Railroad Tie Inventory

This study builds on existing research for forest resources and adds the treating, service use, and disposition stages of creosote-treated wood railroad ties. Previous studies, such as research conducted by the Consortium for Research on Renewable Industrial Materials (CORRIM), have investigated the environmental impacts of wood products. CORRIM's efforts build on a report is-

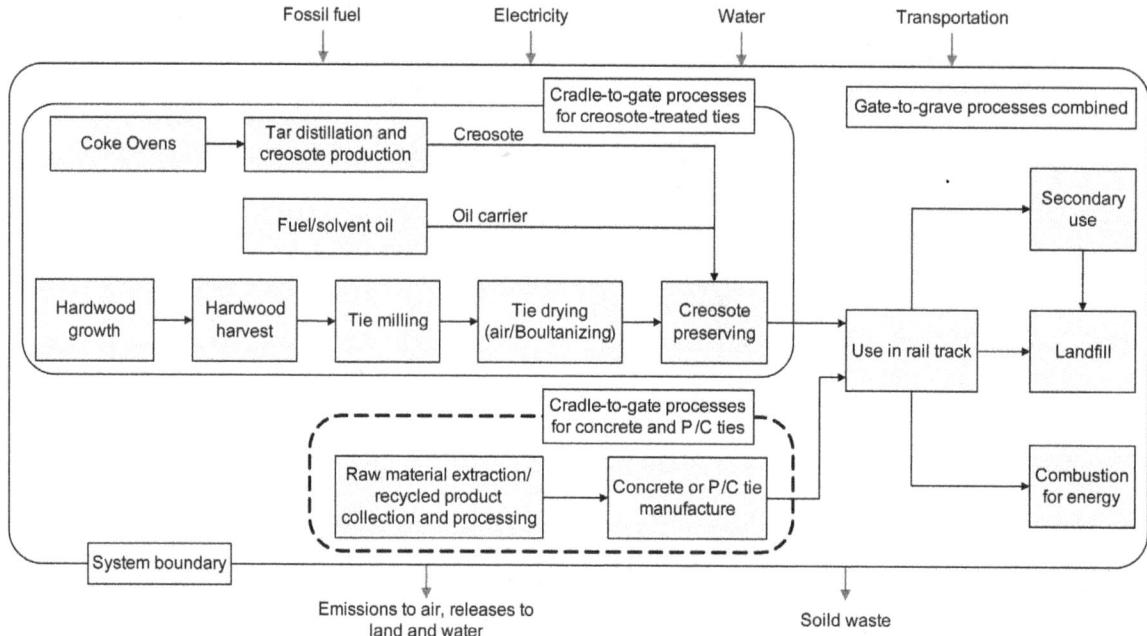

Figure 1. Life cycle stages of railroad ties.

sued under the auspices of the National Academy of Science regarding the energy consumption of renewable materials during production processes [16]. CORRIM's recent efforts [17-19], have focused on an expanded list of environmental aspects necessary to bring wood products to market.

The main source of forest products LCI data used in this study is Oneil et al. [19]. Data include forestry practices applicable to hardwood products from the Northeast and North/Central U.S from the forest (cradle) to the mill (gate). Hardwood trees naturally regenerate and fertilizer usually is not applied, thus, the environmentally relevant inputs are limited to the fuel required to cut, trim, load, and transport logs to mills. Bergman and Bowe [20] completed a gate-to-gate LCI of hardwood lumber milling process inputs and outputs that is adapted in the inventory of this LCA to represent hardwood railroad tie production. Inputs and outputs include electricity and fuel requirements, transportation, water use, and particulate emissions. The data from Oneil et al. and Bergman and Bowe are allocated by volume for a "typical" tie measuring18 cm (7-in) high by 23 cm (9-in) wide by 2.6 m (8.5-ft) long.

Twenty-two (22) creosote treating plants in the U.S. provided primary data responses to a questionnaire covering operations in 2007. The total volume of creosote-treated ties reported in the surveys is approximately 2.0 million cubic meters (71,000 Mcf) of product, including approximately 1.7 million cubic meters (60,000 Mcf) of hardwood crossties. Vlosky [15] estimates US industry total creosote railroad tie treatment in 2007 at approximately 2.0 million cubic meters (71,000 Mcf). Gauntt

[14] estimates that 20,394,000 new wood ties were purchased by North American railroads in 2007, or between 1.9 to 2.1 million cubic meters (67,000 to 75,000 Mcf) at an average volume of 0.093 to 0.11 cubic meters (3.3 to 3.7 cubic feet) per tie. Both estimates support the representation that all, or nearly all, creosote treatment in the US provided input to the primary data used in this study.

The LCI for creosote production considers both coke oven and tar distillation processes. Creosote preservative is produced to meet AWPA standards P1/P13 [21], P2 [22], or P3 [23]. Standard P2 creosote generally is used for crossties. A weighted average of creosote types from the survey data is used as the reference preservative for this LCA. The treaters surveyed as part of this LCA report a weighted average of creosote preservative use of 94% creosote (both P1/P13 and P2) and 6% petroleum oil.

AWPA [24]specifies creosote retention of 112 kg/m^3 (7.0 pounds per cubic foot (pcf)) or refusal for oak and hickory crossties. Retention is based on gauge measurement, meaning that retention is the total weight of creosote injected divided by the total volume of wood treated. The average creosote use rate, as reported in surveys, is approximately 88 kg/m^3 (5.5 pcf). The difference is consistent with the AWPA specifications because a large percentage of the total volume of wooden ties accepts less than specified amounts of preservative ("refusal"). Therefore, survey data are used in this study.

Outputs in the form of solid waste, waste water discharges, chemical releases from process equipment and stored product are primary data. Releases of creosote to

Life Cycle Assessment of Creosote-Treated Wooden Railroad Crossties in the US with Comparisons to Concrete and Plastic Composite Railroad Crossties

151

air are reported under the Toxic Release Inventory (TRI) reporting program and include releases from the process equipment, such as tank vents and treating cylinders. Evaporative losses from the finished ties are estimated at 0.12 kg/m^3 (7.5 pounds/Mcf) for the first 120 days following treatment [25]. Treatment process releases of creosote, used in the LCI, are summarized in **Table 1**.

Creosote-treated railroad ties are installed at 49.5 cm (19.5 inch) spacing, center-to-center, or at a frequency of 3249 ties per 1.61 km (1.0 mile). Service life is a function of quality and species of wood, quality and type of treatment, laying condition, use intensity, and environmental factors. Based on studies of US railways by Zarembski [26] and contact with industry sources, a 35-year average service life is assumed in this LCI for creosote-treated railroad ties, an estimate greater than the 15-year estimate in Japan [27], the 20 to 30-year estimate in Australia [28], and the 24 to 30-year estimate in Switzerland [29]. Extended service life by dual treatment with borate and creosote also is addressed through sensitivity analysis. Maintenance applications of preservative to an installed tie, such as ones containing borate, are considered rare and are not included in this LCA. The amount of steel, including tie plates, spikes, and rail anchors, is calculated and inventoried in the LCA. The system boundary does not include supportive ballast except for concrete tie products that require additional ballast material for stability. Only the ballast that is required in addition to that normally used for wood and P/C ties is considered.

Studies done by Becker *et al.* [30], Brooks [31], Burkhardt *et al.* [32], Chakraborty [33], Gallego *et al.* [34], Geimer [35], Gevao and Jones [36], and Kohler and Kunniger [37] have investigated the releases of "creosote" over time and the release mechanism (*i.e.*, releases by volatilization or leaching).The term "creosote" describes the liquid used to treat wood ties, but is imprecise when applied to environmental releases to air, soil, or water. None of the creosote release studies provide individual chemical constituent information necessary as inputs in this LCI for determining impact indicators; thus,

release estimates were developed for this study. The molecular weights and mass fractions of the numerous chemical components of AWPA Standard P2 creosote are provided by Sparacino [38] and are used to estimate fractional amounts of chemical components released from ties at treatment and during time in service.

Creosote constituents are released in proportion to their pure vapor pressures (VPs) and initial concentrations. Constituent VPs range from approximately 4 kilopascal (KPa) to approximately 2×10^{-6} KPa (6×10^{-1} pounds-force per square inch absolute (psia) to 3×10^{-7} psia). Creosote constituents are sorted into four groups by VP (high, medium-high, medium, and low). For each group, assumptions are made regarding the amount of each constituent released and the fraction of the release emitted into the air, as shown in **Table 2**. These creosote loss factors are multiplied times the constituent mass in creosote and calculated as the amount released. The release times the air fraction value is the amount released to the air. Total releases of creosote are estimated to average approximately 1% per year with releases to air at approximately 0.1% per year of initial treatment mass.

According to the Railway Tie Association, approximately 17.1 million wood ties are removed from active and inactive track in the US per year. Following removal, the ties 1) are recycled to other treated wood uses, such as landscape materials (39%); 2) are beneficially used for energy recovery (56%); or 3) are disposed as waste in landfills (5%).

Table 1. Treating process outputs from creosote-treater surveys.

Source	Amount (kg/m^3)
Creosote contained in storm water runoff	0.00019
Creosote discharged to treatment works	0.00038
Creosote releases (drips) to ground	0.000045
Creosote component emissions to air	0.038
Hazardous waste disposed	0.64
Other waste disposed	3.7

Table 2. Release of creosote constituents by vapor pressure.

Assumption for constituents	Assumed loss/yr	Fraction to air	Fraction to ground	Mass loss fraction/yr	Mass loss to air fraction/yr
For each constituent with a VP of Xe-1	2.0%	30%	70%	0.0019%	0.00057%
For each constituent with a VP of Xe-2	1.5%	15%	85%	0.0037%	0.00056%
For each constituent with a VP of Xe-3	1.0%	5.0%	95%	0.0024%	0.00013%
For each constituent with a VP of Xe-4+	0.50%	0%	100%	0.0014%	0%
Total measured mass fraction	0.86		Sum:	0.0094%	0.0013%
% of total measured:				1.1%	0.15%
Projected release for 35 years				33%	4.4%
Release as fraction of initial treatment				38%	5.1%

Removed ties beneficially used as a fuel are modeled as fuel in a steam-electric power plant and the energy value is calculated assuming 20% moisture and considering the carbon content of the remaining wood, creosote, and carrier petroleum oil. Electricity production is based on 50% thermal efficiency. The amount of electricity produced from the tie fuel is entered as an electricity credit. All wood carbon emitted is inventoried as biogenic carbon dioxide. All creosote and petroleum oil preservative carbon emitted is inventoried as fossil carbon dioxide. Emissions from energy recovery are inventoried and assumed to occur with the use of advanced particulate controls. Credits, from recycling wood ties to energy following use, result in some LCI inputs being less than zero, and thus are environmentally beneficial when summed for the whole product life cycle, as shown in **Table 3**.

In contrast to an LCA done to evaluate GHG emissions from concrete and treated wood sleepers (crossties) by Crawford [28], this LCA accounts for anthropogenic GHG and biogenic GHG as neutral related to global warming. Crawford assumed all wood mass from the forestry product, not used as ties, was burned as waste and that at the end of service life, wood ties fully decayed. The carbon dioxide released from forestry product and ties was counted as a GHG, the same as fossil carbon dioxide, with no accounting for carbon uptake by tree growth and the assumption that no beneficial energy was produced either from forestry biomass or from used ties. This LCA better reflects actual North American practice and develops GHG conclusions that contradict those by Crawford.

Steel tie plates and other parts installed with ties are inventoried in the use stage by mass, assuming production in a blast furnace. Recycled steel is inventoried in the final fate stage both as a negative use offsetting the initial use and as the amount of electricity typically used in an electric arc "mini-mill" to melt and reform steel shapes. In this manner, as recycled steel approaches 100%, the minimum inputs required for steel are those to melt and shape steel in each use cycle. Primary steel manufacture, in a blast furnace, is based on inventory data from NREL. NREL database information assumes 85% of steel is recycled. New steel yield from recycled steel is 95%. Energy input to mini-mills, processing recycled steel, is assumed to be of 0.011 terajoule (TJ) per metric ton (1.33 kiloWatt hours per pound (kWh/lb) of steel) of grid electricity [39].

Landfill-disposed crossties are modeled as if decayed to a point where the primary phase of anaerobic degradation has occurred and 17% of the product's carbon is released as carbon dioxide, 6% is released as methane, and 77% [40] remains in long-term storage in the landfill. Inputs and outputs related to landfill construction and closure are apportioned on a mass disposed basis [41].

Transportation-related inputs and outputs are quanti-

fied for each life cycle process. Distances and transport modes for preservative supply to treaters, inbound untreated ties, and outbound treated ties are based on treater survey weighted averages.

3.2. Concrete Railroad Ties Inventory

The "representative" concrete tie has a weight of 318 kg (700 pounds), and includes eight strands of 9.5 mm (3/8-in) pre-stressed steel cable. Concrete tie placement is assumed to be at 61 cm (24-in), on center. A survey of concrete tie manufacturers was not done for production inputs and outputs; therefore, some inputs and outputs may not be fully identified or quantified. Elastic fasteners and clips, constructed of steel, are included in the inventory. The maintenance frequency of concrete crossties includes clip replacement only once during the crosstie life. No carbonatization of concrete is accounted for in the inventory. This LCI does not account for polymer tie pads, pad replacements, or repairs to concrete tie seat areas, items that might add to indicator impacts.

The Railway Tie Association commissioned a study of concrete tie service life specifically for use in this LCA project [42]. The study concluded: "*It appears that a reasonable estimate for concrete tie service life under North American railroad operating conditions is between 40 and 45 years.*" However, the study noted that while concrete ties were installed by one railroad as early as the 1970s, current concrete ties are a relatively new product within the modern North American railroad system with the average age of in-service ties being approximately 13 years. Life variability is high with projected life from approximately 20 years (using Norfolk Southern data) to 41 years (Canadian National data). Premature concrete tie failures have been documented [43], further supporting a conservative service life estimate. Given the high variability and still unknown long-term performance, an assumption of 40 years is used in this LCI for average concrete tie life.

Concrete ties require additional ballast compared to wood or P/C tie systems. Only the additional ballast required for concrete ties is considered in this LCA. 23 cm (9 in) of additional ballast is assumed for the concrete tie model.

Concrete tie rail systems offer advantages to railroads in select situations. In particular, some, but not all, railroads use concrete ties for heavy-haul, higher-curvature track locations. The greater weight of the concrete ties is thought to reduce rail movement in comparison to the lighter wooden ties. Such special situations are outside the scope of this LCA.

When concrete ties are removed from service, it is assumed that a small fraction (5%) will be reused by railroads while most either will be crushed and reused as

Life Cycle Assessment of Creosote-Treated Wooden Railroad Crossties in the US with Comparisons to Concrete
and Plastic Composite Railroad Crossties

153

Table 3. Creosote-treated, concrete, and P/C railroad tie cradle-to grave life cycle inventory summary (per tie).

Infrastructure process	Units	Creosote-treated (/tie) Service life = 35 yrs	Concrete (/tie) Service life = 40 yrs	P/C(/tie) Service life = 40 yrs
Inputs from technosphere				
Electricity, at grid, US	kWh	−54	128	123
Natural gas, processed, at plant (feedstock)	m^3	−2.8	7.4	18
Natural gas, combusted in industrial boiler	m^3	2.9	0.65	7.6
Diesel fuel, at plant (feedstock)	L	0	0	0
Diesel fuel, combusted in industrial boiler	L	−0.11	0.83	0.31
LPG, combusted in equipment	L	0.0035	0	0.00034
Residual oil, processed (feedstock)	L	0	0	0
Residual fuel oil, combusted in industrial boiler	L	0.71	0.052	0.097
Diesel fuel, combusted in industrial equipment	L	3.1	0.86	0.19
Gasoline, combusted in industrial equipment	L	0.11	0.085	0.041
Hogfuel/biomass (50%MC)	kg	3.2	1.6	1.5
Coal-bituminous & sub. combusted in boiler	kg	12	0.055	0.016
Coal (feedstock)	kg	0	7.8	0
Energy (Unspecified)	MJ	0	21	0
Truck transport, diesel powered	ton-km	65	110	100
Rail transport, diesel powered	ton-km	122	569	131
Barge transport, res. oil powered	ton-km	−1.7	4.2	4.5
Ship transport, res. oil powered	ton-km	20	10	5.5
Diesel use for transportation	L	2.6	6.6	3.6
Residual oil use for transportation	L	0.086	0.088	0.027
Harvested logs	m^3	0.11	0	0
Untreated green ties	m^3	0.11	0	0
Coal tar by-products	kg	10	0	0
Creosote	kg	9.2	0	0
Used treated ties	m^3	0	0	0
Landfill capacity	ton	0.011	0.23	0.090
Inputs from nature				
Water	L	26	320	315
Unprocessed coal	kg	4.5	44	37
Unprocessed U_3O_8	kg	−0.000033	0.000091	0.000083
Unprocessed crude oil	L	7.6	11	5.35
Unprocessed natural gas	m^3	3.0	2.0	19
Biomass/wood energy	MJ	0.0000016	0.0000051	0.0000024
Hydropower	MJ	−14	38	35
Other renewable energy	MJ	−1.1	2.6	2.5
Biogenic carbon (from air)	kg	4.9	0	−1.3
Other mined mineral resources	kg	7.3	717	11
Outputs to nature (to air unless otherwise stated)				
CO_2-Fossil	kg	20	207	133
CO_2-Non-fossil	kg	−35	1.7	1.6
Carbon monoxide	kg	0.37	0.55	0.25
Ammonia	kg	0.00014	0.00044	0.00012
Hydrochloric acid	kg	−0.0026	0.024	0.019
Hydrofluoric acid	kg	−0.0010	0.0025	0.0024
Nitrogen oxides (NOx)	kg	0.20	0.57	0.17
Nitrous oxide (N_2O)	kg	0.00019	0.00049	0.00027
Nitric oxide (NO)	kg	0.0037	0	0
Sulfur dioxide	kg	−0.19	0.76	0.93
Sulfur oxides	kg	0.043	0.09	0.044
Particulates (PM10)	kg	0.10	0.077	0.0093
VOC	kg	0.0074	0.026	0.027
Methane	kg	0.29	0.27	0.83
Creosote	kg	0.42	0	0
Creosote to soil	kg	2.6	0	0
Solid wastes to landfill	kg	11	6.7	4.1
Solid wastes to recycle	kg	1.6	5.5	2.3
Process solid & hazardous waste to landfill	kg	0.068	0	0

aggregate (25%) or disposed in landfills (70%). The low fraction of recycled to aggregate reflects the difficulty and expense of grinding high-strength reinforced concrete. Steel from embedded fasteners and reinforcement from recycled ties is assumed to be recycled and inventoried in the same way as with creosote-treated ties.

Concrete railroad ties disposed in landfills have inputs and outputs associated with landfill construction and closure proportional to the mass of disposed ties. No releases or emissions are modeled from concrete ties once disposed in a landfill.

3.3. P/C Railroad Tie Inventory

P/C ties can be made of recycled plastics, generally polyethylene, but often include other materials such as steel fiber, steel reinforcing bar, shredded used tires, mineral filler, virgin plastic, or concrete. The modeled representative P/C tie is assumed to be 8% virgin HDPE plastic, 7% talc (mineral filler), and the balance a mixture of post-consumer recycled milk bottles, grocery bags, and tires [44]. Electric energy is required to process the mixture and extrude the P/C product. The spacing of the P/C product and the required steel used to fasten the P/C tie to the track are assumed the same as creosote-treated crossties.

While recycled plastic does not carry the inputs and outputs of virgin material, post-consumer plastic use requires collection and processing inputs [45]. Inputs and outputs included in the inventory are similar to those for thermoplastics recycling by Garrain *et al.* [46].

P/C ties have not yet developed enough history to accurately predict service life. This LCA assumes that P/C ties will provide an average service life of 40 years, similar to concrete ties. Also, the P/C tie market is not yet sufficiently mature to know how ties will be handled when removed from service. For this LCA it is assumed that following removal from railroad use, 5% of P/C ties will be reused by a railroad for another purpose, 20% will be recycled to the plastic reuse market, and 75% will be disposed in landfills. Steel, attached to P/C ties, is assumed to be recycled (75%) and is inventoried in the same way as for creosote-treated ties.

A summary of selected inventory inputs and outputs for creosote-treated, concrete, and P/C ties is provided in **Table 3**.

4. Life Cycle Impact Assessment

4.1. Selection of the Impact Indicators

The impact assessment phase of the LCA uses the LCI results to calculate impact indicators. The environmental impact indicators are considered at "mid-point" rather than at "end-point" in that, for example, the amount of greenhouse gas (GHG) emission in mass of carbon diox-

ide equivalent (CO_2-eq) at mid-point is provided rather than estimating end-points of global temperature or sea level increases. The life cycle impact assessment is performed using USEPA's Tool for the Reduction and Assessment of Chemical and Other Environmental Impacts, Version 2002 ((TRACI [47] and [48]) to assess GHG, acidification, ecotoxicity, eutrophication, and smog impacts potentially resulting from life cycle air emissions. Other indicators of interest also are tracked, such as biogenic and anthropogenic contributions to net GHG emissions, fossil fuel use and water use.

4.2. Impact Indicators Considered but Not Presented

The TRACI model, a product of USEPA, and the USEtox model [49] a product of the Life Cycle Initiative (a joint program of the United Nations Environmental Program (UNEP) and the Society for Environmental Toxicology and Chemistry (SETAC)), offer several additional impact indicators that were considered during the development of the LCA, such as human health impacts and impacts to various impact indicators from releases to soil and water. The decision was made not to include these impact indicators because of limited and/or insufficient-data and concerns regarding misinterpretation. Thus, the life cycle inventory includes releases of chemicals associated with impacts (such as human health and land and water ecological impacts), but impact indicators for these categories are not calculated. Land use impacts are beyond the scope of this LCA.

5. Life Cycle Interpretation

5.1. Findings

Impact indicator values are totaled at two stages for creosote-treated, concrete, and P/C crosstie products: 1) the new tie at the manufacturing facility after production, and 2) after service and final disposition. A summary of impact indicator values for all three crosstie products is provided in **Table 4**. Negative-value impacts are recognized as credits or beneficial to environmental conditions. Comparisons are made per year and per 1.61 km (1.0 mi) of railroad track to account for differences in service life expectancy and spacing.

Impact indicator values are normalized to the product (creosote-treated tie, concrete tie, or P/C tie) having the highest cradle-to-grave value, allowing relative comparison of indicators between products on **Figure 2**. The product with the highest value at final disposition receives a value of one, and the other products are fractions of one.

According to the Research and Innovative Technology Administration (RITA [50]), the Class 1 railroad total freight volume was approximately 2.58 trillion ton-

Table 4. Summary of impact indicator totals at life cycle stages for creosote, concrete, and P/C ties (per year of use and per 1.6 km (1mile) of railroad track).

Impact indicator	Units	Creosote-treated ties Service life = 35 yrs Spacing = 49.5 cm (19.5 in)		Concrete ties Service life = 40 yrs Spacing = 61 cm (24 in)		P/C ties Service life = 40 yrs Spacing = 49.5 cm (19.5 in)	
		Cradle-to-gate	Cradle-to grave	Cradle-to-gate	Cradle-to grave	Cradle-to-gate	Cradle-to grave
Greenhouse gas	kg-CO_2-eq	2700	2400	8400	14,000	9200	12,000
Net GHG	kg-CO_2-eq	−7500	−800	8400	14,000	9200	12,000
Fossil fuel use	TJ	0.12	0.093	0.11	0.16	0.23	0.23
Acidification	kg-mole H+	830	65	1800	4400	3200	4,700
Water use	L	1600	2400	22,000	21,000	28,000	26,000
Smog	g NOx-eq/m	12	25	24	58	22	29
Eutrophication	kg-N-eq	0.36	0.84	0.74	1.7	0.54	0.62
Ecotoxicity	kg-2,4-D-eq	6.4	−3.1	65	85	13	30

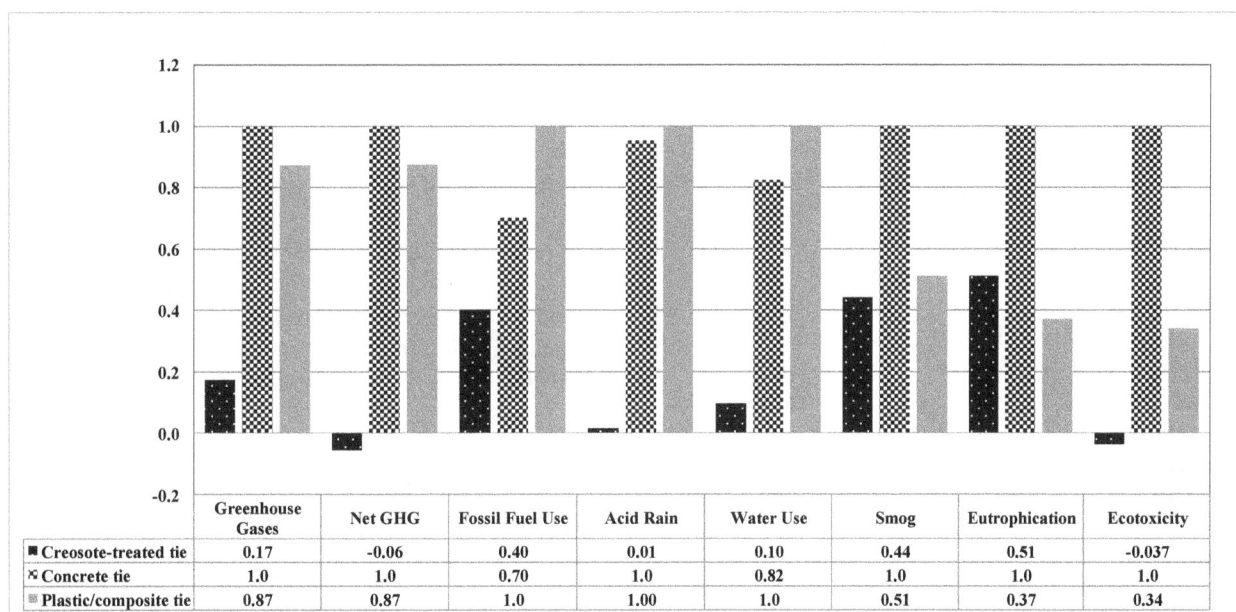

	Greenhouse Gases	Net GHG	Fossil Fuel Use	Acid Rain	Water Use	Smog	Eutrophication	Ecotoxicity
■ Creosote-treated tie	0.17	-0.06	0.40	0.01	0.10	0.44	0.51	-0.037
✕ Concrete tie	1.0	1.0	0.70	1.0	0.82	1.0	1.0	1.0
▦ Plastic/composite tie	0.87	0.87	1.0	1.00	1.0	0.51	0.37	0.34

Figure 2. Creosote-treated wood, concrete, and P/C ties normalized impact comparisons (normalized to maximum impact = 1).

kilometers (1.77 trillion ton-miles) in 2008. The annual impacts attributable to all ties in the US, if made of the same material, are compared as a percentage of the annual Class 1 freight related impacts in **Table 5** (*i.e.*, the impacts calculated for tie manufacture, use, and disposition of each tie material are compared to railroad freight impacts).

5.2. Data Quality Analyses

Data quality analyses per ISO 14044 [13] includes a gravity analysis, uncertainty analysis, and sensitivity analysis.

5.2.1. Gravity Analysis

A gravity analysis is done to identify the creosote-treated

crosstie manufacture, use, and disposition processes most significant to the impact indicator values. This gravity analysis only considers creosote treated ties. Significant contributing processes to the gravity of each impact indicator are described below.

- Anthropogenic GHG emissions are most notably impacted by steel plates and spikes, but also by green tie production and tie treatment. GHG emissions are reduced or offset from steel recycling and for fossil energy offset by producing electricity from recycled used ties.
- Net GHG emissions demonstrate the environmental benefit of wood products that first remove carbon dioxide as the forestry product grows. The net result is

Table 5. National normalized cradle-to-grave impact per year for total miles of track ties in the US as a fraction of total class 1 RR freight transport impacts per RITA [50].

Impact Indicator	Units	Creosote wood ties	Concrete ties	P/C ties
GHG	lb-CO_2-eq	2.4%	6.6%	5.4%
Fossil fuel use	MMBTU	3.1%	6.0%	7.8%
Acidification	lb-mole H+	0.025%	2.0%	2.0%
Smog	g NOx-eq/m	0.33%	0.80%	0.39%
Eutrophication	lb-N-eq	0.32%	0.66%	0.24%

an overall reduction in GHG when using creosote-treated wood ties, whereas ties made of extracted materials only emit GHG in all stages. This difference between wood and products of other materials is clear in **Figure 3**.

- Fossil fuel use is most notably impacted by preservative manufacture and use, the manufacture of steel plates and spikes, and fossil fuel offsets from the energy recovery of used creosote-treated ties.
- Acidification is most notably impacted by the manufacture of steel plates and spikes, and credits from the offset of electricity from energy recovery. The credits received from the beneficial use of ties for energy recovery are large enough to result in an overall cradle-to-grave impact of acidification near zero.
- Water use is most notably impacted by use at the treating facility and in the manufacture of steel plates and spikes.
- The potential to impact smog is most notably impacted by the manufacture of steel plates and spikes, creosote releases from ties in service, and transport of the ties throughout the life cycle; however, a credit also is recognized from steel recycling and the offset of electricity from energy recovery.
- Eutrophication is most notably impacted by the manufacture of steel plates and spikes and the combustion of used ties in the energy recovery stage. Emissions related to transportation also are significant to eutrophication. Eutrophication is reduced by steel recycling.
- Ecotoxicity impact is largely a result of the manufacture of steel plates, creosote releases from ties in-service, and combustion of used ties for energy recovery. The ecotoxicity impact indicator is reduced by steel recycling and the offset from the combustion of used ties for energy recovery, resulting in an overall credit for ecotoxicity.
- As more steel is recycled, lower fossil fuel use, water use, and eutrophication result, but with increases in acidification, and ecotoxicity. These changes occur as less energy is derived from primary sources, such as coal to fuel primary steel production, and more energy is derived from the electric grid for electric arc mills

converting recycled steel.

5.2.2. Uncertainty Analysis
Areas of uncertainty identified in this LCA include:

- The creosote preservative producers did not provide detailed LCI input and output data for creosote production; therefore, industry experts provided estimates for the creosote manufacture model.
- Creosote release estimates, during treatment, storage, use, and disposal in landfills, are guided by research and assumptions. Creosote constituent releases are a function of site- and product-specific factors resulting in uncertainty.
- End-of-life disposition methods employed by railroads vary by operator, based on corporate policies, geographic locations, and economics.
- Landfill fate and release models are based on USEPA GHG emission inventory data [40] and modeled assumptions result in variability of impact indicator values, especially GHG. In this LCA, creosote-treated crossties are conservatively assumed to degrade to the same degree and at the same rate as round wood limbs disposed in a landfill.
- The comparative analysis phase of this LCA includes the assembly of LCIs for concrete and P/C railroad ties. The cradle-to-grave LCIs of concrete and P/C ties include data inputs that involve professional judgment. No survey of manufacturers of the concrete or P/C products was done.

5.2.3. Sensitivity Analysis
Sensitivity analysis was completed to determine the magnitude of changes to impact indicators resulting from assumptions and uncertainties differing from those identified in the LCI and the impact on LCA conclusions. The sensitivity of the creosote-treated railroad tie model was analyzed after variations in: preservative use (retention of preservative in the treated product), releases to environment, service life, post-use disposition, landfill decay percentages, and the addition of borate to creosote treatment as a dual-treatment application. The concrete tie model was investigated using sensitivity analysis and

Life Cycle Assessment of Creosote-Treated Wooden Railroad Crossties in the US with Comparisons to Concrete
and Plastic Composite Railroad Crossties

157

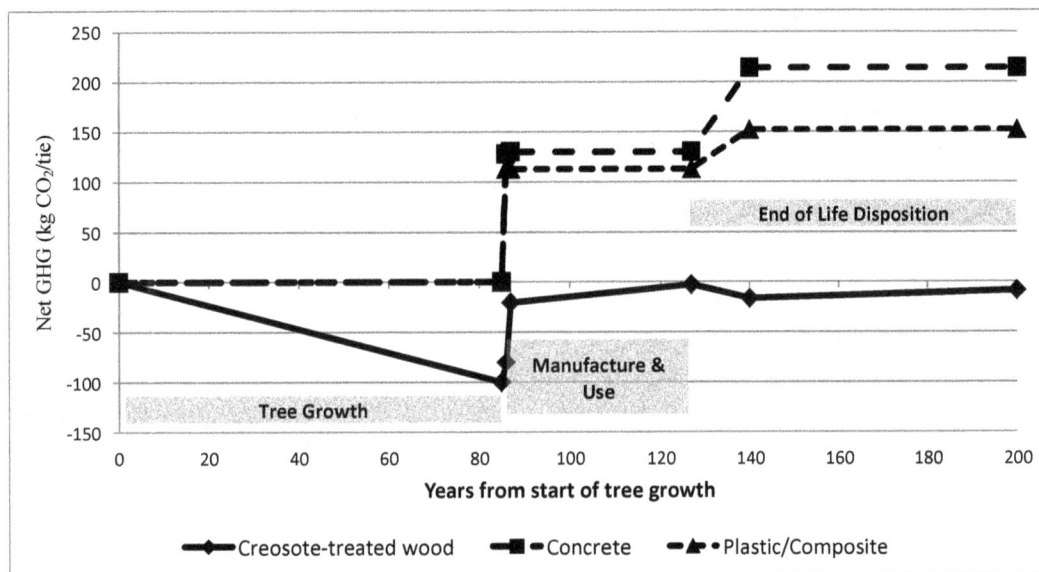

Figure 3. GHG uptake and release by manufacture, use, and disposition of ties over time.

included variations in: post-use fate, service life, impact of rolling resistance, and ballast requirements. The P/C tie model was investigated using sensitivity analysis and included variations in: service life and post-use fate. Items deemed most notable from the sensitivity analysis are further discussed below.

Altering the estimated average service life (35 years) of creosote-treated railroad ties to 20 years results in notable increases to the impact indicators GHG, fossil fuel use, water use, smog, and eutrophication. In this scenario, more ties are recycled for energy recovery and indicators improve (net GHG, acidification, and ecotoxicity) because of additionalfossil fuel offsets. However, even with a shortened service life, many of the impact indicators, including GHG, net GHG, acidification, water use, and ecotoxicity, compare favorably to both alternatives.

Dual treatment with borate followed by creosote has been shown to increase the service life of ties [51], but at the cost of increasing the inputs and outputs of treatment. Addition of treatment with borate, before consideration of extended service life, has minimal impact on the indicators. Assuming a 10-year (30%) increase in service life because of dual-treatment, impact indicators decrease between 10% and 25% for GHG, fossil fuel use, smog, and eutrophication while increasing for other indicators. Changes result from fewer wood ties being used and recycled to energy. Comparisons with alternative products do not change. The use of borates in dual-treatment systems has given rise to railroads experimenting with reducing the amount of creosote used in dual-treated ties by as much as 30% to 40%. If experience proves that these reductions can be maintained without negatively impacting service life, improved treated wood tie performance, in most impact indicator categories, is expected.

Theoretical evaluation indicates that rail transport over concrete tie systems may result is lower fuel consumption than occurs with traditional wood tie track, because the concrete system is "stiffer", thus resulting in less rolling resistance. Modeling indicates that up to 0.19 liter (0.05 gallons) of diesel fuel are saved per 1459 ton-kilometers (1000 ton-miles) of freight [52]. A sensitivity case assuming 10% of fuel saving is attributable to the use of concrete ties, found reductions to GHG (−11%), fossil fuel use (−25%), acidification (−35%), smog (−83%), and eutrophication (−100%). This does not include increased impacts due to increased wear and damage to the trains. Under this scenario, concrete offers lower impacts in comparison to creosote-treated wood for smog and eutrophication, but higher than wood for the other indicators.

A sensitivity test considers less than 10% of creosote-treated ties recycled for energy and over 50% landfilled. Shifting used ties from beneficial energy recovery to landfilling notably impacts indicators for GHG (4-times more) fossil fuel (2-times more) and smog (+55%). Net GHG, acid rain, and ecotoxicity impacts were increased from negative or near zero values to values similar to the alternative products.

5.3. Limitations

The life cycle inventory completed for both concrete and P/C ties was designed to be representative of a product category, and therefore by design, likely will not be accurate for a specific product brand. A survey of manufacturers of concrete and P/C railroad ties was not done; therefore, inputs such as fuel and electricity use, water use, and solid waste generation at the manufacturing facilities are estimated.

6. Conclusions and Recommendations

6.1. Conclusions

The use of creosote-treated railroad ties offers lower fossil fuel and water use and lower environmental impacts than similar products manufactured of concrete and P/C, except for the eutrophication impact indicator for P/C ties.

Compared to creosote-treated railroad ties, and using the assumptions of this LCA with the understanding that actual values can vary from the assumptions, the use of concrete railroad ties results in 1.8 times more fossil fuel use and 8.7 times more water use, and results in emissions with the potential to cause approximately 5.8 times more GHG, 68 times more acid rain, 2.3 times more smog, and 2.0 times more eutrophication.

Compared to creosote-treated railroad ties, the use of P/C ties results in 2.5 times more fossil fuel use and 11 times more water use, and results in emissions with the potential to cause 5.0 times more GHG, 72 times more acid rain, and 1.1 times more smog. Creosote railroad ties result in approximately 1.4 times more eutrophication impact than P/C railroad ties.

The life cycle of creosote-treated ties results in credits (or environmental benefits) for the net GHG and ecotoxicity impact indicators.

Reuse of wood ties for energy improves the environmental life cycle performance.

This study includes the comparison of creosote-treated railroad ties to concrete and P/C ties. The results conform with the ISO 14040 and ISO 14044 standards and are suitable for public disclosure. A detailed, peer-reviewed Procedures and Findings Report can be requested by contacting the TWC at www.treated-wood.org/contactus.html. This LCA covers one treated wood product in a series of LCAs commissioned by the Treated Wood Council (TWC). The series of treated wood product LCAs covers alkaline copper quaternary (ACQ)-treated lumber [53], borate-treated lumber [54], pentachlorophenol-treated utility poles [55], chromated copper arsenate (CCA)-treated marine pilings [56], and CCA-treated guard rail systems [57].

6.2. Recommendations

Recycling of ties to energy production should be supported and increased. The LCA shows clear benefits to the impact indicators considered, particularly fossil energy, GHG, acidification, and ecotoxicity with the use of used ties as an energy source and potential remains for increased reuse. The fuel offset gained by recycling creosote-treated ties for energy recovery is 20 times greater than energy recovery from landfill disposal. Furthermore, offsets result in a significant decrease in GHG emissions when ties are recycled for energy compared to a slight increase in GHG emissions when landfilled.

Each tie recycled for energy represents approximately 0.5% of the annual U.S. per capita GHG emissions and fossil fuel usage. Thus, approximately 200 ties recycled for energy will offset the GHG and fossil fuel impacts of one typical US resident. If all ties replaced annually in the US (approximately 20 million ties) are recycled for energy, their use would offset the GHG and fossil fuel use equivalent to nearly 100,000 residents.

Utilization of dual treatment of ties should be promoted and increased in high decay regions. A service life extension of 10 to 15 years is expected when using borate/creosote dual treatment. The resulting impacts from the use of borate/creosote dual treatment are more than offset by the reduced impacts resulting from a longer service life. Wider application of dual treatment of ties in high decay regions will result in lower overall life cycle impacts.

Research evaluating the use of biodiesels as carriers of oil-borne preservatives, such as creosote and pentachlorophenol, should be continued. Data are needed to demonstrate both that biodiesel has lower impact indicators than fossil diesel, and that its use does not impact service life of treated products. If supported by these data, substitution of biodiesel for fossil oil may decrease the need for fossil oil in preservative.

Landfill disposal should be minimized. The treated wood industry and utilities should seek to minimize releases of methane resulting from disposal of wood in landfills in two ways: minimize disposal in landfills and if disposal is necessary, encourage disposal in landfills equipped with methane collection systems. Minimizing disposal is especially beneficial, because it reduces use of landfill capacity, reduces release of methane from landfills, and offsets fossil fuel use and GHG emissions with renewable biogenic fuel use.

Production facilities should continue to strive to reduce energy inputs through conservation and innovation, including sourcing materials from locations close to point of treatment and use. Also, the use of biomass as an alternate energy source can reduce some impact category values compared to the use of fossil fuel energy or electricity off the grid.

7. Acknowledgements

The authors wish to thank the TWC for their funding of this project. The TWC members and its Executive Director, Mr. Jeffrey Miller, have been integral in its completion. The authors and TWC thank the Railway Tie Association for additional studies conducted to support this LCA. We also thank the internal reviewers, James Clark, Craig McIntyre, and Maureen Puettmann, and the independent external reviewers, Mary Ann Curran, Paul

Cooper, and Yurika Nishioka for their support, patience, and perseverance in seeing this project through to completion.

REFERENCES

[1] D. J. Forkenbrock, "Comparison of External Costs of Rail and Truck Freight Transportation," *Transportation Research Part A*, Vol. 35, No. 4, 2001, pp. 321-337.

[2] G. Gould and D. Niemeier, "Review of Regional Locomotive Emission Modeling and the Constraints Posed by Activity Data," *Transportation Research Record: Journal of the Transportation Research Board*, Vol. 2117, 2009, pp. 24-32.

[3] C. Fracanha and A. Horvath, "Evaluation of Life-Cycle Air Emission Factors of Freight Transportation," *Environmental Science and Technology*, Vol. 41, No. 20, 2007, pp. 7138-7144.

[4] E. Garshick, *et al.*, "Lung Cancer in Railroad Workers Exposed to Diesel Exhaust," *Environmental Health Perspectives*, Vol. 112, No. 15, 2004, pp. 1539-1543.

[5] P. Qiao, J. Davalos, and M. Zipfel, "Modeling and Optimal Design of Composite-Reinforced Wood Railroad Crossties," *Composite Structures*, Vol. 41, No. 1, 1998, pp. 87-96.

[6] R. Resor, A. Zarembski and P. Pradeep, "Estimation of Investment in Track and Structures Needed to Handle 129,844-kg (286,000-lb) Railcars on Short-Line Railroads," *Transportation Research Record: Journal of the Transportation Research Board*, Vol. 1742, 2001, pp. 54-60.

[7] R. Ibach, "Wood Handbook-Wood as an Engineering Material. General Technical Report. FPL-GTR-113," Forest Service, Forest Products Laboratory, Madison, 1999.

[8] J. Morrell, "Disposal of Treated Wood," *Proceedings for the Environmental Impacts of Preservative-Treated Wood Conference*, Gainesville, 8-11 February 2004, pp. 196-209.

[9] C. C. Schnatterbeck, "Handbook on Wood Preservation," American Wood Preservers' Association, Baltimore, 1916.

[10] J. Bigelow, S. Lebow, C. Clausen, L. Greimann and T. Wipf, "Preservation Treatment for Wood Bridge Application," *Transportation Research Record: Journal of the Transportation Research Board*, No. 2108, 2009, pp. 77-85.

[11] K. Andersson, M. Eide, U. Lundqvist and B. Mattsson, "The Feasibility of Including Sustainability in LCA for Product Development," *Journal of Cleaner Production*, Vol. 6, No. 3-4, 1998, pp. 289-298.

[12] International Organization for Standardization (ISO), "Environmental Management-Life Cycle Assessment-Principles and Framework," Switzerland, 2006.

[13] International Organization for Standardization (ISO), "Environmental Management-Life Cycle Assessment-Requirements and Guidelines," Switzerland, 2006.

[14] J. Gauntt, "Welcome to the Future and What Will They Think of Next?" *Crossties*, Vol. 89, No. 4, 2008, pp. 13-17.

[15] R. Vlosky, "Statistical Overview of the U.S. Wood Preserving Industry: 2007," Louisiana State University Agricultural Center, Los Angeles, 2009.

[16] C. Boyd, *et al.*, "Wood for Structural and Architectural Purposes. Committee on Renewable Resources for Industrial Resources: Panel II," *Wood and Fiber*, Vol. 8, No. 1, 1976, pp. 3-72.

[17] L. Johnson, B. Lippke, J. Marshall and J. Comnick, "Forest Resources—Pacific Northwest and Southwest. CORRIM Phase I Final Report Module A. Life-Cycle Environmental Performance of Renewable Building Materials in the Context of Residential Building Construction," Seattle, 2004.

[18] L. Johnson, B. Lippke, E. Oneil, J. Comnick and L. Mason, "Forest Resources—Inland West. CORRIM Phase II Report Module A. Environmental Performance Measures for Renewable Building Materials with Alternatives for Improved Performance," Seattle, 2008.

[19] E. Oneil, *et al.*, "Life-Cycle Impacts of Inland Northwest and Northeast/North Central Forest Resources," *Wood and Fiber Science*, Vol. 42, 2010, pp. 29-51.

[20] R. Bergman and B. Bowe, "Environmental Impact of Producing Hardwood Lumber Using Life-Cycle Inventory," *Wood and Fiber Science*, Vol. 40, No. 3, 2008, pp. 448-458.

[21] American Wood Protection Association, "Standard P1/P13-09. Standard for Creosote Preservative," In: 2010 *Book of Standards*, Birmingham, 2010, p. 109.

[22] American Wood Protection Association, "Standard P2-09. Standard for Creosote Solution," In: 2010 *Book of Standards*, Birmingham, 2010, p. 110.

[23] American Wood Protection Association, "Standard P3-09. Standard for Creosote-Petroleum Solution," In: 2010 *Book of Standards*, Birmingham, 2010, p. 111.

[24] American Wood Protection Association, "Standard U1-10 Use Category System: User Specification for Treated Wood," In: 2010 *AWPA Book of Standards*, Birmingham, 2010, pp. 5-71.

[25] American Wood Preservers' Institute, "Clean Air Act Title V Guidance Manual for Wood Preserving Facilities," Fairfax, 1995.

[26] A. Zarembski, "Development of Comparative Crosstie Unit Costs and Values," *Crossties*, Vol. 87, No. 6, 2007, pp. 17-18.

[27] M. Emoto, H. Takai, T. Tsujimura and H. Ueda, "Fundamental Investigation of LCA of Cross Tie," *Railway Technical Research Institute*, Vol. 40, No. 4, 1999, pp. 210-213.

[28] R. Crawford, "Greenhouse Gas Emissions Embodied in Reinforced Cncrete and Timber Railway Sleepers," *Environmental Science & Technology*, Vol. 43, No. 10, 2009, pp. 3885-3890.

[29] T. Kunniger and K. Richter, "Comparative Life Cycle Assessment of Swiss Railroad Sleepers, IRG/WP 98-50117," *Paper prepared for the 29th Annual Meeting,*

Maastricht, 1998.

[30] L. Becker, G. Matuschek, D. Lenoir and A. Kettrup, "Leaching Behavior of Wood Treated with Creosote," *Chemosphere*, Vol. 42, No. 3, 2001, pp. 301-308.

[31] K. Brooks, "Polycyclic Aromatic Hydrocarbon Migration from Creosote-Treated Railway Ties into Ballast and Adjacent Wetlands. Research Paper FLP-RP-617," Department of Agriculture, Forest Service, Forest Products Laboratory, Madison, 2004.

[32] M. Burkhardt, L. Rossi and M. Boller, "Diffuse Release of Environmental Hazards by Railways," *Desalination*, Vol. 226, No. 1-3, 2008, pp. 106-113.

[33] A. Chakraborty, "Investigation of the Loss of Creosote Components from Railroad Ties," University of Toronto, Toronto, 2001.

[34] E. Gallego, F. Roca, J. Perales, X. Guardino and M. Berenguer, "VOCs and PAHs Emissions from Creosote-Treated Wood in a Field Storage Area," *Science of the Total Environment*, Vol. 402, No. 1, 2008, pp. 130-138.

[35] R. Geimer, "Feasibility of Producing Reconstituted Railroad Ties on a Commercial Scale: Research Paper FPL 411," United States Department of Agriculture Forest Service, Forest Products Laboratory, Madison, 1982.

[36] B. Gevao and K. Jones, "Kinetics and Potential Significance of Polycyclic Aromatic Hydrocarbon Desorption from Creosote-Treated Wood," *Environmental Science and Technology*, Vol. 32, No. 5, 1998, pp. 640-646.

[37] M. Kohler and T. Kunninger, "Emission of Polycyclic Aromatic Hydrocarbon (PAH) from Creosoted Railroad Ties and Their Relevance for Life Cycle Assessment," *Springer*, Vol. 61, No. 2, 2003, pp. 117-124.

[38] C. Sparacino, "Final Report—Preliminary Analysis for North American CTM Creosote P2," Research Triangle Institute, Research Triangle Park, 1999.

[39] M. D. Fenton, "Mineral Commodity Profiles—Iron and Steel," US Geologic Survey, US Department of Interior, Reston, 2005.

[40] USEPA, "Inventory of U.S. Greenhouse Gas Emissions and Sinks: 1990-2007: Report No: EPA 430-R-09-004," Washington DC, 2009.

[41] J. Menard, *et al.*, "Life Cycle Assessment of a Bio-Reactor and an Engineered Landfill for Municipal Solid Waste Treatment," 2003. www.lcacenter.org/InLCA-LCM03/Menard-presentation.ppt

[42] A. Zarembski, "Assessment of Concrete Tie Life on US Freight Railroads," Report Submitted to the Railway Tie Association, 2010.

[43] Crossties, "UP Makes Claim Against CXT Inc. for Failing Ties," *Crossties*, Vol. 93, No. 1, 2012, p. 1.

[44] S. Morris, "Market Watch," 2008. http://www.marketwatch.com/news/story/tieteck-llc-sells-over-one/story.aspx

[45] U. Arena, M. Mastellone and F. Perugini, "Life Cycle Assessment of a Plastic Packaging Recycling System," *International Journal of Life Cycle Assessment*, Vol. 8, No. 2, 2003, pp. 92-98.

[46] D. Garrain, P. Martinez, R. Vidal and M. Belles, "LCA of Thermoplastics Recycling," 2009. http://www.lcm2007.org/paper/168.pdf

[47] J. Bare, G. Norris, D. Pennington, and T. McKone, "TRACI—The Tool for the Reduction and Assessment of Chemical and Other Environmental Impacts," *Journal of Industrial Ecology*, Vol. 6, No. 3-4, 2003, pp. 49-78.

[48] USEPA, "Tool for the Reduction and Assessment of Chemical and Other Environmental Impacts (TRACI)," 2009. http://www.epa.gov/nrml/std/traci/traci.html

[49] R. Rosenbaum, *et al.*, "USEtox—The UNEP-SETAC Toxicity Model: Recommended Characterization Factors for Human Toxicity and Freshwater Ecotoxicity in Life Cycle Impact Assessment," *The international Journal of Life Cycle Assessment*, Vol. 13, No. 7, 2008, pp. 532-546.

[50] Research and Innovative Technology Administration (RITA), Bureau of Transportation Statistics, 2010. http://www.bts.gov/publications/national_transportation_statistics/html/table_01_46a.html

[51] M. G. Sanders and T. L. Amburgey, "Tie Dual Treatments with TimBor and Creosote or Copper Naphthenate —20 Years of Exposure in AWPA Hazard Zone 4," *Crossties*, Vol. 90, No. 5, 2009, pp. 20-22.

[52] AREMA, "Section 2.1 Resistance to Movement," In: *AREMA Manual for Railway Engineering*, American Railway Engineering and Maintenance-of-Way Association, Lanham, 1999.

[53] C. Bolin and S. Smith, "Life Cycle Assessment of ACQ-Treated Lumber with Comparison to Wood Plastic Composite Decking," *The Journal of Cleaner Production*, Vol. 19, No. 6-7, 2011, pp. 620-629.

[54] C. A. Bolin and S. T. Smith, "Life Cycle Assessment of Borate-Treated Lumber with Comparison to Galvanized Steel Framing," *The Journal of Cleaner Production*, Vol. 19, No. 6-7, 2011, pp. 630-639.

[55] C. Bolin and S. Smith, "Life Cycle Assessment of Pentachlorophenol-Treated Wooden Utility Poles with Comparesons to Steel and Concrete Utility Poles," *Renewable and Sustainable Energy Reviews*, Vol. 15, No. 5, 2011, pp. 2475-2486.

[56] C. Bolin and S. Smith, "Life Cycle Assessment of CCA-Treated Wood Marine Piles in the US with Comparisons to Concrete, Galvanized Steel, and Plastic Marine Piles," *Journal of Marine Environmental Engineering*, Vol. 9, No. 3, 2012, pp. 239-260.

[57] C. Bolin and S. Smith, "Life Cycle Assessment of CCA-Treated Wood Highway Guard Rail Posts in the US with Comparisons to Galvanized Steel Guard Rail Posts," *Journal of Transportation Technologies*, Vol. 3, No. 1, 2013, pp. 58-67.

16

Integrating Strategic and Tactical Rolling Stock Models with Cyclical Demand

Michael F. Gorman
Department of MIS, Operations and Decision Sciences, University of Dayton,
Dayton, USA

ABSTRACT

In the transportation industry, companies position rolling stock where it is likely to be needed in the face of a pronounced weekly cyclical demand pattern in orders. Strategic policies based on assumptions of repetition of cyclical weekly patterns set rolling stock targets; during tactical execution, a myriad dynamic influences cause deviations from strategically set targets. We find that optimal strategic plans do not agree with results of tactical modeling; strategic results are in fact suboptimal in many tactical situations. We discuss managerial implications of this finding and how the two modeling paradigms can be reconciled.

Keywords: Rolling Stock; Network Management; Strategic; Tactical

1. Introduction

Many freight transportation companies managing rolling stock fleets (e.g., containers, trailers, truck tractors, railcars, locomotives, etc.) face highly regular weekly cycles in supply of and demand patterns for these resources. For examples, supply and demand for rail locomotives may depend on the number of train terminations and originations, or in trucking, delivered loads contribute to container supply and historical order patterns indicate likely demand. These supply and demand vectors are heavily influenced by day of week (e.g., weekday versus weekend patterns). In this paper, we refer to "strategic" models as those based on this regular repeating patterns.

At a more tactical level, a transportation company must establish the best levels of rolling stock assets each day to support these highly cyclical and uncoordinated supply and demand patterns; transportation companies often keep non-zero levels of rolling stock capacity at locations in their network in anticipation of future demand because of the costs and time constraints of repositioning rolling stock. Because of the strong repeating weekly patterns of supply and demand, a company might develop target rolling stock levels based on strategic planning models that assume a regular weekly pattern to maximize the return on its rolling stock asset.

Although these regular patterns can be used for strategic planning, each week actual supply and demand levels

vary around those patterns because of the stochastic nature of supply and demand, resulting in a deviation from the strategic plan. In this "tactical" setting, actual rolling stock inventory varies from the strategic targets; tactical models are deployed based on a starting condition for rolling stock levels.

In the tactical setting, in order to resolve such deviations and return to strategic target rolling stock levels, a company might make efforts to return to the optimal strategic inventory capacity levels such as increased or decreased allocation of the asset. However, the recovery or adjustment path often carries its own costs, so the company must assess if and when to adjust back to strategic targets.

Both strategic and tactical models have problems in implementation. The strategic model is difficult to manage in real time environment because of the assumption of cyclical repetition. The strategic model gives no indication how to react to deviations from the long run strategic optimum. On the other hand, strictly tactical modeling reflects current conditions in the network given prior events, and doesn't necessarily lead to any long run goals or targets. One intuitive solution in the tactical model paradigm is to start with current conditions, but at the end of the cycle, "recover", or return to the strategic target levels.

This research evaluates recovery strategies from a deviation from the strategic target rolling stock levels and the appropriate integration of strategic and tactical models. We find that in some cases deviations from strategic optima are in fact advantageous; that is, the strategic optima may not be optimal in a tactical setting, calling into question the utility of strategic modeling of problems of this type. Managing to a target rolling stock level can be misdirected effort, creating additional costs with questionable incremental benefits. We evaluate conditions that give rise to this situation and make recommendations on how to reconcile the approaches. Based on these results, we make recommendations on the trade-offs between short run and long run rolling stock management.

2. Literature

The use of the words "strategic" and "tactical" require some definition. In some cases (e.g., [1]), strategic models focus on design of the transportation system, where tactical models focus on its operations. This is not the intended definition in this paper. Rather, we define a strategic model as one that assumes cyclicality, and a constraint on the ending time period ties it back to the starting time period. In this paper, "tactical" models start with a given starting condition, and may or may not have a constraint on the final period. In short, tactical models do not assume or require cyclicality. Because the two models have the same structure, and deal with the same issues in asset management, but differ only in assumption, their juxtaposition is warranted.

There are numerous examples of strategic and tactical models as defined above juxtaposed in the transportation literature. Similar to the strategic planning horizon described in [2] we define strategic patterns as those assuming a cyclic, repeating pattern that can be sustained in the long run, supporting a regular cycle and more strategic plan. An alternative to strategic planning is a more tactical orientation, which we call the tactical paradigm. Similar to the "daily" horizon of [2], we define tactical planning tools as have some initial (time = t_0) rolling stock inventory levels. In a similar way, [3], discusses various planning horizons in passenger rail: strategic, tactical, operational and short term. The tactical models of [3] are equivalent to the strategic models discussed here, and the operational models and short term planning models look at daily deviations as do these tactical models.

As depicted in **Figure 1(a)**, strategic models include an arc from the end of the planning horizon, back to the start, imposing a repeatable cycle. In strategic models, ending period ending stock variables must equal the beginning period's starting value, thus creating continuity and consistency in the strategic model. Just as a circle must tie back to itself, so does a cyclic model's starting and ending inventory.

In the tactical model paradigm, the models are tied to a starting inventory condition (given all prior patterns of supply and demand and management allocation decisions, including unanticipated supply and demand shocks). Given a starting rolling stock inventory and anticipated cyclic supply and demand, what is the best course of action for managing these critical assets? Simply, tactical models react to current conditions which are a result of past known and exogenous events; Strategic models plan for them by viewing yesterday's events are next week's future events. Tactical models are necessary for dealing with a starting condition that are the result of prior events; strategic models are useful for establishing what the optimal conditions would be in the long run. The question addressed here is how to align these two modeling paradigms.

As depicted in **Figure 1(b)**, tactical model might be specified with or without a constraint on the final state at the end of the horizon. On one hand, if there is a single deviation from the strategic target levels (say due to an unexpected supply or demand shock), the goal might be to manage from the current disequilibrium towards the "strategic target" level of inventory ($I_n = I_n^s$; dubbed "recovery" mode). Alternatively, the tactical setting might be more open-ended, with no constraint on ending inventory (dubbed tactical "reactionary" mode). The question remains, in the stochastic environment, how fast should adjustment take place if at all; how much weight should be given to strategic considerations? Secondarily, how close to strategic targets is "close enough", and what is the cost of deviating from this target? What is the cost of adjustment to the strategic optimum?

It is common in the literature to take either a strategic or tactical perspective on the problem without considering the alternative. For examples, [4-6] look at the challenge of managing railcars in the face of uncertain demand in the tactical setting, but do not consider a longer-run, strategic allocation of rolling stock or their optimal stocking levels. On the other hand, [7] considers fleet sizing under strategic assumptions, but do not discuss deviations from the strategic plans brought on by stochastic elements in a tactical, real time setting.

Similarly, [8] creates weekly repeating cycles of (strategic) locomotive to train assignments (and accompanying "ground arcs", or inventory decisions). On the other hand, [9] describes in-plant tactical locomotive manage-

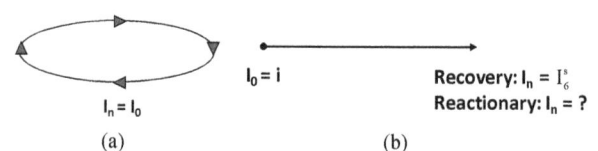

Figure 1. Conceptualization of strategic and tactical modeling paradigms. (a) Strategic modeling paradigm; (b) Tactical modeling paradigm.

ment over a two-hour window with no consideration for longer-term considerations.

Similar to rolling stock, manpower capacity also must be managed by location. Reference [10] plans drayage operations and [11] plans drivers in a tactical setting, with a starting location for drivers and tractors specified, but because of short planning and order visibility horizons, make no consideration for repeatability or consistency of scheduling. Reference [12] discusses a tactical rail crew planning model given an initial starting location, and claims the model can be used for strategic manpower planning over a longer horizon, but does not show how to account for different starting conditions and stochastic train schedules.

A similar schism in focus can be found in the capacity pricing and yield management literature in freight transportation. For examples, [13] shows the importance of rolling stock balance in rail intermodal in a strategic setting, but does not talk about adjustment mechanisms in a tactical setting. Similarly, [14] builds a strategic logistics queueing network model that creates prices to place rolling stock capacity where and when it has the highest value. Reference [15], on the other hand, present a similar problem viewed from a tactical setting with a given starting inventory. There is no discussion in these articles on how to transition from a tactical yield management situation to the long run strategic pricing strategy, or how to translate strategic recommendations to a tactical implementation environment.

Reference [3] discusses decision horizon tradeoffs in passenger rail, which is considerably less volatile than freight rail systems, so short term planning models face similar constraints as air passenger service. However, [3] describes the different modeling paradigms, this research does not detail their differences nor try to reconcile them. It might be added that this research investigates the handoffs between the tactical and operational environments in the more uncertain freight transportation environment, where the differences of the two paradigms is more pronounced and the decisions on how to reconcile more difficult to address.

We should differentiate this research from the literature on "refleeting" or disruption management and recovery in the airline literature. This literature focuses on building robust cyclic schedules with respect to disruptions [16] and getting back on schedule in a least-cost way given a disruption [17]. In the airline case, there is a fixed schedule that must be followed per customer expectations and industry norms. This situation implies a required fleet capacity and mandatory and expedient recovery to the strategic condition; in the freight transportation case, service provision depends on a stochastic order pattern without customer reservations with no obligation to follow specific schedule or level of capacity

provision. The reason this distinction is important is that where airlines view the schedule as a constraint and recovery is mandatory, freight providers are not constrained to return to the strategic plan, but must choose both whether to recover, and how fast to return to the longer run strategic target levels.

In general, there is a dearth of literature which tries to bridge the gap in the planning process between strategic rolling stock planning and tactical or real time execution in freight transportation. A notable counter example is an early attempt to meld tactical and strategic models in [18], which mitigates end effects in tactical vehicle allocation by proposing a "transient" (tactical) portion and a "stationary" (strategic) portion of the problem. With a discount factor, the repeating (looping) portion of the stationary portion represents the net present value of future flows. Reference [15] points out the managerial need for establishing target container inventory levels in intermodal, and identifies the problem surrounding the reconciliation of these two modeling paradigms, but does not resolve or make recommendations on how to resolve them. The contribution of this research is derived from its focus on the cost of deviation from the strategic plan, the cost of recovery, and derivation of appropriate recovery strategies. This research is the first to identify and quantify a contradiction between optimal strategies in strategic and tactical paradigms.

3. Mathematical Modeling

Below we provide a modeling construct which allows us to capture both strategic and tactical paradigms for comparative analysis. For simplicity of exposition and modeling, we will focus on the allocation of a single rolling stock inventory in a single location, but the results apply directly to the full multi commodity time space network. The model could be expanded to incorporate all locations in a transportation network, but for the purpose of this research, a single location model adequately demonstrates the point. Further, as noted in Gorman [15], the single node view matches the managerial focus of transportation companies managing tactical rolling stock inventories.

The decision variable, denoted A_t, is the allocation of capacity of various types (tractors, drivers, locomotives, railcars, containers) to demands, D_t, of different types (trains, orders) in any period, t. The source of the allocated capacity in any period is based on the inventory of the resource carried from the previous period, I_t, and that are made available from the supply process in that period, S_t. In the tactical setting, supply and demand are considered exogenous. Demand is exogenous based on customer order patterns. Supply is exogenous because it is the result of terminated usage from past allocation decisions in the tactical setting and because it is the result of allocation decisions made in other geographic locations in both the tactical and strategic modeling paradigms.

The allocation of each asset depends on its cost and revenue profiles in various uses. The explicit cost of excess inventory is higher inventory carrying costs. The opportunity cost of excess rolling stock is the acceptance of lower profitability business (higher cost or lower revenue) in order to utilize the asset.

The cost of high inventory must be balanced against the opportunity cost of low inventory levels. In the single asset case, the primary opportunity cost of a rolling stock inventory shortage is lost revenue. In the multiple asset case, a shortage of a preferred asset requires the use of a less preferred alternative—either a lower revenue or higher cost asset. Different asset classes which are imperfect substitutes, with "preferred" and "less preferred" assignment which constitute varying cost profiles and capabilities which govern the feasibility of their assignment. For examples, locomotives of 4 and 6 axles have different fuel efficiencies, tractive effort potential, or consist interoperability, making them have different efficiency levels for different train types [8]. Different railcars might have different equipment rent if they are foreign owned or not [15]. Drivers may be different distances from an order, requiring varying dray costs. Containers of different ownership have different cost structures and rail routing options [15]. In each case, there is some incremental cost of using the imperfect substitute for a given demand. From a revenue perspective, different sizes of containers or equipment on railcars might have diminishing values to customers, or from a safety

stock perspective, additional inventory of any kind has a diminishing but positive probability of use, so incremental units of capacity have diminishing expected revenue [15].

As discussed in [15] such opportunity costs are likely to be increasing as the level of surplus or shortage grows. Both the incremental costs and diminishing revenues contribute to a decreasing valuation of any asset at a location as the quantity of that asset grows. We define the "profit advantage function", as $y * (x - A_t/D_t)^z$, of one class of rolling stock inventory over another to capture this diminishing profitability relationship. (This is simply the profit function in the case with only one asset class.) A_t/D_t is the percentage of total demand on day t allocated to asset A on day t $(0 \leq A_t \leq D_t)$. A generalized diminishing profitability function with parameters x, y, and z is specified in order to test for sensitivity of our results for different functional forms of the profit equation: y specifies the highest valuation of the most appropriate allocation, x determines the percentage of total allocation at which the expected contribution becomes negative, and z indicates the concavity (z > 1), convexity (z < 1) or linearity (z = 1) of the functional form. Such a relationship is easily estimated from historical data on assignments and profitability. Examples of each functional form are demonstrated in **Figure 2**.

4. Optimization Model

The optimization model that serves as the basis of our study is described in Equations (1)-(4). The profit advan-

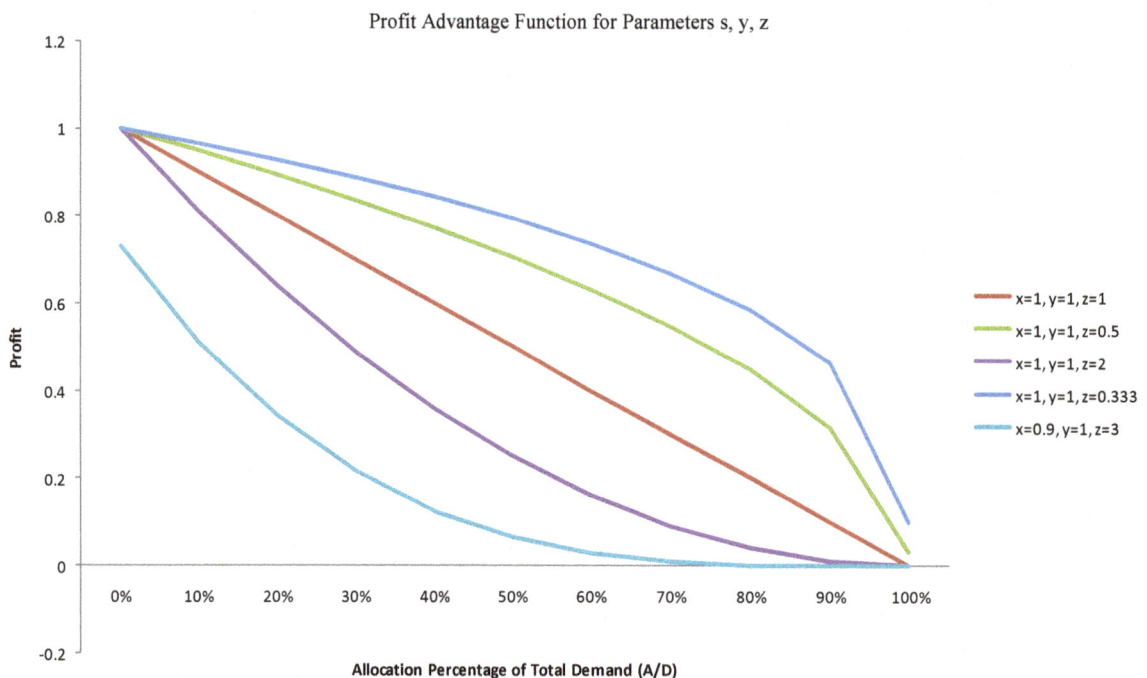

Figure 2. Profit advantage function for parameters s, y, z.

tage function is the first component of the objective function (Equation (1)). A secondary disadvantage of carrying inventory, I_t, of some asset class is its holding cost, HC, which is subtracted from the expected profitability of each assignment level, the second component of Equation (1).

The constraints governing assignments are given in Equations (2)-(4). Allocations of an asset must be less than demand (2), inventory in any period equals the inventory of the prior period, plus new supply, less allocation in this period (3), and inventory must be non-negative.

Maximize

$$p^* = y\left(x - D_t / A_t\right)^z - \sum_{t=1}^{7} HC I_t \qquad (1)$$

Subject to:

$$0 \leq A_t \leq D_t \forall t \qquad (2)$$

$$I_{t+1} = S_{t+1} - A_{t+1} + I_t \forall t \qquad (3)$$

$$I_t \geq 0 \forall t \qquad (4)$$

Without loss of generality, we focus on the weekly supply and demand paper with seven daily time periods, t. Constraints 5s and 5d differentiate the strategic and tactical modeling paradigms. In the strategic paradigm, starting inventory is related ending inventory to assure cyclic repeatability:

$$I_1 = I_7 + S_1 - A_1 \qquad (5s).$$

We define the resulting profit to the strategic problem given in Equations (1)-(5s) as p^s, optimal allocations vector as A_t^s, and resulting inventory vector as I_t^s.

In the tactical model, constraint 5s is replaced with 5d:

$$I_1^d = I_0 0 + S_1 - A_1 \qquad (5d),$$

where I_0 is some initial, exogenous inventory value in the tactical setting given past supply and demand shocks.

Within the tactical paradigm, management might not constrain ending inventory, I_7, following a reactionary, short term tactical (t) strategy, choosing to react to supply and demand perturbations with a short term focus, disregarding the strategic optimum. We denote objective function values, inventory levels and allocations as p^t, I^t and A^t. Alternatively, management might pursue what we will call a "recovery" (r) strategy that allows them to regain strategic inventory levels by constraining ending inventory to equal that of the strategic model as in Equation (6).

$$I_7^r = I_7^s. \qquad (6)$$

We denote objective function values, inventory levels and allocations as p^r, I^r and A^r. For any given D and S arrays, each of these three models lead to different values of the decision variables, inventory and total objective function values.

5. Numerical Example

We illustrate the optimization models with a numerical example. In this illustration, key input parameters are: HC = 0.1, x = y = z = 1. We select a random demand (D) and supply (S) arrays as depicted in **Table 1**, and solving the strategic, reactionary and tactical optimization models given in Equations (1)-(6), results in optimal profitability (p), fleet allocations (A) and inventory levels (I) under each paradigm displayed in the columns as labeled. In the strategic model, I_0 is based on I_7 at the end of the week, thus is endogenously determined. In this example, the optimal end-of-week inventory is 7 units to achieve a strategic sustainable profit of $72.74 per week. If a manager was myopic or did not put emphasis on the somewhat uncertain future supply and demand patters, he would ignore end-of-week inventory on the subsequent week's profits (the constraint that $I_0 = I_7$ is removed). As a result, $74.67 could be earned in a week. However, the ending inventory of 0 units ($I_7 = I_0 = 0$) in reactionary mode drives the subsequent week's profit down to $69.29 if the goal is to "recover"—return to the strategic optimum inventory ($I_7 = 7$)—by the end of the second week (as shown in the last column of **Table 1**, in which starting inventory is zero and ending inventory is seven). Average profits fall from a strategic expectation of $72.74, to an average of $71.98 = ($74.67 + 69.29)/2. Thus, unsustainable short-term profit is gained at the expense of future recovery costs. It seems reasonable, then, to strive to achieve strategic target inventories.

However, deviations from the strategic optimum could be for exogenous reasons no fault of the manager, such as an unanticipated supply or demand shock. Let us assume a single a priori exogenous supply shock leads to some deviation from strategic optimum inventory at the end of Day 0. In this case, the manager optimizes given some starting inventory level. The manager has a choice to try to recover to the strategic target inventory or not.

We solved the tactical model to optimality for reasonable levels of starting inventory ranging from zero to 59 results in a quadratic shaped profit curve as indicated in **Figure 3**. While the strategic optimum (and the recovery tactical target ending inventory) starting inventory is seven, profitability in a given week is maximized at a starting inventory of 25 units, resulting in $4.82 (6.6%) in higher profits from the presence of that inventory with no recovery to strategic targets, and $2.89 (2.6%) increased profits if recovery to strategic is completed by day seven. Although such profit is not sustainable in the long run if the strategic supply and demand processes are representative of normal patterns, these results call into

Table 1. Number example of state optimal inventory, and dynamic reactionary and dynamic recovery profit levels.

Day	Input Data: Cyclical Order and Release Data — Demand Process (D)	Supply Process (S)	Steady State Model ($I_0 = I_7$; I_7 is a decision variable) — Decision Variable: Fleet Allocation (A)	Resultant: Non-Fleet Demand Coverage (D-A)	Resultant: Fleet End of Day Inventory ($I_t = I_{t-1} + S_t - A_t$)	Resultant: Fleet Demand Coverage Percent (A/D)	Dynamic Reactionary Model ($I_0 = I'_7 = 7$; I_7 unconstrained) — Decision Variable: Fleet Allocation (A)	Resultant: Non-Fleet Demand Coverage (D-A)	Resultant: Fleet End of Day Inventory ($I_t = I_{t-1} + S_t - A_t$)	Resultant: Fleet Demand Coverage Percent (A/D)	Dynamic Recovery Model ($I_0 = I'_7 = 0$; $I'_7 = 7$) — Decision Variable: Fleet Allocation (A)	Resultant: Non-Fleet Demand Coverage (D-A)	Resultant: Fleet End of Day Inventory ($I_t = I_{t-1} + S_t - A_t$)	Resultant: Fleet Demand Coverage Percent (A/D)	Optimal Starting Tactical Inventory ($I_0 = 0$; $I'_7 =$ Unconstrained) — Decision Variable: Fleet Allocation (A)	Resultant: Non-Fleet Demand Coverage (D-A)	Resultant: Fleet End of Day Inventory ($I_t = I_{t-1} + S_t - A_t$)	Resultant: Fleet Demand Coverage Percent (A/D)
End Day 0					7				7				-				25	
1	54	10	17	37	-	31%	17	37	-	31%	10	44	-	19%	27	27	8	50%
2	51	16	16	35	-	31%	16	35	-	31%	16	35	-	31%	23	28	1	45%
3	20	18	8	12	10	40%	8	12	10	40%	8	12	10	40%	8	12	11	40%
4	87	20	30	57	-	35%	30	57	-	35%	30	57	-	35%	31	56	-	35%
5	15	21	7	8	14	46%	7	8	14	49%	7	8	14	46%	7	8	14	49%
6	76	20	31	45	3	41%	34	42	-	44%	31	45	3	41%	34	42	-	44%
7	30	15	11	19	7	36%	15	15	-	50%	11	19	7	36%	15	15	-	50%

Input Fleet Profit Parameters		Steady State		Dynamic Reactionary		Dynamic Recovery		Optimal Starting Tactical
		Fleet Profitability (p)	$76.12	Fleet Profitability (p)	$77.04	Fleet Profitability (p)	$72.67	Fleet Profitability (p) $80.81
x	1	Less Fleet inventory cost	$3.37	Less Fleet inventory cost	$2.37	Less Fleet inventory cost	$3.37	Less Fleet inventory cost $3.29
x	1	Net Fleet Benefit	$72.74	Net Fleet Benefit	$74.67	Net Fleet Benefit	$69.29	Net Fleet Benefit $77.52
z	1							
HC	0.1							

question whether deviations from strategic targets are in fact bad, and if strenuous efforts should be made to return to strategic targets. Similar unanticipated supply and demand shocks over the planning horizon could render striving for strategic targets both infeasible and unprofitable. In a highly stochastic environment, the short-sighted manager who discounts the future states might actually achieve superior results.

6. Monte Carlo Simulation

Of course, this example could be a special case. To investigate how wide spread this result is, we conducted numerical experiments based on a Monte Carlo simulation. The algorithm below compares the strategic opti-

mum profit with the profit level consistent with the optimal tactical inventory level. To compare the cost of deviation from target levels, we solve both models repeatedly in n Monte Carlo replications of weekly S and D patterns. We then constrain I_0 for all reasonable levels of inventory (I_{max}) to evaluate the change in cost from the optimal strategic level p^s from dynamic model objective function values p^r and p^t. Importantly, we compare the deviation from optimal p_s profit levels for each starting level of I.

The approach takes three general steps:

1) For any random generation of S and D arrays, solve the strategic model to establish long run targets (s);

2) Perform sensitivity analysis with respect to devia-

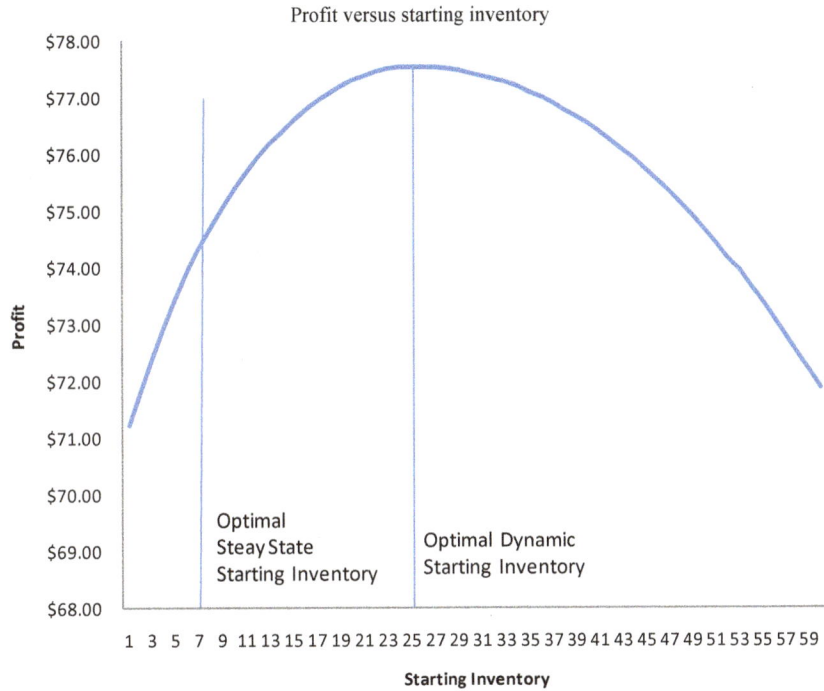

Figure 3. Profit versus starting inventory.

tions from optimum starting inventory. For all reasonable starting inventory levels, I_0, solve tactical model twice, once in recovery mode (r), once in tactical reactive mode (t), comparing tactical model objective function values with that of strategic model.

3) The difference between the optimal based on strategic target and optimal based on current tactical inventory is the objective function loss or gain from deviating from the strategic optimum.

The algorithm below describes the steps in more detail.

For j = 1 to n
   Generate random replications for S and D processes
   Solve strategic model for p^s, I^s, A^s
   For i = 0 to I_{max}
      $I_0 = i$
      Solve the tactical model for p^r, I^r, A^r
      DeviationCost^r = p^s– p^r
      Solve the tactical model for p^t, I^t, A^t
      DeviationCost^t = p^s- p^t
   Next i
Next j

6.1. Experimental Design

We set up a balanced experiment with 12 scenarios: 2 variance levels (High, Low), 3 supply/demand ratio levels (High, Medium and Low supply), and 3 functional forms (linear, concave and convex). In the structured experiments, we held three parameters constant: x = 1, y = 33 * HC. For the high supply and demand variance, a

uniform distribution was used to generate supply and demand vectors, D ~ U(1,100); for the low supply and demand variance, a Poisson distribution was used, D ~ Poisson(50). For high supply, the expected value of S = 0.5 D, for medium supply, S = 0.4 D, and for low supply, S = 0.33 D. Strategic model results reflect a single model run; tactical model results reflect the tactical model run with the best starting inventory (minimum deviation cost from strategic optimum). We conducted 40 randomly generated replications for each scenario. For the functional form of the profit advantage function, we set z = 2 (convex), 1 (linear) and 0.5 (concave). Descriptive statistics are presented in **Table 2**.

We can see from **Table 2**, no matter what the functional form, variance or supply demand ratio, the average objective based on ideal tactical starting inventory is always higher than the strategic optimum. That is, there is some level of inventory other than the strategic optimum which makes the objective increase, over the decision horizon. This effect is generally more pronounced in the high variance scenarios. Simply, the strategic, strategic optimum is not optimal in a tactical setting.

Further, by comparing the tactical-reactionary objective function value to the tactical-recovery objective, we see that recovery has a cost; by working to return to some steady-state ideal inventory by the end of the week, profit potential is lost. It is worth mentioning that the profit performance of the reactionary modeling paradigm may not be sustainable because the end of the week inventory levels may not support future business patterns well, but

Table 2. Comparison of objective function values in stratified random experiments.

Linear		Objective Function Values					
		Strategic		Tactical-Reactionary		Tactical-Recovery	
		Mean	Std Dev	Mean	Std Dev	Mean	Std Dev
HSHVLIN	High Supply, High Var, Linear	$ 29,701	$ 11,478	$ 32,591	$ 8282	$ 31,359	$ 10,188
MSHVLIN	Medium Supply, High Var, Linear	$ 37,084	$ 7555	$ 39,131	$ 7557	$ 38,200	$ 7770
LSHVLIN	Low Supply, High Var, Linear	$ 42,821	$ 8560	$ 46,200	$ 8211	$ 44,792	$ 8536
HSLVLIN	High Supply, Low Var, Linear	$ 38,449	$ 3125	$ 38,721	$ 3194	$ 38,328	$ 3338
MSLVLIN	Medium Supply, Low Var, Linear	$ 47,333	$ 2311	$ 48,595	$ 2582	$ 48,365	$ 2533
LSLVLIN	Low Supply, Low Var, Linear	$ 52,093	$ 2688	$ 55,429	$ 3012	$ 55,007	$ 3037
Average	Average	$ 41,247	$ 9962	$ 43,445	$ 9595	$ 42,675	$ 10,130

Concave		Objective Function Values					
		Strategic		Tactical-Reactionary		Tactical-Recovery	
		Mean	Std Dev	Mean	Std Dev	Mean	Std Dev
HSHVCAVE	High Supply, High Var, Concave	$ 44,996	$ 16,983	$ 49,366	$ 14,501	$ 47,814	$ 15,396
MSHVCAVE	Medium Supply, High Var, Concave	$ 53,432	$ 10,964	$ 59,134	$ 10,648	$ 56,805	$ 10,903
LSHVCAVE	Low Supply, High Var, Concave	$ 58,131	$ 14,223	$ 64,313	$ 12,215	$ 62,401	$ 13,334
HSLVCAVE	High Supply, Low Var, Concave	$ 66,596	$ 3,451	$ 71,860	$ 4,156	$ 71,202	$ 4,039
MSLVCAVE	Medium Supply, Low Var, Concave	$ 62,281	$ 3,048	$ 65,348	$ 3,383	$ 64,780	$ 3,286
LSLVCAVE	Low Supply, Low Var, Concave	$ 58,267	$ 4,223	$ 59,898	$ 4,418	$ 59,542	$ 4,477
Grand Total	Average	$ 57,284	$ 12,361	$ 61,653	$ 11,523	$ 60,424	$ 12,124

Convex		Objective Function Values					
		Strategic		Tactical-Reactionary		Tactical-Recovery	
		Mean	Std Dev	Mean	Std Dev	Mean	Std Dev
HSHVCAVE	High Supply, High Var, Convex	$ 14,510.63	$ 10,279.24	$ 16,246.75	$ 7789.73	$ 15,464.76	$ 8856.95
MSHVCAVE	Medium Supply, High Var, Convex	$ 25,155.82	$ 5988.79	$ 26,010.81	$ 6377.54	$ 25,848.05	$ 6309.42
LSHVCAVE	Low Supply, High Var, Convex	$ 26,019.86	$ 7343.24	$ 26,973.27	$ 7779.07	$ 26,594.44	$ 7634.80
HSLVCAVE	High Supply, Low Var, Convex	$ 22,919.27	$ 1593.77	$ 22,983.77	$ 1573.92	$ 22,940.86	$ 1594.72
MSLVCAVE	Medium Supply, Low Var, Convex	$ 25,722.16	$ 1766.67	$ 25,763.86	$ 1777.73	$ 25,750.90	$ 1776.19
LSLVCAVE	Low Supply, Low Var, Convex	$ 28,307.34	$ 1698.89	$ 28,434.25	$ 1736.51	$ 28,395.70	$ 1729.95
Grand Total	Average	$ 23,958.98	$ 6682.89	$ 24,436.24	$ 5915.60	$ 24,246.38	$ 6259.72

this risk must be balanced with the reward of enhanced high probability short term profit.

A more direct comparison of the three models can be conducted by looking at the foregone profit under each random demand generation. In each replication, we calculated the absolute increase in profits between the modeling paradigms objective function values. Over 90% of the time, the differences were non-zero, demonstrating the modeling paradigm and resulting operating policy does make a difference to profits under most supply and demand conditions. **Table 3** presents the differences in mean profit levels under each modeling paradigm and scenario. We see that there is a starting inventory level in the tactical setting that leads to an average of 5.5% higher profits that the inventory suggested by the strategic model. Further, an average of 1.8% profits are fore-

gone when striving to recover to the strategically identified optimum rather than simply focusing on the near term. In total, 7.3% profits are lost by focusing on a strategically identified optimum that is inappropriate in a tactical, execution setting.

6.2. Completely Randomized Experimental Design

To ensure robustness, we also ran a completely randomized design experiment that varied x, y, z and HC: x ~ U(0.5,1), y ~ U(100, 500), z = U(1, 3), HC ~ U (5,30) with both the medium and high supply, and with low and high variance. We generated 250 replications in this completely randomized design. Although the standard deviation of the profit differentials was higher due to the more varied input data, the results from the stratified

Table 3. Difference in mean profit levels between modeling paradigms.

Linear

		Tactical Recov-Strategic		Tactical React-Recovery	
		Abs. Diff	Pct. Diff	Abs. Diff	Pct. Diff
HSHVLIN	High Supply, High Var, Linear	$ 2891	8.9%	$ 1232	3.9%
MSHVLIN	Medium Supply, High Var, Linear	$ 2048	5.2%	$ 931	2.4%
LSHVLIN	Low Supply, High Var, Linear	$ 3379	7.3%	$ 1408	3.1%
HSLVLIN	High Supply, Low Var, Linear	$ 272	0.7%	$ 393	1.0%
MSLVLIN	Medium Supply, Low Var, Linear	$ 1262	2.6%	$ 230	0.5%
LSLVLIN	Low Supply, Low Var, Linear	$ 3336	6.0%	$ 423	0.8%
Average	Average	$ 2198	5.1%	$ 769	2.0%

Concave

		Tactical Recov-Strategic		Tactical React-Recovery	
		Abs. Diff	Pct. Diff	Abs. Diff	Pct. Diff
HSHVCAVE	High Supply, High Var, Concave	$ 4370	8.9%	$ 1553	3.2%
MSHVCAVE	Medium Supply, High Var, Concave	$ 5702	9.6%	$ 2329	4.1%
LSHVCAVE	Low Supply, High Var, Concave	$ 6181	9.6%	$ 1912	3.1%
HSLVCAVE	High Supply, Low Var, Concave	$ 5264	7.3%	$ 657	0.9%
MSLVCAVE	Medium Supply, Low Var, Concave	$ 3068	4.7%	$ 569	0.9%
LSLVCAVE	Low Supply, Low Var, Concave	$ 1631	2.7%	$ 357	0.6%
Grand Total	Average	$ 4369	7.1%	$ 1229	2.1%

Convex

		Tactical Recov-Strategic		Tactical React-Recovery	
		Abs. Diff	Pct. Diff	Abs. Diff	Pct. Diff
HSHVCAVE	High Supply, High Var, Convex	$ 1736	10.7%	$ 782	5.1%
MSHVCAVE	Medium Supply, High Var, Convex	$ 855	3.3%	$ 163	0.6%
LSHVCAVE	Low Supply, High Var, Convex	$ 953	3.5%	$ 379	1.4%
HSLVCAVE	High Supply, Low Var, Convex	$ 64	0.3%	$ 43	0.2%
MSLVCAVE	Medium Supply, Low Var, Convex	$ 42	0.2%	$ 13	0.1%
LSLVCAVE	Low Supply, Low Var, Convex	$ 127	0.4%	$ 39	0.1%
Grand Total	Average	$ 630	3.1%	$ 236	1.2%

experiment were supported; in every case there was an inventory level in a tactical setting that was preferred to the strategic, stead state optimum, and any effort of inventory recovery to the strategic levels incurred a cost to the objective function. The mean profit improvement of fortuitous tactical inventory deviations from strategic levels is $491.67 (2.0%) with a standard deviation of 1239.3. Inventory level recovery costs on average $544.14 (2.2%) with a standard deviation of 1426.23.

6.3. Optimal Recovery Time

To the extent that the reactionary model is inconsistent with the strategic cycle, it is not sustainable in the long run; it is benefited by the lack of a constraint on the ending inventory each week tying it to the start-of-week inventory. The fact that there is a cost of recovery to strategic, but fortuitous deviations from long run optima is

not sustainable gives rise to the question, "What is the optimal recovery path?" We conducted a sensitive analysis with respect to one dimension of this problem; given an initial surplus of inventory, how quickly should the inventory be reduced to the strategic optimum under different costs of inventory? We reformulated the single-week model as a 6-week model with repeating demand pattern, thus we could remove the end of week constraint in the recovery model, allowing the number of days to return to strategic targets to be endogenous, rather than imposed by constraint by week's end.

Given an initial inventory of 100 units (in this example a shock which causes an excess inventory of 57 units over the strategic target of 43), we evaluated how many days pass before inventory returns to the optimal strategic targets. The answer depends on the cost of the inventory excess, and the opportunity cost of recovery. **Figure**

4 shows the path to recovery under the various scenarios. In the low cost of inventory case (Invcost ≤ 5), the optimal recovery interval was 15 days; for higher inventory costs, the optimal recovery interval was 8 days. In every case, despite having the same total days to recovery, over the recovery interval, the deviation from the long run target inventory was smaller as inventory costs were higher.

6.4. Multiple Supply Shocks

To this point, we have considered a single deviation from strategic optimum, and the managerial options for adjusting to it. The far more common case is for repeated, daily deviations from planned inventory levels that result from regular deviations from planned supply of and demand for resources.

We designed and experiment with Poisson arrivals of demand around a daily mean demand. Similar to the optimal recovery path experiment, we generated instances of supply and demand over a six week period. We compared the repeating, strategic optimum inventories to the tactical level, reactive inventories in the face of random demand, as shown in **Figure 5**.

Because daily arrivals were random, optimal tactical deviations from planned inventory were pervasive and regular. More importantly, any cost incurred to regain the

strategic target is likely in vain; subsequent supply and demand shocks essentially decimate any anticipated benefits of being on target. In this case, the strategic optimum targets provides even less value as a managerial target. Because a manager never knows what tomorrow will bring, efforts to manage to a target based on expectations prove costly and unrewarding. In short, in the face of regular and pervasive supply and demand shocks, strategic targets have little or no role in tactical decision making.

7. Managerial Implications

The managerial implications from this research are palpable. Strategic models like those described in the literature review such as [8,13,14] propose strategic models as the basis for managing various fleets. However.front line decision makers who tend to be "short sighted"—maximizing current profits while eschewing future opportunities—may be more rational than the strategic modeling results would imply. Because of the real explicit and opportunity costs of managing to an inventory target, and the uncertainty of future conditions, a manager might rationally sacrifice uncertain future benefits for near term gain. In the case of multiple supply and demand shocks, not only is there a cost of recovery, but a successful recovery is not likely to improve future profit expectation.

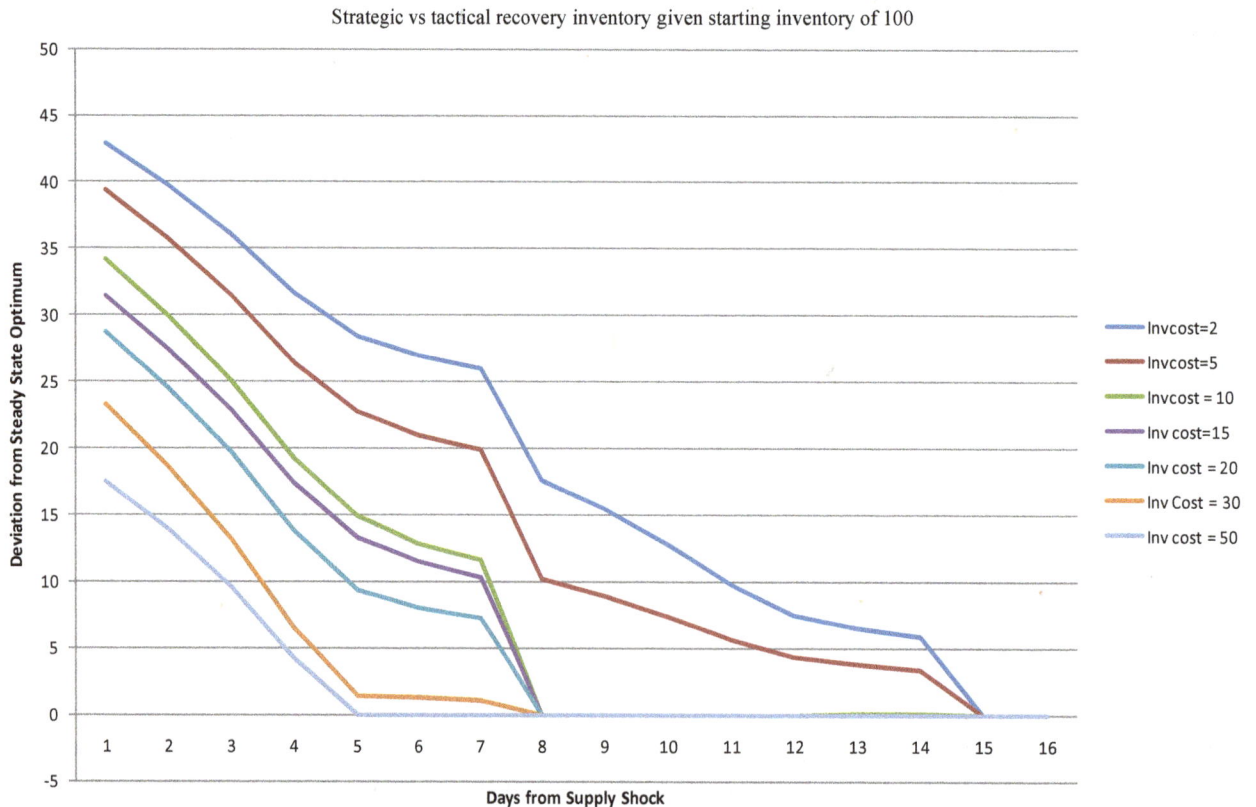

Figure 4. Strategic vs tactical recovery inventory given starting inventory of 100.

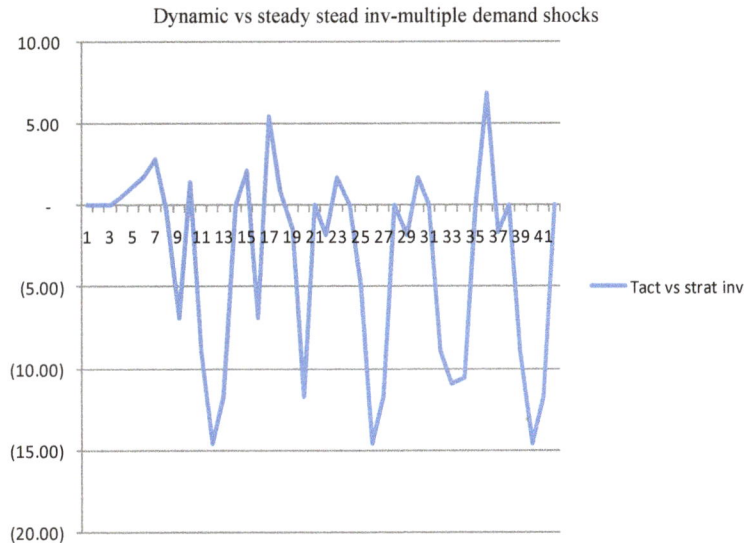

Figure 5. Dynamic vs steady stead inv-multiple demand shocks.

Simply, a manager should not be held accountable for managing fleet inventory relative to some long-term ideal or target.

7. A Strategy for Combining Tactical and Strategic Models

Given our finding that strategic model targets are not relevant in a tactical setting, and tactical model results may not be sustainable on a continued basis, what is the appropriate course of action? Simply, the strategic and tactical optimum inventories represent bounds on the optimal inventory. We suggest the optimal target inventory level rests in the interval between tactical and strategic optima; the actual optimum depends on the cost of inventory and the cost of adjustment. We suggest that rather than a target inventory level, a target inventory range is a better goal; the range is determined by the interval between the tactical optimum for each day (given current conditions), and the strategic optimum (given common long run conditions).

Alternatively, replicating the modeling horizon with tactical values in the first interval and strategic values in subsequent intervals allows the model to endogenously arrive at an optimal recovery period. Any use of either strategic or tactical modeling paradigm in isolation will likely lead to errant managerial action. It should be noted that any managerial action geared towards managing to that target is tempered by future uncertainty of supply and demand.

8. Conclusions

In this research, we evaluate the differences in strategic and tactical modeling paradigms with highly cyclical supply and demand patterns. While these modeling approaches have both been widely used in the literature, the difference in the model recommendations and managerial implications have not been explored.

We find that in some cases, deviations from strategic optima may in fact be advantageous. That is, the strategic optimal target stock levels are not optimal in an execution setting. Thus, care should be taken when "managing to a target" that is well meant and derived optimally, but can be misdirected effort, actually creating additional costs and foregone profits.

Simply, we are faced with a paradox that strategic and tactical model recommendations do not necessarily match. While having and idea of strategically "where you want to be" is important to long run profitability and operational feasibility, managing too strictly to these targets can be shown to be suboptimal. Thus, a coordinated blend of the two approaches is required. We recommend an appropriate mix of the two model regimes: Each modeling paradigm sets a boundary on optimal operational parameters. Strategic models set long policy, tactical models to set optimal behavior given current conditions and the long run strategic targets. Any achieved value in this range is acceptable; which target to pursue more aggressively depends on the relative costs of adjustment and opportunity costs of straying from a strategic target.

This research focused on a single perturbation that drives a deviation from strategic targets; future research might examine more fully the managerial implications of persistent supply and demand shocks on the role strategic modeling in a highly stochastic setting.

REFERENCES

[1] J. Roy and T. Crainic, "Improving Intercity Freight Routing with a Tactical Planning Model," *Interfaces*, Vol. 22,

No. 3, 1992, pp. 31-44.

[2]   J. Cordeau, P. Toth and D. Vigo, "A Survey of Optimiza-
      tion Models for Train Routing and Scheduling," *Trans-
      portation Science*, Vol. 32, No. 4, 1998, pp. 380-404.

[3]   D. Huisman, L. Kroon, R. Lentink and M. Vromans, "Op-
      erations Research in Passenger Railway Transportation,"
      Statistica Neerlandica, Vol. 59, No. 4, 2005, pp. 467-497.

[4]   M. F. Gorman, D. Sellers and D. Acharya, "CSX Railway
      Cashes in on Optimization of Empty Equipment Distribu-
      tion," *Interfaces*, Vol. 40, No. 1, 2010, pp. 5-16.

[5]   M. F. Gorman, K. Crook and D. Sellers, "North American
      Freight Rail Industry Real-Time Optimized Equipment
      Distribution Systems: State of the Practice," *Transporta-
      tion Research Part C*, Vol. 19, 2011, pp. 103-114.

[6]   W. B. Powell and T. A. Carvalho, "Real-Time Optimiza-
      tion of Containers and Flatcars for Intermodal Opera-
      tions," *Transportation Science*, Vol. 32, 1998, pp. 110-
      126.

[7]   H. Sherali, E. Bish and Z. Xiaomei, "Polyhedral Analysis
      and Algorithms for a Demand-Driven Refleeting Model
      for Aircraft Assignment," *Transportation Science*, Vol.
      39, No. 3, 2005, pp. 349-366.

[8]   R. K. Ahuja, J. Liu, J. B. Orlin, D. Sharma and L. A.
      Shughart, "Solving Real-Life Locomotive-Scheduling
      Problems," *Transportation Science*, Vol. 39, No. 4, 2005,
      pp. 503-517.

[9]   M. Lübbecke and U. Zimmermann, "Engine Routing and
      Scheduling at Industrial In-Plant Railroads," *Transporta-
      tion Science*, Vol. 37, No. 2, 2003, pp. 183-197.

[10] Y. Ileri, M. Bazaraa, T. Gifford, G. Nemhauser, J. Sokol,
      and E. Wikum, "An Optimization Approach for Planning
      Daily Drayage Operations," *Central European Journal of
      Operations Research*, Vol. 14, No. 2, 2006, pp. 141-156.

[11] A. Erera, B. Karacık and M. Savelsbergh, "A Dynamic
      Driver Management Scheme for Less-than-Truckload Carr-
      iers," *Computers& Operations Research*, Vol. 35, No. 11,
      2008, pp. 3397-3411.

[12] B. Vaidyanathan, K. Jha and R. Ahuja, "Multicommodity
      Network Flow Approach to the Railroad Crew-Schedul-
      ing Problem," *IBM Journal of Research & Development*,
      Vol. 51, No. 3-4, 2007, pp. 325-344.

[13] M. F. Gorman, "Intermodal Pricing Model Creates a Net-
      work Pricing Perspective at BNSF," *Interfaces*, Vol. 31,
      No. 4, 2001, pp. 37-49.

[14] D. Adelman, "Price-Directed Control of a Closed Logis-
      tics Queueing Network," *Operations Research*, Vol. 55,
      No. 6, 2007, pp. 1022-1038.

[15] M. F. Gorman, "Hub Group Implements a Suite of OR
      Tools to Improve Operations," *Interfaces*, Vol. 40, No. 5,
      2010, pp. 368-384.

[16] A. Schaefer, E. Johnson, A. Kleywegt and G. Nemhauser,
      "Airline Crew Scheduling Under Uncertainty," *Trans-
      portation Science*, Vol. 39, No. 3, 2005, pp. 340-348.

[17] A. Jarrah, J. Goodstein and R. Narasimhan, "An Efficient
      Airline Re-Fleeting Model for the Incremental Modifica-
      tion of Planned Fleet Assignments," *Transportation Sci-
      ence*, Vol. 34, No. 4, 2000, pp. 349-363.

[18] R. E. Hughes and W. B. Powell, "Mitigating End Effects
      in the Dynamic Vehicle Allocation Model," *Management
      Science*, Vol. 34, No. 7, 1988, pp. 859-879.

17

Tracking National Household Vehicle Usage by Type, Age, and Area in Support of Market Assessments for Plug-In Hybrid Electric Vehicles

Yan Zhou, Anant Vyas, Danilo Santini
Argonne National Laboratory, Argonne, USA

ABSTRACT

Plug-in electric vehicle (PHEV) technology is seen as promising technology for reducing oil use, improving local air quality, and/or possibly reducing GHG emissions to support a sustainable transportation system. This paper examines the usage of household vehicles to support assessment of the market potential of plug-in hybrid electric vehicles (PHEVs), the higher purchase price of which requires high usage rates to pay off the investment in the technology. According to the 2009 National Household Travel Survey (NHTS), about 40% of household vehicles were not used on the survey travel day [1]. This study analyzed household vehicle use and non-use by vehicle type, age, area type (metropolitan statistical area [MSA] and non-MSA), and population density. Vehicles used on survey day with or without a reported travel time and distance in the survey are considered "vehicles used". All others are referred to as "vehicles not used". We divided the "vehicles not used" into three categories: 1) left at home while other household vehicles were used; 2) not used because travelers used other modes; and 3) no household trips. The "vehicle used" consists of two categories: 1) those with distance and time data and 2) those with no travel data. Within these five categories, vehicles were subdivided according to four vehicle types: car, van, SUV, and pickup. Each vehicle type was further subdivided in two age groups: 10 years or less ($\leq$10) and more than 10 years (>10). In addition, vehicle usage was compared in both MSAs and non-MSAs and during weekdays and weekends. Results indicate that most vehicles—especially pickups—are not used because the households own and use other vehicles. Moreover, SUVs—especially newer SUVs ($\leq$10 years)—are the most utilized vehicle type and should be strongly considered as a primary vehicle type for PHEVs, in addition to cars.

Keywords: Vehicle Usage; PHEV; NHTS

1. Introduction

According to the Energy Information Administration (EIA), in 2009 the transportation sector was responsible for 70% of petroleum consumption and 33% of GHG emissions in the United States [2]. Within the transportation sector, light-duty vehicles account for nearly 60% of its petroleum consumption [3]. By reducing fossil fuel use per mile of service delivered, sustainable transport leads to greater energy security and reduced greenhouse gas (GHG) emissions. Among several initiatives supported by the US government, one is to diversify transportation energy sources by using electricity to drive light-duty vehicles.

Although specifics are important, at a broad conceptual level, the technology for plug-in hybrid electric vehicles (PHEVs) is similar to that in regular hybrid electric vehicles (HEVs), except that they employ bigger batteries, which are recharged through electric vehicle supply equipment by drawing electricity from the grid.

Regarding specifics, there are three basic powertrain configurations and operational capabilities: series, parallel, and series-parallel (power-split) [4]. When the grid-to-vehicle series operational capability is implemented in a PHEV, stored grid-supplied battery energy propels the vehicle initially. This phase of operation is called charge-depleting (CD) operation. The more power and energy that are available in the battery pack, the greater the vehicle's ability to operate all electrically during CD. All PHEVs considered in this paper have some degree of ability to operate all electrically with grid electricity. The parallel and series-parallel tend to have much less power than the series configuration and are more likely to operate with both the engine and battery simultaneously providing power during CD operation. The three designs are distinguished from one another with respect to their operational capabilities when the engine comes on.

In a series PHEV, only the electric motor directly drives the wheels; the engine does not. There is a second electric machine, which operates as a generator (an electric machine can be reversed in rotational direction and can operate either as a generator or motor). The series configuration is an engine-to-generator-to-motor-to-wheels pathway. After the battery energy is depleted to a predetermined level, an internal combustion engine (ICE) turns the generator, which supplies current to the electric motor, which then rotates the vehicle's drive wheels. When excess electric energy is available, the generator recharges the battery pack. Since it must do all the work of moving the vehicle, the electric motor of a series PHEV must be larger than that of the other two types.

By adding the conventional engine-to-transmission-to-wheels pathway, the parallel design can simultaneously transmit power to the drive wheels from both the internal combustion engine and the battery. Compared to the series PHEV, the parallel operational capability primarily uses the conventional mechanical link from the engine to the wheels, eliminating the ability of the engine to support simultaneous series operation. This restriction allows the parallel PHEV to use only one electric machine (motor/generator) of less power than in the series configuration, thereby reducing cost. By using two electric machines, the series-parallel design has the flexibility to operate the onboard engine to support either series or parallel mode, or both simultaneously. A specialized mechanical step splits engine power into two pathways: one to the wheels and another to the generator. For this reason, the series-parallel system is called a "power split". Compared to the series PHEV, both the parallel and series-parallel are able to use much less electric machine and battery pack power, thus cutting costs. For a given level of acceleration capability, the series-parallel is more expensive than the parallel, but it is more efficient.

A PHEV travels its initial miles by making use of energy from the grid, which has been stored in the battery. If the power of the battery pack and electric machines is sufficient, propulsion may happen all electrically. However, as electric power capability is reduced to cut PHEV cost (as in parallel or parallel-series configurations), the battery power must often be supplemented by engine power during CD operation. Thus, PHEV designs will vary with respect to the share of electricity and fuel used as the battery pack is discharged. All PHEVs operate in "hybrid mode" (as an HEV) on fossil fuels once the battery is depleted, although they can differ in the way they do so, according to powertrain configuration, as previously discussed.

Because electricity is generated through the use of coal, nuclear, natural gas, hydro, and wind sources, widespread acceptance of PHEVs could diversify energy sources used in the transportation system. It is fairly well understood that the reduction in petroleum use by PHEVs increases with a corresponding increase in their onboard energy storage, which increases nonlinearly (less well known) with the size of the employed batteries. In many (but not all) cases, the increase in vehicle weight associated with bigger batteries partially offsets the potential reduction in petroleum use by PHEVs during engine operation. Another poorly understood attribute is that the power of the battery pack and electric machines is an important factor in the ability of PHEVs to electrify miles. Also in a nonlinear fashion, the higher the electrical power, the more electricity is used per mile of CD operation. More power means fewer miles until depletion of the pack and lower fossil fuel use during depletion, which translates into greater electrification potential. Clearly, the PHEV technology can cover a wide variety of options with respect to technical attributes, such as the battery chemistry, the amount of grid electricity that can be stored in the battery, and the powertrain and fuel choices. In addition, the driving behavior of consumers, such as driving aggressiveness and daily travel distance, could also significantly affect the energy use and the GHG effects of PHEVs.

Plug-in electric drive's ability to eliminate oil use has become increasingly attractive since 2007 as 1) technical and economic feasibility has improved; 2) oil prices have increased significantly on average; and 3) oil prices have become more volatile. Starting in 2010, almost all of the major vehicle manufacturers offered—or planned to soon offer—PHEVs for sale to the mass market. Although the broad PHEV technology offers great promise, many questions about details remain unanswered. This paper examines the use of household vehicles to support assessment of the market potential of the many different PHEV technology options.

2. Contribution

To our best knowledge, no similar study has been conducted to assess the vehicle utilization by demographic factors. We have searched Google Scholar and TRID (http://www.trid.trb.org/) by using the following key words: vehicle utilization rates and vehicle usage rates.

New vehicle technologies are expensive at the early stage of implementation and require high usage to pay off. This paper helps decision makers and manufacturers identify the proper market niche of early vehicle models on the basis of the usage rate. Moreover, this paper details the reasons why vehicles were not used on travel days, which provides alternative perspectives to identify

Tracking National Household Vehicle Usage by Type, Age, and Area in Support of Market Assessments for
Plug-In Hybrid Electric Vehicles

175

potential markets for PHEV powertrains.

3. Searching for High-Usage Vehicles

This paper examines usage of household vehicles—by type, age, and area—to support assessment of the market potential of PHEVs. However, the information obtained in the study is applicable to any costly powertrain that sharply reduces fuel costs, whether by use of a less-expensive fuel or by higher efficiency. High usage rates are needed for such technologies to pay off. The paper complements a paper presented in 2011 [5], in which the 2001 National Household Transportation Survey (NHTS) was used to examine vehicle records, separating those records into groups of vehicles 1) older and newer than 10 years of age, and 2) above and below 50 miles of use per day.

One issue for the purchaser of such vehicles is the warranted life of the battery pack, which is at present eight years for the Nissan Leaf and Chevrolet Volt. If a pack replacement were necessary, and diminished rates of use were anticipated after the warranty period, the costs might be prohibitive and lead to a need to scrap the vehicle. Accordingly, the target market was drivers who used their vehicles intensively (high miles per day and many days per year), so that such vehicle would otherwise have its end of useful life at about the same time as the end of the useful life of the PHEV or EV battery (*i.e.*, about 8 years). Another consideration was battery "cycle" life, which is believed to be about 3000 cycles. However, Vyas *et al.* assumed 5000 cycles to be possible by the year 2020 [6]. An assumption of charging overnight once for 90% of days for 10 years would lead to 3285 cycles, and so vehicles with a pack cycle life of 5000 cycles could be charged more than once per day on average, but not twice each day. The issue of charging a second time during the day has been addressed in Vyas *et al.* 2009 [7] and Elgowainy *et al.* 2012 [8]. Because calendar life (years of pack life) and cycle life have different causal mechanisms, a capability for greater cycle life may not lead to longer calendar life. Thus, there may be an incentive to use expensive PHEVs and EVs as intensively as possible, particularly if 5000 cycles or more can be obtained within a 10-year calendar life.

One aspect of use that has not been investigated is the proportion of days that a vehicle is in operation. We are aware that the 90% assumption used in the computations described above is optimistic and probably not typical. This paper is intended to address that assumption in the context of the 10-year life break point assumptions made previously [5]. In this paper, we analyze the probability of daily use to enhance our understanding of the market for personal-use PHEVs and EVs.

According to the 2009 NHTS [1], about 40% of vehicles on the survey travel day are reported as "not used".

At first glance, it appears that many households do not travel by personal vehicle. As shown in **Table 1**, of the 60.9% of vehicles used, 19.5% are vehicles greater than 10 years old, while 41.4% are relatively new vehicles (≤10 years old). However, of the 39.4% vehicles not used, percentages of old vehicles and new vehicles are almost identical. Besides vehicle age, there are many other factors that can affect vehicle usage, such as vehicle type, residential area type, and travel day.

Emerging new vehicle technologies offer opportunities to reduce the US transportation sector's dependence on petroleum and possibly reduce greenhouse gas emissions. Prior experience with hybrid electric vehicle technology has shown that new technologies, such as PHEVs, will first be introduced in the passenger car [9]. Once the technologies are successfully introduced in passenger cars, they may be made available in other vehicle types. Analysis of the types of vehicles households use more frequently is needed to assist transportation analysts and decision makers. Knowing the vehicle usage by type, location, travel day, and population density would be helpful in estimating the benefits of new-technology vehicles in terms of the energy use and emission reductions associated with daily travel.

Many factors influence daily household vehicle usage; this study focuses on the following: vehicle type, vehicle age, travel day, residential location, and population density. Four vehicle types—car, van, SUV, and pickup—were selected. Vehicle age is divided into two groups: less than or equal to 10 years old (≤10 years) and greater than 10 years old (>10 years), while residential location is subdivided into metropolitan statistical area (MSA) and non-MSA. Population density in square kilometers includes five groups, ≤386, 387 - 1544, 1545 - 3860, 3861 - 9650, and >9650.

4. Identifying Unique Used/Non-Used Vehicle Records

The 2009 NHTS was conducted from March 2008 through May 2009. Information relating to sampled households, household members, vehicles owned, and travel during one day was collected. The survey was designed to collect travel data during a typical year and on all seven days of the week, including all holidays. The household sample consisted of a random national sample

Table 1. Pattern of household vehicle usage.

Category	≤10 Years	>10 Years
Used (%)	41.4	19.5
Not Used (%)	19.7	19.4
Percent of Used	68.0	32.0
Percent of Total	61.6	38.9

and several add-on samples (comprising additional households in selected areas where local planning entities paid to have the sample size expanded). Care was taken to assign weights such that the final sample would provide estimates representative of the national population. The NHTS dataset contains data for 150,147 households that own 309,163 vehicles. Various files provide detailed data relating to households, persons, vehicles, and daily (travel day) trips. This study utilized travel day trip and vehicle files, which also contain data related to characteristics of households and household members. Vehicles in the travel day file are sampled for only one day, making it impossible to track the weekly, monthly, or seasonal behavior of any single vehicle.

In this analysis, vehicles with or without reported travel time and distance are considered "vehicles used". All others are called "vehicles not used". We subdivided the "vehicles not used" into three sets: 1) left at home; 2) used other modes; and 3) no trips. The first "vehicle not used" set represents vehicles that were left at home while residents drove other household vehicles. The second set, "used other modes", represents the vehicles left at home while household members used other travel modes, such as public transit, carpooling, bicycle, walking, or traveling as passengers in someone else's vehicle. The last set, "no trips", represents vehicles left at home because the household members did not make any trips. Within the "vehicles used" group, vehicles were subdivided as "with travel data" and "without travel data". The "without travel data" set represents vehicles that were used for travel, but because the respondents did not report travel distance or time, these vehicles are often excluded from travel-related analysis. For these five usage sets, vehicles were further subdivided into four vehicle types: car, van, SUV, and pickup. Each vehicle type was further subdivided into two age groups: 10 years or less (≤10) and over 10 years (>10). Finally, the vehicle usage was compared by household location: MSA and non-MSA.

We first created three subsets of day trip file records based on the trip information: *driver-set* 1, *driver-set* 2, and *other*. *Driver-set* 1 includes the driver trip records with reported travel distance and time, while *driver-set* 2 includes driver trip records without travel distance and/or time. The driver trip means the survey responder is the driver of this particular trip. The *other* subset contains all the non-driver trips. Next, we created unique files out of the first two files containing one record by household identification code (HOUSEID) and vehicle number (VEHID). Because a household may report many daily trips with detailed information for the same vehicle but may not do so for some trips, a few vehicles ended up in both of the *driver* files. We deleted the duplicates in *driver-set* 1 and *driver-set* 2 so that vehicle numbers are not duplicated.

Figure 1 shows the procedures used to match the vehicle file with the three trip files to identify unique "used" or "non-used" vehicle records. First, we matched the vehicle file with a trip file *driver-set* 1, without any duplicate records, by HOUSEID and VEHID. The matched records are the vehicles used with reported travel data. Care was taken to separate the non-matched vehicle records. Next, the first non-matched vehicle file was further matched with *driver-set* 2 by HOUSEID and VEHID. The matched records for this step are the vehicles used without reported trip distance and/or time, while the non-matched file includes the vehicles not used. The non-matched file from these two steps was matched with the combined file (*driver-set* 1 and *driver-set* 2) by unique HOUSEID only; the matched records are the vehicles left at home while household members used other vehicles to travel. Next, the non-matched file generated in this step was matched with the *other* file by HOUSEID; the matched records are the vehicles owned by the household members who traveled by using other modes (e.g., public transit, bicycle). Finally, the non-matched file of the last step includes vehicles that were not used because the household members did not travel at all.

5. Identifying Patterns in Household Vehicle Usage Data Records

The total weighted number of vehicles included in the 2009 NHTS vehicle file is 211,501,318. **Table 2** shows the percentage distribution by the five usage/non-usage categories. As the table shows, "left at home" is the largest "not used" vehicle group at 28.8%. All percentages were calculated by using weighted NHTS numbers.

Further examination of the age of the vehicles "left at home" was conducted; the distribution of vehicles ac-

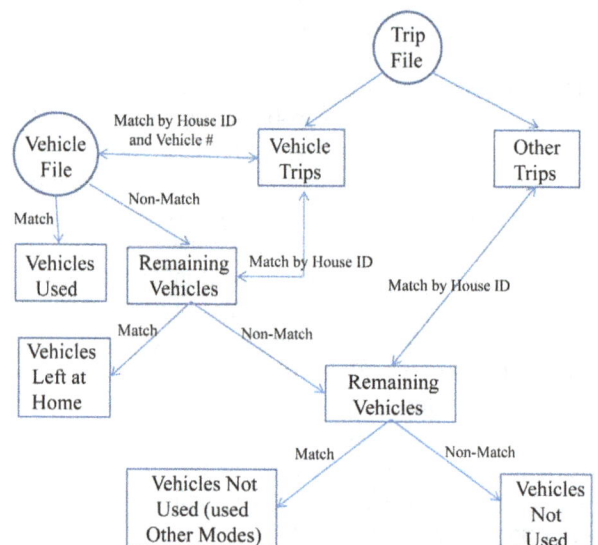

Figure 1. Procedures for vehicle usage identification.

Tracking National Household Vehicle Usage by Type, Age, and Area in Support of Market Assessments for Plug-In Hybrid Electric Vehicles

177

Table 2. Use and non-used household vehicles by vehicle age.

Category	≤10 Years	>10 Years	All Ages
Used (%)	67.8	50.1	60.9
Left at Home (%)	23.0	36.9	28.4
Other Modes (%)	1.7	1.8	1.7
No HH Trip (%)	7.5	11.1	8.9
Total (%)	100	100	100

cording to the five categories within the two age groups is also shown in **Table 2**. We are not surprised to find that older vehicles (>10 years) are more likely to be "left at home" than newer vehicles (≤10 years). More importantly, the probability of use of a vehicle drops very sharply from 68% for the ≤10-year age group to 50% for the >10-year age group.

Vehicle age might not be the only reason that the vehicles were not used. For the same vehicle type within a similar age group, residential location might also affect vehicle usage. Therefore, the vehicle use/non-use pattern was analyzed by age group, vehicle type, *and* residential location type. The residential location was subdivided as in an MSA and in a non-MSA. **Table 3** shows the resulting distribution; percentages of each location type add up to 100%. As shown in **Table 3**, the passenger car has the largest utilization rates, both in and out of an MSA, especially the newer cars (≤10 years). The usage rate drops more sharply from the ≤10 years to >10 years groups in an MSA. Newer (≤10 years) SUVs are more frequently used than older (>10 years) SUVs—almost triple the use rate for both MSA and non-MSA. No other vehicle type shows this kind of usage pattern. This is partly because SUVs have been the most rapidly growing class of vehicle.

Figure 2 shows vehicle use and non-use by vehicle and area type such that all percentages add up to 100%. Compared with those in non-MSAs, people in MSAs tend to own more newer (≤10 years) cars than other vehicle types. In rural areas, the difference between cars and other vehicle types is not that significant. SUVs are the second most common type of vehicle, far ahead of pickups in MSAs, although there are slightly fewer SUVs outside MSAs. Note that a significant percentage of pickups are left at home, especially the older ones. Cars constitute the most common type of vehicle: 105,595,553 of the 211,501,318 vehicles in the survey. As stated previously, SUVs are the second most prevalent vehicles at 41,116,312 units, pickups are third at 37,738,450, and vans are the least prevalent at 17,356,299. Considering that the total number of pickups is much lower than the number of cars in the survey, their share of vehicles "left at home" is surprisingly high. Pickups are clearly the most likely vehicle type to be "left at home" when other

vehicle types are available for travel. There are more old (>10 years) pickups in rural areas; however, they are more likely to be among the vehicles "left at home". This shows that people in rural areas find the load-carrying attributes of pickups valuable, but use them only when needed.

Although SUVs and vans have high usage rates, the total number of these two vehicle types is much smaller than the number of cars. Since the car is the most prevalent vehicle, more of them are used than any other vehicle class. However, the car is not the most used type of vehicle. We analyzed the pattern of use and non-use by individual vehicle type; the ≤10-year age group is shown graphically in **Figure 3**. The total usage percentage for each vehicle type adds up to 100%. Among the newer (≤10 years) vehicles, 71.4% of cars, 73.7% of SUVs, 57.7% of pickups, and 73.8% of vans are used. SUVs and vans are the most-used vehicle type with a usage rate of almost 74%, while the usage rate for pick- ups is the lowest. One possible reason for higher usage rates for vans and SUVs is that they may be used more frequently for carrying family members on weekends and vacations. Even so, the difference is not large.

The distribution of "used" and "not used" newer (≤10 years) vehicles inside and outside a MSA is shown in **Table 4**. Different from **Figure 3**, the sum of percentages for each area type (MSA or Non-MSA) adds up to 100%in **Table 4**. The overall use share for newer cars (**Table 4** column total) is much higher in an MSA than in a non-MSA. Conversely, pickup trucks are used about half as much by MSA residents as non-MSA residents. MSA and non-MSA use shares of newer vans and SUVs are very similar.

Besides residential location, we also examined the vehicle usage pattern by population density, as shown in **Figure 4**. Population density per square kilometer was divided into five groups: ≤386, 387 - 1544, 1545 - 3860, 3861 - 9650, >9650. For newer vehicles (≤10 years old), the utilization rate of all vehicle types decreases as the density increases, as does ownership. SUVs, vans, and pickup trucks have higher shares of ownership by density category than cars only in areas with a population density of less than $386/km^2$. Pickup ownership in areas of less than $386/km^2$ is over 50% of total pickup truck ownership. Another finding to note is that more pickups are left at home than used for all of the population density groups, except in the highest-density areas.

Compared to the MSA vs. non-MSA breakdown, the relative probability of ownership of SUVs and minivans (*share* of total ownership of the type) appears to be greater than the relative probability of ownership of cars in the two lowest-density categories. Either way it is examined (**Table 4** or **Figure 4**) the SUV is a strong second to the car, with its proportional share relative to cars

Table 3. Percentages use and non-use of household vehicles by vehicle age and type.

Location	Vehicle Type*	Age (years)	Used-Data (%)	Used-No Data (%)	Left at Home (%)	Other Modes (%)	No HH Trip (%)
MSA	Car	≤10	22.8	0.1	5.8	0.5	2.6
		>10	13.0	0.2	6.7	0.4	2.8
	SUV	≤10	11.5	0	2.7	0.3	1.0
		>10	2.8	0	1.8	0.1	0.5
	Pickup	≤10	4.7	0.1	2.6	0.1	0.5
		>10	3.1	0	3.4	0.1	0.7
	Van	≤10	4.0	0	0.9	0.1	0.3
		>10	2.0	0	1.1	0.1	0.3
	All	All	63.9	0.4	25.1	1.8	8.9
Non MSA	Car	≤10	15.6	0	4.8	0.3	1.7
		>10	10.3	0.1	6.6	0.5	2.9
	SUV	≤10	9.2	0	2.5	0.2	1.0
		>10	3.2	0	2.5	0.1	0.6
	Pickup	≤10	7.3	0	4.9	0.2	1.1
		>10	6.2	0	8.5	0.3	1.7
	Van	≤10	3.1	0	0.9	0.1	0.3
		>10	1.6	0	1.1	0.1	0.4
	All	All	56.5	0.1	31.7	1.9	9.7
All Type	Car	≤10	21.3	0.1	5.7	0.4	2.4
		>10	12.4	0.1	6.7	0.5	2.8
	SUV	≤10	10.9	0	2.6	0.3	1.1
		>10	2.8	0	1.9	0.1	0.5
	Pickup	≤10	5.3	0	3.0	0.1	0.6
		>10	3.8	0	4.5	0.2	0.9
	Van	≤10	3.8	0	0.9	0.1	0.3
		>10	1.9	0	1.2	0.1	0.3
	All	All	62.3	0.2	26.6	1.9	9.0

*Vehicle type "Other" is not included in the calculation.

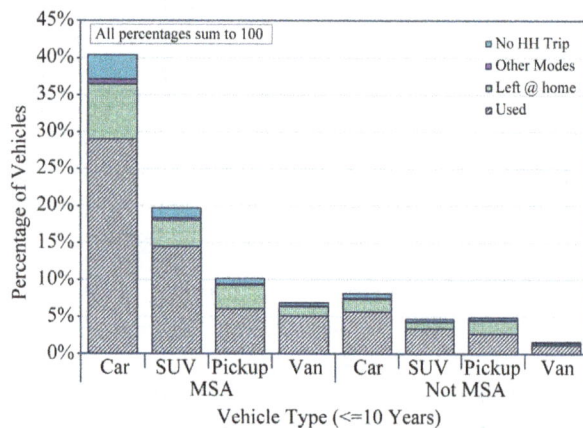

Figure 2. Use and non-use of household vehicles (≤10 years) by vehicle type in MSA and non-MSA.

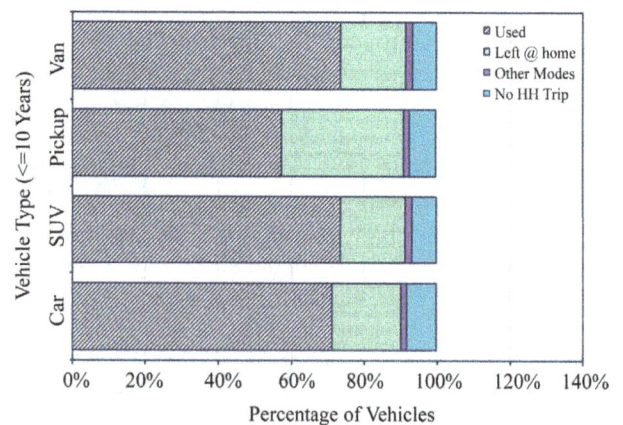

Figure 3. Pattern of use and non-use of individual vehicle type (≤10 years).

Tracking National Household Vehicle Usage by Type, Age, and Area in Support of Market Assessments for Plug-In Hybrid Electric Vehicles

179

Table 4. Percentage use and non-use of ≤10-year-old vehicles by type and location.

	MSA*				Non-MSA*			
	Car	SUV	Pickup	Van	Car	SUV	Pickup	Van
Used	37.8	18.9	7.9	6.5	29.3	17.4	13.7	5.8
Left at Home	9.6	4.5	4.3	1.6	9.1	4.6	9.2	1.6
Other Modes	0.8	0.4	0.2	0.2	0.5	0.4	0.5	0.1
No HH Trip	4.3	1.7	0.9	0.5	3.3	1.8	2.0	0.6
Column Total	52.5	25.4	13.3	8.8	42.3	24.2	25.4	8.2
Percent Used	71.9	74.3	59.4	74.1	69.4	71.9	53.8	71.4

*Vehicle type "Other" is not included in the calculation.

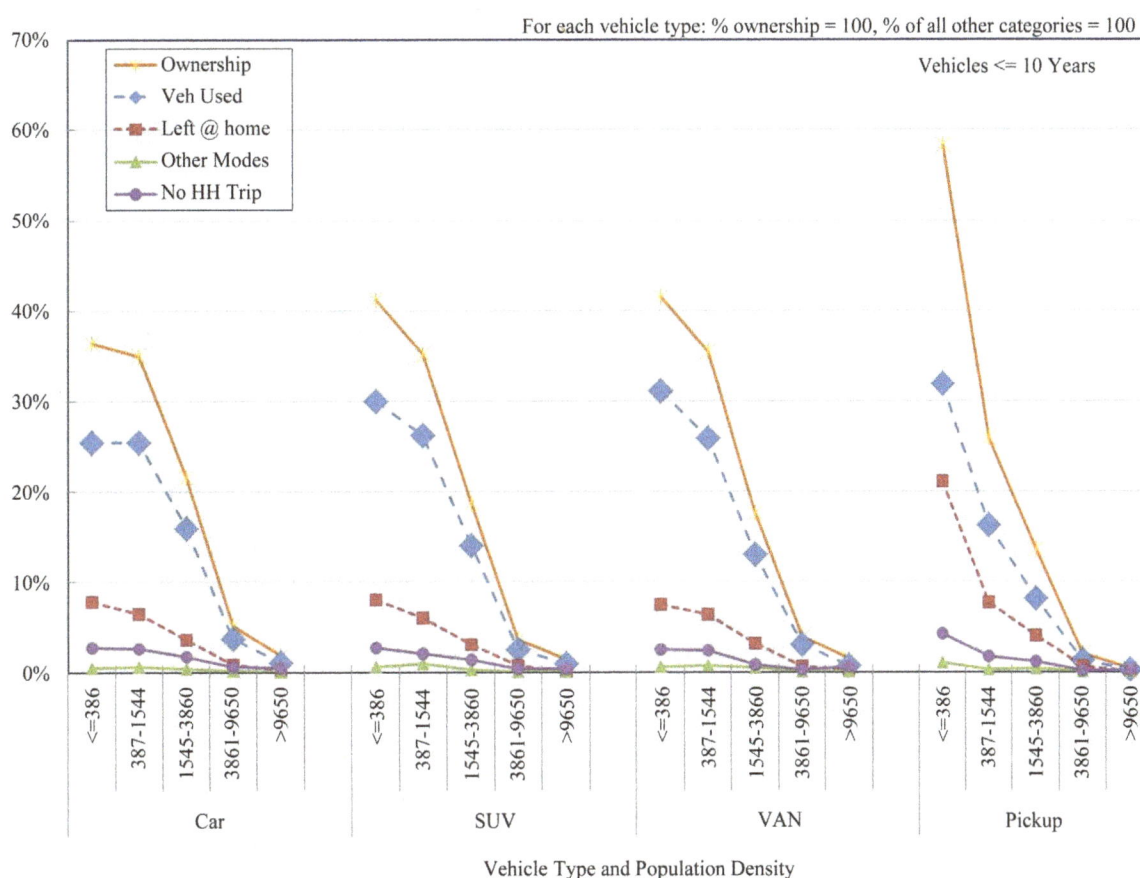

Figure 4. Vehicle usage pattern by population density.

much higher in non-MSAs and in the two lowest-density categories that we have chosen.

Earlier research done by the co-authors of this paper found that all upcoming electric-drive vehicles (including PHEVs) will be considerably more fiscally sound investments than conventional vehicles for vehicles driven more than 80 km/day at gasoline prices higher than $4/gallon [5]. Other authors have recently also reached the conclusion that high rates of vehicle use are necessary for the financial investment in PHEVs to pay off [10,

11,12]. **Figure 5** shows the regional vehicle use rates of newer vehicles (≤10 years old) for a daily travel distance of more than 80 km by type in areas with different population density. The percentages within each population density group sum to 100. The use share for cars increases monotonically with increasing population density and then drops when the density is over 9650/km². Conversely, the share of SUV use first decreases when density increases, but it increases in the highest-density group. The shares of van use are the most stable across

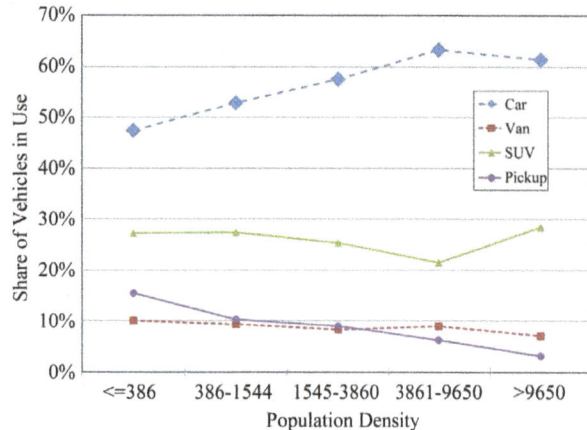

Figure 5. Vehicle (≤10 years old) usage by type and population density.

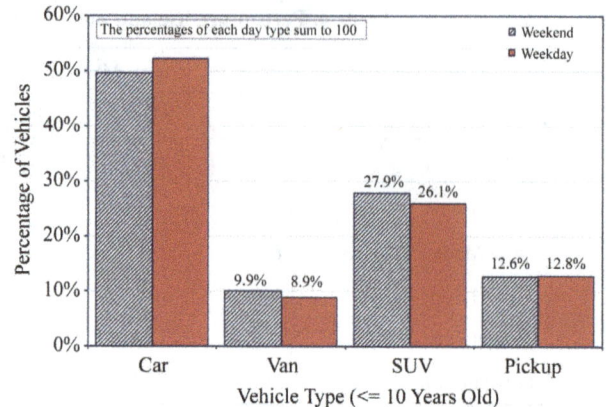

Figure 6. Vehicle usages by vehicle type on weekdays and weekend.

density categories.

As a result of the increase in share of SUVs in the highest population density category, we developed a behavioral hypothesis. We noted that there is a significant increase in the use of public transit only in this density category, which represents only a small share of US households. However, there are many cities in the world that have populations of this density. We wondered if the use pattern of vehicles might be significantly different in such places. Perhaps SUVs are purchased preferentially for this use pattern. Such thinking also led us to wonder if multiple-vehicle households that owned cars and SUVs might generally be more likely to use the SUV on the weekend, leaving the car parked.

To address these questions, we first examined the impact of travel day in terms of use/non-use on weekdays or weekends without location and population density factors. Although the direction of change for cars and SUVs was consistent with the general hypothesis, our analysis indicated that there is no statistically significant difference (through t-test) in vehicle use/non-use patterns between weekdays and weekends, for either new vehicles or old vehicles. **Figure 6** shows the usage of newer (≤10 years) vehicles by vehicle type on weekdays and weekends. A similar trend was observed for older vehicles (>10 years).

However, when considering population density, noticeable differences were found between weekday and weekend usage pattern for all four types of vehicles. Because we think potential PHEV markets will first be for cars and SUVs. **Figure 7** uses these two vehicle types as examples to demonstrate the used and non-used rates on weekends and weekdays by population density. More cars are used on weekdays than weekends among all of the population density groups. This is also true for SUVs, except for in dense urban areas. For the SUV, only 60% are used on weekdays in the highest-density category.

However, this category has a much higher SUV usage rate, about 80%, on weekends. This finding supports the hypothesis that many of the persons who purchase SUVs in dense urban areas intend to use them on the weekends, but much less during the week. Conversely, those who purchase cars consistently intend to use them during the week more frequently than on the weekend. Within the three lowest-density-use categories, the greater overall use rate of SUVs appears to be due to the greater probability that they will be used on the weekend than will cars. So, since consistent daily use is an important factor in the financial viability of PHEVs, the SUV is slightly favored over the car in this respect in the three lower-density areas, but not in very densely populated areas with public transit. For both cars and SUVs in the United States, the probability of daily use in the most densely populated zones is much lower than in the rest of the country. If this pattern is also prevalent elsewhere in the world, it has negative implications for cost-effective implementation of zero-emissions driving capability within the densest metro areas, where this feature is generally thought of as most desirable.

6. Conclusions

We analyzed 2009 NHTS data to more accurately predict usage rates for four major types of vehicles (cars, SUVs, vans, pickup trucks) that might adopt PHEV powertrains. To assist in investigating the types of vehicles in which the plug-in feature would be most utilized —in terms both of gross number of vehicles and rates of use per vehicle—we controlled for effects on use/non-use pattern by vehicle type, vehicle age group, household location type (MSA vs. non-MSA), and travel day (weekday vs. weekend).

Study conclusions include the following:

• Most vehicles not used in the survey are "left at home" because household members own other vehicles or because multiple household drivers ride to-

Tracking National Household Vehicle Usage by Type, Age, and Area in Support of Market Assessments for
Plug-In Hybrid Electric Vehicles

181

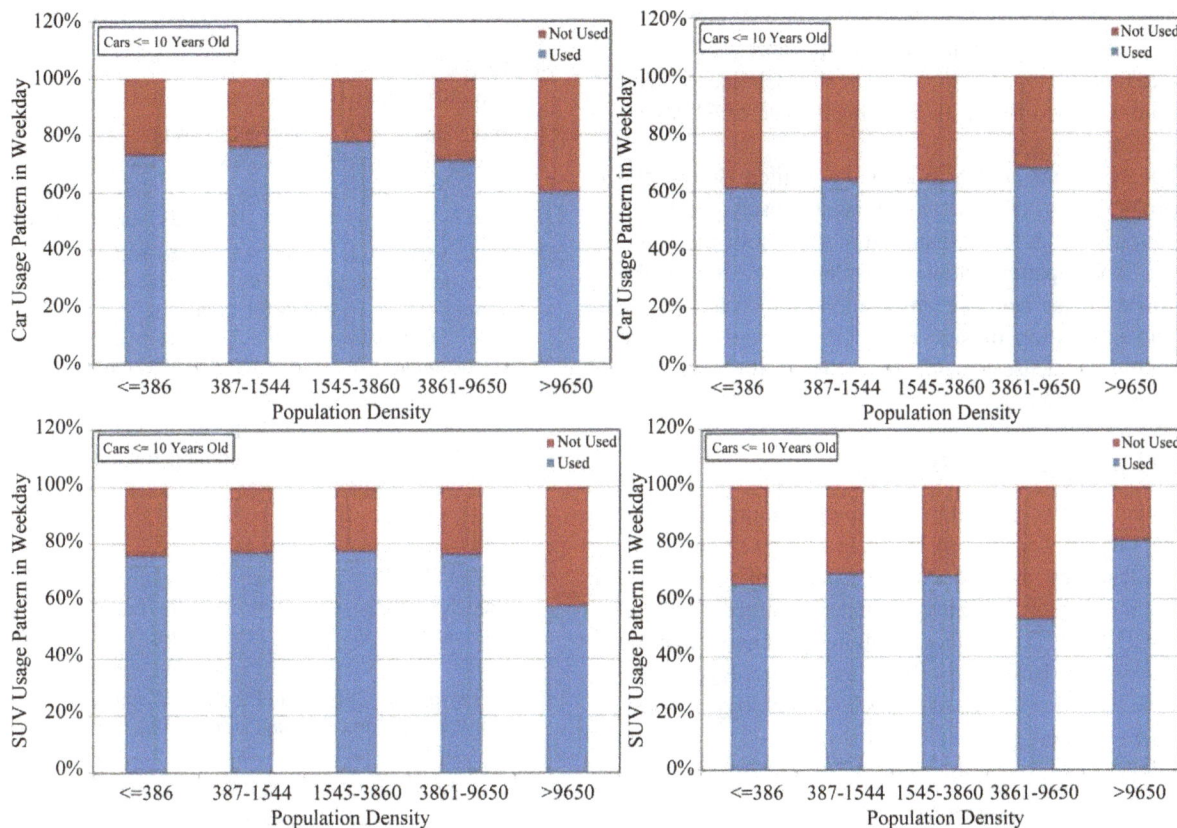

Figure 7. Car and SUV usage pattern on weekends and weekday.

gether and use another vehicle on a given day.

- Pickup trucks are most commonly an extra vehicle;
- Considering all four vehicle types together, the probability of use of a vehicle drops very sharply from 68% for the ≤10-year age set to 50% for the >10-year age set. For cars and SUVs in MSAs, the mean rates for the ≤10-year group are 72% and 74%, respectively, which are slightly higher than those in non-MSAs.
- For any of the four vehicle types, usage rates within MSAs are higher than those outside MSAs. By a slight margin over cars, SUVs and vans are the most frequently used vehicles.
- Cars, especially those ≤10 years in age, are the most-owned vehicle type, because of much higher ownership rates in MSAs.
- Pickup trucks are used much less intensively and last much longer. Their longer lifetime raises their share of the >10-year age group. Pickups are also far more likely to be "left at home" while household members use other vehicles to travel, particularly in rural areas. Thus, period required to pay off a battery pack is the longest for pickup trucks, and pickup trucks are the most likely vehicle type to outlast the calendar life of and warranty period for PHEV battery packs. Pickup truck longevity makes the odds of needing a battery replacement much higher than those of any other

class of vehicle.

- Although vans, SUVs, and cars do not differ significantly in their usage rates, the size of the car market is the largest, followed by the SUV market, which significantly exceeds the size of the van market.
- Given the findings of this analysis, considering only probability of vehicle use each day, the greatest market potential for PHEVs lies in cars and SUVs in low-density areas of MSAs (*i.e.*, suburbs). Interpreting the density results, at locations where high-rise multi-family residences result in high average population density (and the possible viability of public transit), the daily use rates of vehicles are much lower than elsewhere, reducing the probability of payback of battery packs through frequent charging.
- Since plug-in vehicles are relatively more economically attractive in MSAs, the small share of pickup trucks in MSAs shows that pickup trucks are a considerably smaller market than the more numerous cars and SUVs. Design priorities for pickups would logically carry a higher priority based on the operation characteristics outside of MSAs. Even when pickup trucks are found in MSAs, their rate of use is much less than that for other vehicle types.
- For vehicles <10 years old, SUVs in the three lowest population density categories are more consistently

used over the full week than are cars. For these density categories, car use drops off more rapidly on weekends, presumably as some households owning both cars and SUVs shift to family use of SUVs on weekends.

- For both cars and SUVs in the United States, the probability of daily use in the most densely populated zones is much lower than that in the rest of the country. If this pattern is also prevalent elsewhere in the world, it has negative implications for cost-effective implementation of zero-emissions driving capability within the densest metro areas, where this feature is generally thought of as most desirable.

7. Acknowledgements

This work was supported by the Energy Storage Program of the Vehicle Technology Program of the Office of Energy Efficiency and Renewable Energy of the United States Department of Energy, under contract # DEAC 02-06CH11357. The authors gratefully acknowledge the sponsorship of David Howell, Team Leader, Hybrid and Electric Systems, Office of Vehicle Technology, US Department of Energy.

REFERENCES

[1]   Federal Highway Administration, "2009 National Household Travel Survey," US Department of Transportation, Washington DC. http://nhts.ornl.gov/download.shtml

[2]   Energy Information Agency, Annual Energy Review, "Energy Information Administration," US Department of Energy, Washington DC, 2011.

[3]   S. C. Davis, S. W. Diegel and R. G. Boundy, "Transportation Energy Data Book," 30th Edition, Oak Ridge National Laboratory, Oak Ridge, 2011.

[4]   A. Elgowainy, J. Han, L. Poch, M. Wang, A. Vyas, M. Mahalik and A. Rousseau, "Well-to-Wheels Analysis of Energy Use and Greenhouse Gas Emissions of Plug-In Hybrid Electric Vehicles," Center for Transportation Research, Argonne National Laboratory Report ANL/ESD/10-1. http://www.transportation.anl.gov/pdfs/TA/629.PDF

[5]   D. J. Santini and A. D. Vyas, "Where Are the Market Niches for Electric Drive Passenger Cars?" *Proceedings of the 90th Annual Meeting of the Transportation Research Board*, Washington DC, 23-27 January 2011.

[6]   A. D. Vyas, D. J. Santini, M. Duoba and M. Alexander, "Plug-In Hybrid Electric Vehicles: How Does One Determine Their Potential for Reducing US Oil Dependence?" *Proceedings of the 23rd Electric Drive Vehicle Symposium* (*EVS* 23), Anaheim, 2-5 December 2007.

[7]   A. D. Vyas, D. J. Santini and L. R. Johnson, "Potential of Plug-In Hybrid Electric Vehicles to Reduce Petroleum Use: Issues Involved in Developing Reliable Estimates," *Transportation Research Record*, 2009, pp. 55-63.

[8]   A. Elgowainy, Y. Zhou, A. D. Vyas, M. Mahalik, D. Santini and M. Wang, "Impacts of Plug-In Hybrid Electric Vehicles Charging Choices in 2030," *Proceedings of the 91st Annual Meeting of the Transportation Research Board*, Washington DC, 23-27 January 2012.

[9]   A. Taylor III, "The Birth of the Prius," *Fortune*, 2006. http://money.cnn.com/magazines/fortune/fortune_archive/2006/03/06/8370702/

[10]   T. Stephens, J. Sullivan and G. A. Keoleian "A Microsimulation of Energy Demand and Greenhouse Gas Emissions from Plug-In Hybrid Electric Vehicle Use," *Proceedings of the Electric Vehicle Symposium*, Los Angeles, 6-9 May 2010.

[11]   P. Propfe, *et al.*, "Cost Analysis of Plug-In Hybrid Electric Vehicles Including Maintenance & Repair Costs and Resale Values," *Proceedings of Electric Vehicle Symposium* 26, Los Angeles, 2012.

[12]   A. Rousseau, M. Badin, N. Redelbach, A. Kim, D. Da Costa, D. Santini, A. Vyas, F. Le Berr and H. Friedrich, "Comparison of Energy Consumption and Costs of Different HEVs and PHEVs in European and American Context," European Electric Vehicle Congress, Brussels, 19-22 November 2012.

18

Coupled Vibration Analysis of Vehicle-Bridge System Based on Multi-Body Dynamics

Deshan Shan, Shengai Cui, Zhen Huang

Bridge Engineering Department, Southwest Jiaotong University, Chengdu, China

ABSTRACT

For establishing the refined numerical simulation model for coupled vibration between vehicle and bridge, the refined three-dimensional vehicle model is setup by multi-body system dynamics method, and finite element method of dynamic model is adopted to model the bridge. Taking Yujiang River Bridge on Nanning-Guangzhou railway line in China as study background, the refined numerical simulation model of whole vehicle and whole bridge system for coupled vibration analysis is set up. The dynamic analysis model of the cable-stayed bridge is established by finite element method, and the natural vibration properties of the bridge are analyzed. The German ICE Electric Multiple Unit (EMU) train refined three-dimensional space vehicle model is set up by multi-system dynamics software SIMPACK, and the multiple non-linear properties are considered. The space vibration responses are calculated by co-simulation based on multi-body system dynamics and finite element method when the ICE EMU train passes the long span cable-stayed bridge at different speeds. In order to test if the bridge has the sufficient lateral or vertical rigidity and the operation stability is fine. The calculation results show: The operation safety can be guaranteed, and comfort index is "excellent". The bridge has sufficient rigidity, and vibration is in good condition.

Keywords: Cable-Stayed Bridge; Coupled Vibration; Co-Simulation; Multi-Body System Dynamics; Finite Element Method

1. Introduction

Yujiang River Bridge is a large-span double-pylon steel truss cable-stayed railway bridge with double cable plane and double track which is located on Nanning-Guangzhou high-speed railway line. It is the most convenient railway from Guangxi Province to Perl River Delta according to the Eleventh Five Year Plan of China Railways. The span combination of main bridge is 36 m + 96 m + 228 m + 96 m + 36 m, and the total length is 492 m. The maximum designed speed is 250 kilometers per hour, and the distance between track centers is 4.6 m. And the design live load is C-live load. One hand, the higher running speed of vehicles makes the coupled vibration between high speeded trains and bridge structure is more noticeable. On the other hand, because of the high flexibility, the cable-stayed bridge has lower rigidity compared to common bridge [1,2]. So it is necessary to have a coupled vibration analysis of vehicle-bridge system for Yujiang River Bridge, and then the safety and comfort of vehicle running is evaluated based on the analysis results. For meet the current requirement of the refined analysis

for the vehicle-bridge coupled vibration system, the refined dynamic model of the whole vehicle is set up by the multi-body system dynamics methods. And the dynamic analysis model of the bridge is established by finite element method. Finally, the co-simulation method based on multi-body system dynamics and finite element method is adopted to calculate the 3-D vibration responses of the vehicle-bridge system d by when the ICE EMU train moving on the long span cable-stayed bridge under different speeds and the dynamic index of bridge and structure are evaluated as well.

2. Finite Element Model and Natural Vibration Properties of Bridge

The main girder of this cable-stayed bridge is steel truss girder with triangle shape, and is composed of two-piece main truss girder with 14 m height , the 15 m truss spacing and 12 m panel length. Welded integral node with 50 mm maximum plate thickness is adopted in the main trusses. Box section with 1000 mm width is applied on both upper and lower chords of the main trusses. The sectional dimension of upper chords is 1000 × 1260 mm, and the sectional dimension of lower is 1000 × 1400 mm except 1000 × 2000 mm for overburdened zone. Box and H shape sections are choose in the web member respectively according to the different loading conditions, the section size of the former is 1000 × 1040 mm and 1000 × 940 mm, while the latter is 1000 × 900 mm. The pylon is diamond shape. There are 8 pairs of cables on each side span and 16 on middle span, in a total the number of cable pairs is 32. Distance between the adjacent cables on the main girder is 12 meters while 2 meters on the pylons. The bridge is discretized by the 3-D frame finite element method. The spatial beam element is adopted to simulate the main girder, pylons, piers and bridge deck, and the cables are simulated by 3-D bar element. Tapered cross-sections of the pylons are considered in the model as well. The sectional area of cables and initial cable-tension are

provided by the design units. The calculation model of the cable-stayed bridge with 1906 nodes and 1623 elements is carried out finally.

According to the dynamic analysis model of the cable-stayed bridge established, the natural vibration characters of the bridge are analyzed. The first ten natural frequencies and respective mode shapes can are shown in **Table 1**. And **Figure 1** shows some mode shapes.

Because of the floating-type of main girder, the first order natural mode shape is longitudinal drift, and the frequency is 0.386 Hz; the fundamental lateral modal shape is the second order modal of the bridge, its frequency is 0.426 Hz, and the modal shape is symmetric bending of main girder; the fundamental vertical modal shape is the third order modal of the bridge with the frequency is 0.842 and symmetric vertical bending of main girder; and the fundamental torsion modal shape is

Table 1. List table of first ten natural frequencies and mode shapes.

Order	Frequency (Hz)	Mode Shape
1	0.386	Longitudinal drift of main girder
2	0.429	Symmetric lateral bending of main girder
3	0.824	Symmetric vertical bending of main girder
4	0.893	Antisymmetric lateral bending of main girder with opposite direction bending of pylon
5	0.951	symmetric lateral bending of main girder with same direction bending of pylon
6	1.079	Antisymmetric lateral bending of main girder with opposite direction bending of pylon
7	1.280	Symmetric lateral bending of right side span
8	1.296	Symmetric lateral bending of left side span
9	1.650	Antisymmetric lateral bending of main girder
10	1.663	Symmetric lateral bending of main girder with torsion of main girder

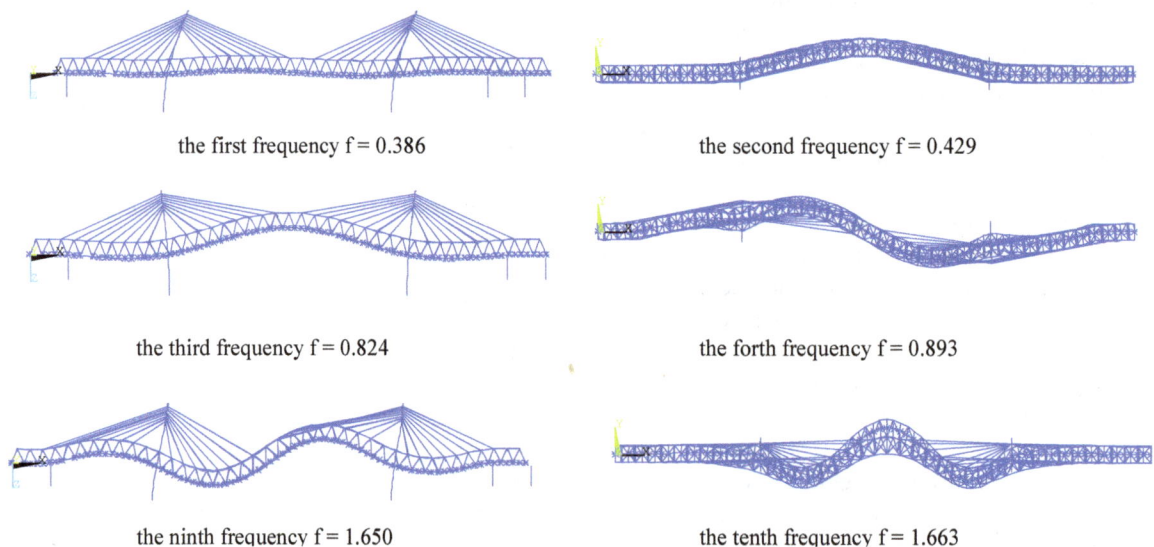

the first frequency f = 0.386

the second frequency f = 0.429

the third frequency f = 0.824

the forth frequency f = 0.893

the ninth frequency f = 1.650

the tenth frequency f = 1.663

Figure 1. Part space modes of the cable-stayed.

the tenth order of the bridge, its frequency is 1.663 Hz with the symmetric torsion of main girder.

3. Realization of Vehicle-Bridge Coupling Vibration in Multi-Body System

The behavior of the vehicle-bridge system is a complex coupled time-varying dynamic problem. Such a problem is generally solved by a numerical simulation based on a dynamic interaction model for the whole vehicle -bridge system. Theoretically, the analysis model for the vehicle-bridge system is composed of two subsystems, the moving vehicle subsystem and the bridge subsystem, which are simulated as two elastic substructures, each of them characterized by some vibration patterns. The two subsystems interact with each other through the contact forces between the wheels and the rail surface.

3.1. Multi-Body Dynamics Model of Vehicle

Multi-body system dynamics is adopted to establish the vehicle system 3-D model [3]. The inter-relationship between each body is realized by the force element which reflects complex features and motion constraint hinge for realizing the stylized modeling, and multi-body dynamics equations are formed automatically. The defects of traditional derivation method are corrected by the proposal modeling method, so it is considered to be powerful evidence of a breakthrough in the vehicle dynamics.

In the multi-body dynamics modeling and simulation of the vehicle system, the characteristics and connection of each body in the vehicle can be confirmed by the definition of rigid body, hinge, constraint, force element, wheel-rail contact model and so on, and then a series of dynamic governing equations are formed [4], the vehicle dynamics model can be divided into three parts from the viewpoint of spring suspension system, car body, boogie and wheel-set. The German ICE Electric Multiple Unit (EMU) train calculation model is set up by multi-system dynamics software SIMPACK. Wheel-set is linked to the bogie frames through the primary suspension, and the bogie frames is linked to the car body through the secondary suspension. Not only two dampers, two snakelike movement dampers and two vertical dampers, but also elastic lateral stoppers are assembled between the car body and bogie. There are 42 degree of freedom in the vehicle model with 34 single hinges and 8 constraints. And the Multi-body dynamics model of locomotive is shown in **Figure 2**. The composition of the train is 2 × (locomotive + trailer + 3 × locomotive + trailer + locomotive), and the total number of cars is 16, the multi-body dynamics model of train is shown in **Figure 3**.

Based on the theory of multi-body dynamic system, the equations of motion for the ith vehicle can be derived as follows:

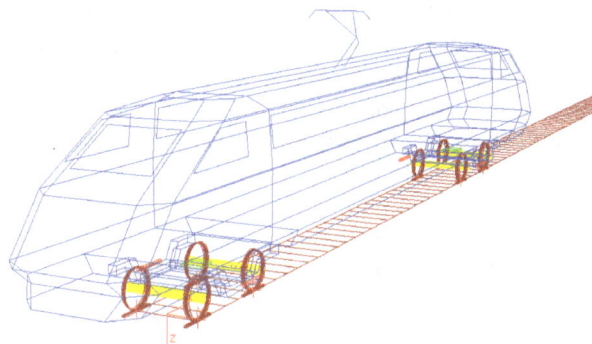

Figure 2. Multi-body dynamics model of locomotive.

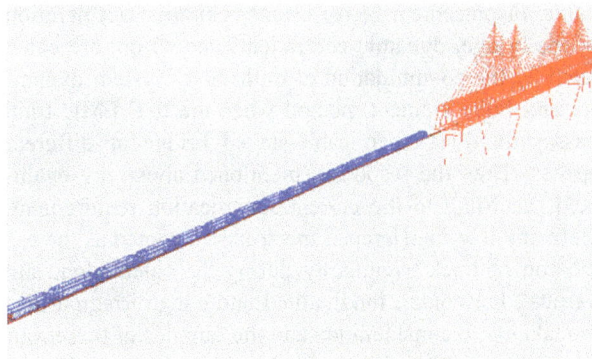

Figure 3. Simulation model of coupled vibration.

$$[M_v]\{\ddot{x}_v\}+[C_v]\{\dot{x}_v\}+[K_v]\{x_v\}+\{F\} \qquad (1)$$

where $[M_v]$, $[C_v]$ and $[K_v]$ denote the mass, damp and stiffness matrices of vehicle respectively; $\{x_v\}$ denotes the column vector of general unknown displacement, $\{\dot{x}_v\}=\frac{\partial\{x\}}{\partial t}$ and $\{\ddot{x}_v\}=\frac{\partial^2\{x\}}{\partial t^2}$. $\{F\}$ denotes the column vector of exciting forces.

3.2. Co-Simulation of SIMPACK and ANSYS

An effective strategy has been provided by co-simulation to solve the coupled vibration problem of vehicle-bridge system. And in the light of the principle of co-simulation, the initial value of the vehicle and bridge are independently solved through their appropriate methods. Finite element method is a powerful tool to analyze dynamic behavior of the bridge accurately, and the multi-body dynamics method is adopted to achieve the vehicle dynamics behavior which contains the complex wheel-rail relationship, then the co-simulation of coupled vehicle-bridge vibration is realized through the wheel-rail data exchange preprocessing program at the discrete information points of wheel-rail contact surface [5,6]. The data exchange process of co-simulation based on SIMPACK and ANSYS is detail described in the reference

[7]. The normal force between wheel and rail is determined by Hertz nonlinear contact theory, and the vertical and horizontal creep forces is obtained aided by the equivalent Hertz contact characteristics and Kalker's simplified nonlinear rolling contact theory-FASTSIM Algorithm [8].

4. Calculation and Analysis of Vehicle-Bridge Coupled Vibration

The 3-D dynamic responses, such as the maximum vertical and lateral vibration acceleration of the vehicle, comfort index, rate of wheel load reduction, derailment coefficient, axle lateral force, vertical and horizontal dynamic displacement of the bridge, vibration acceleration of the bridge, dynamic coefficients and so on, are calculated by the co-simulation of multi-body system dynamics and finite element method when the ICE EMU train passes the long span cable-stayed bridge at different speeds. Then the responses mentioned above are evaluated according to the current specification requirement. Germany low-interference spectrum is adopted as the excitation of track irregularity during the calculation, and vertical, horizontal, longitudinal and gauge irregularities are taken into consideration and the samples of these four kind irregularities are obtained by the trigonometric series. There are four speed cases, 250, 270, 290 and 300 km per hour respectively. The responses of locomotive and trailer are shown in **Table 2**, and the responses of the bridge under different train operating speeds are shown

in **Table 3**. Comfort index is evaluated by Sperling index Wz. The reduction rate of wheel weight in the **Table 2** is defined as $\Delta P/P$, ΔP is the reduction magnitude of wheel weight at the decrease side of the wheel weight, and P is the static wheel weight.

As shown in **Tables 2** and **3**, generally speaking the vibration responses of the vehicle and bridge gradually increase along with the speed increase. In all 4 calculation cases, the maximum of derailment coefficient is 0.211, and the maximum of reduction rate of wheel weigh is 0.262. According to the Specification GB-559985, both of these two indexes are less than the second limit value 1.0 and 0.6 respectively. So the train running safety can be guaranteed. The maximum vertical and lateral acceleration of vehicle are 1.066 m/s^2 and 0.681 m/s^2 respectively, both of these two indexes are less than their corresponding limits values, 1.3 and 1.0 m/s^2. Both the vertical and lateral comfort index is less than 2.5, which means the comfort index is "excellent". The maximum of vertical displacement in the mid-span of main span is 123.688 mm, the vertical deflection-span ratio is 1/1843 correspondingly, and the maximum of lateral displacement in the mid-span of main span is 2.749 mm, the lateral deflection-span ratio is 1/83,000 correspondingly. The maximum vertical acceleration of bridge is 0.386 m/s^2 while lateral acceleration is 0.107 m/s^2. The maximum of dynamic coefficient is 1.200. All of these results show that the bridge has a good vibration performance.

Table 2. Responses of locomotive and trailer.

	Speed cases [km/h]	250	270	290	300
Locomotive	Lateral acceleration [m/s^2]	0.547	0.591	0.646	0.681
	Vertical acceleration [m/s^2]	0.720	0.840	0.996	1.066
	Lateral force of wheel axle [kn]	22.831	26.877	31.166	33.672
	Reduction rate of wheel weight	0.184	0.212	0.243	0.262
	Derailment coefficient	0.146	0.170	0.196	0.211
	Lateral comfort index SP	2.101	2.109	2.122	2.128
	Vertical comfort index SP	2.179	2.224	2.312	2.310
Trailer	Lateral acceleration [m/s^2]	0.436	0.432	0.438	0.441
	Vertical acceleration [m/s^2]	0.687	0.746	0.853	0.933
	Lateral force of wheel axle [kn]	19.134	21.432	24.201	25.871
	Reduction rate of wheel weight	0.191	0.219	0.247	0.261
	Derailment coefficient	0.133	0.150	0.173	0.184
	Lateral comfort index SP	2.145	2.120	2.098	2.088
	Vertical comfort index SP	2.121	2.153	2.195	2.224

Table 3. Responses of the bridge under different train running speeds.

		Speed cases [km/h]	250	270	290	300
Maximum displacement [mm]	Vertical	Mid-span of left side span	13.795	13.731	13.761	13.753
		Mid-span of main span	109.345	114.489	120.599	123.688
		Mid-span of right side span	13.511	13.763	14.398	14.550
	Later	Mid-span of left side span	0.574	0.505	0.485	0.495
		Mid-span of main span	2.422	2.519	2.653	2.749
		Mid-span of right side span	0.520	0.544	0.541	0.595
Maximum acceleration [m/s²]	Vertical	Mid-span of left side span	0.162	0.246	0.254	0.278
		Mid-span of main span	0.309	0.311	0.319	0.386
		Mid-span of right side span	0.135	0.211	0.254	0.343
	Later	Mid-span of left side span	0.083	0.081	0.081	0.107
		Mid-span of main span	0.053	0.043	0.051	0.074
		Mid-span of right side span	0.062	0.081	0.076	0.083
Maximum vertical acceleration [m/s²]			0.309	0.311	0.319	0.386
Maximum lateral acceleration [m/s²]			0.083	0.081	0.081	0.107
Maximum vertical deflection-span ratio [1/x]			2085	1991	1891	1843
Maximum lateral deflection-span ratio [1/x]			94137	90512	85940	83000
Dynamic coefficient [1+μ]			1.061	1.111	1.1700	1.200

5. Conclusion

The refined simulation model of the Germany ICE Electric Multiple Unit (EMU) train is set up by multi-system dynamics methods. Then the dynamic analysis model of the cable-stayed bridge is established by finite element method. Finally the dynamic responses of vehicle-bridge coupled vibration are analyzed by co-simulation based on multi-body system dynamics and finite element method when the ICE EMU train passes the long span cable-stayed bridge. From the co-simulation analysis, the derailment coefficient, reduction rate of wheel weight, vertical and lateral acceleration of the vehicle are met the specification requirements, and both the vertical and lateral comfort index are "excellent" when the ICE EMU train passes the bridge at different working conditions. The results also show that it is safe when the train moving through this bridge at the design speed of 250 km per hour with enough running safety and comfort margin for the vehicle, and the bridge structure owns good dynamic performance.

6. Acknowledgements

The research reported herein has been conducted as part of the result of a series of research projects granted by the Chinese National Science Foundation with 51078316, Chinese Railway Ministry Science and Technology Research and Development Program with 2011G026-E & 2012G013-C, and Sichuan Province Science and Technology Project with 11JC0318.

REFERENCES

[1] W. H. Guo, X. R. Guo and Q. Y. Zeng, "Vibration Analysis of Train-Bridge System for Cable-Stayed Bridge Scheme of Nanjing Yangtse Bridge on Beijing-Shanghai High Speed Railway [J]," *China Civil Engineering Journal*, Vol. 32, No. 3, 1999, pp. 23-26.

[2] X. Z. Li and S. Z. Qiang, "Vehicle-Bridge Dynamic Analysis for Long Span Highway and Railway Bi-Purpose Cable-Stayed Bridge [J]," *Journal of Vibration and Shock*, Vol. 22, No. 1, 2003, pp. 6-26.

[3] Z. S. Chen and C. G. Wang, "Railway Vehicle Dynamoics and Control [M]," China Railway Press, Beijing, 2004.

[4] B. R. Miu, W. H. Zhang, S. N. Xiao, *et al.*, "Car-Body Fatigue Life Simulation Based on Multi-Body Dynamics and FEM [J]," *Journal of the China Railway Society*, Vol. 29, No. 4, 2007, pp. 38-42.

[5] S. Dietz, G. Hippmann and G. Schupp, "Interaction of Vehicles and Flexible Tracks by Co-Simulation of Multi-Body Vehicle Systems and Finite Element Track Models [J]," *Vehicle System Dynamics Supplement*, Vol. 37, 2003,

pp. 372-384.

[6] S. G. Cui, "Refined Simulation Research of Vehicle-Bridge
     Coupled Vibration Based on the Multi-Body System Dy-
     namics and Finite Element Method [D]," Southwest Jiao-
     tong University, Chengdu, 2009.

[7] S. G. Cui, B. Zhu and Z. T. Huang, "Comparative Analy-
     sis of Different Wheel/Rail Contact Models in Vehicle

and Bridge Coupled Vibration [J]," *Chinese Journal of Applied Mechanics*, Vol. 27, No. 1, 2010, pp. 63-67.

[8] J. J. Kalker, "A Fast Algorithm for the Simplified Theory
     of Rolling Contact," *Vehicle System Dynamics*, Vol. 11,
     No. 1, 1982, pp. 1-13.

19

How Cities Influenced by High Speed Rail Development: A Case Study in China

Jing Shi, Nian Zhou

Institute of Transportation Engineering, Tsinghua University, Beijing, China

ABSTRACT

This study aims at analyzing various extents of transportation equity change and the accessibility change of cities along High Speed Rail (HSR) line in China. To evaluate the accessibility change and transportation equity, certain cities along the transportation corridor which the successfully operating WuGuang HSR lies in are selected as samples of influenced ones first. Cities' connection with HSR and their sizes are considered as choosing criteria. Weighted mean travel time and Generalized weighed travel time, which integrated with generalized cost, are both calculated in the two different scenarios to indicate accessibility changes of each city. The two scenarios are with or without the advent of WuGuang HSR. Accessibilities in different scenarios were carefully contrasted. It was found out that there is no significant change in the mean accessibility aspect of the region between the two scenarios. Then transportation equity issue of the transportation corridor is discussed based on the results of Gini coefficient using the data of Passenger Transportation Balance index. Passenger Transportation Balance index is proposed in this article to indicate the how much is one city's transportation demand satisfied. The research found that cities had HSR stations built-in all attained a more accessibility improvement comparing with the others. And the gap between the accessibilities of the cities along the HSR line and those which are not is relatively wider. However, the equity evaluation using Transportation Balance Gini Coefficient reveals that massive investment in HSR has no potent evidence for equity improvement.

Keywords: WuGuang HSR; Accessibility Evaluation; Weighted Mean Travel Time; Generalized Weighted Travel Time; Transportation Equity; Gini Coefficient

1. Introduction

This research is interested in how transportation equity and cities' accessibility in a transport corridor would change after the advent of High Speed Rail (HSR). HSR differs from other transport modes mainly in three aspects, the higher speed comparing with conventional rail, the better quality of service and a relative high capacity among passenger transport modes. As a production of huge investment to the local transportation system, the influence HSR may have in changing transportation equity has been noticed as in [1]. Transportation equity is usually evaluated in four aspects: equity among different areas, different social groups, different traffic modes, and different generations [2]. The four all refer to the fairness and justice of the distribution of transportation resources on two or more units. Thus, there is no doubt that accessibility can be used in illustrating transportation equity, as accessibility is an important indicator of transportation resources. Accessibility has already been used in trans-

portation equity evaluation by using the indicators such as Gini coefficient and Theil index [3]. This research will use the data of accessibility in transportation equity evaluation.

HSR was considered playing an important role in evaluating interregional accessibility early in 1997 [4]. For the meaning of accessibility varies as it is defined and operationalised in different scenarios [5], researchers who were interested in the accessibility changes caused by the operation of HSR have delved into this topic from different angles. Usually accessibility issue of HSR is analyzed together with competition and cooperation of HSR and other transport modes [6-12]. Aware of the fact that HSR mainly impacts on passenger transport, accessibility is defined as the extent to which transport systems enable groups of individuals to reach out for opportunities or destinations in this research. Travel time and transportation fare are the factors considered in this article.

This research is focusing on city level. They are acting as centripetal as well as centrifugal forces in the whole region [13].The importance of cities in the spatial organization as well as their roles as important locations of production, consumption, exchange and control is recognized by researchers years ago [14]. A large part of the literature on HSR, which basically linking cities together, focused on impacts it brings to cities in aspects of economic development [15,16], labor market [17] and other relevant aspects [18-22]. The reason why cities were studies instead of regions in those researches is that most of economic activities occur in cities and they provide most of the attractive workplaces. That is to say the accessibility of cities is more suitable for revealing the impact of HSR.

Although there are plenty of studies on relative issues on accessibility and city development caused by the advent of HSR, little attention was allocated on the differences among cities' transportation equity change and accessibility change in a region. Studying changes of accessibilities in the view of the system means concentrating more on whether or not city is more accessible comparing with others. China started establishing its HSR network since decades ago. While there is still a long period of time before the plan is accomplished, some HSR lines have already been put into use and the operation goes fairly well. How HSR impacts the accessibility of the systems of cities along the transportation corridor has attracted lots of attentions in Chinese scholars [23-25]. However, as the country with the longest HSR lines in operation and going to have the largest HSR network in the world, empirical study on the accessibility change is inadequate [26,27]. This paper aims at finding out how cities are influenced by HSR in China, basing on the case study of WuGuang HSR.

2. Methodology Approach

2.1. Accessibility Evaluation

To find out how accessible the cities are and then to evaluate the transportation equity，accessibility evaluation needs to be taken first. In this study, accessibility indicator is a reflection of the impact of the advent of HSR on the passengers transferring and then the economic development. This means that accessibility should relate to the role of transport systems which will help groups of individuals to attain opportunities in different cities. Thus, accessibility is defined as the extent to which transport systems enable groups of individuals to reach out for opportunities or destinations. The selection of indicator is a very important issue for measuring changes in the accessibility, since the results of evaluations vary with indicators used. In this paper, we will analyze the variation of accessibility changes by the clas-

sic accessibility indicator weighted mean travel time first. This measure does not relate to the short distance and expresses the extent to which a changed link modifies the location by reducing access times to other urban areas [28]. It can be represented as Equation (1):

$$A_i = \left(\sum_{j=1}^{n} M_j T_{ij} \right) \bigg/ \sum_{j=1}^{n} M_j \tag{1}$$

where A_i is the accessibility of a node (city) i. T_{ij} is the shortest travel time between i and j. M_j is the weight (for example, population or gross domestic product) of the node j. The factor M_j is used to value the importance of the minimum-time T_{ij} [11]. n is the number of cities studied.

Weighted mean travel time considers all relationships within the area of study, but routes on short distances do not contribute more than other factors in the calculations since there is no distance decay. Thus it picks up impacts of new link more dispersal in location than other accessibility indicators, like daily accessibility indicator and economic potential indicator [28]. In this study, the weight M_j is specialized by using GDP, for economy is mainly concerned. M_j is calculated using the Equation (2):

$$M_j = \sqrt{GDP_j \times P_j} \tag{2}$$

where GDP_j is the gross domestic product, and P_j is the number of resident population of city j. The number of resident population is used to indicate the scale of the labor market as well as an indicator for job opportunities.

The weighted mean travel time was used in lots of researches [23,24]. In this article, the accessibility will be evaluated by this indicator as well. The result will show us how accessibilities changed if considering travel time only.

Shorten travel time is on reason why HSR caught much attention. However, HSR is also known for its ticket price beside of its high speed. The price of second class ticket is about 0.43 Yuan/km, while the traditional rail is 0.13 Yuan/km. A transportation mode beyond affordance is not an available choice for passengers and will not really help to improve accessibility. Thus, it is necessary to consider ticket price in evaluating accessibility. The Generalized weighed travel time, which integrated time and fare, is considered in this article.

$$a_{ij} = \min \left[\frac{M_j \times T_{ij,k}}{\sum_{j=1}^{n} M_j} + \frac{F_{ij,k} \big/ TV_j}{\sum_{j=1}^{n} \left(1 / TV_j \right)} \right] \tag{3}$$

where a_{ij} is the extent of accessibility between node (city) i and j. $T_{ij,k}$ is the shortest travel time by available transport mode k. M_j is the weight of the node i as in Equation (1). n represents the number of cities studied. TV_j is the generalized travel time value of city j. $T_{ij,k}$ is the travel fare due to transport mode k.

It is a common belief that people from different social groups have different travel time values. There are plenty of methods to calculate TV_j. And different methods usually lead to different results. The following Equation is one of the most widely used methods in travel time evaluation in China [29].

$$TV_j' = GDP_j / (P_j \times WH) \qquad (4)$$

In Equation (4), WH is the number of legal working hours of a year in China, which is 2000. TV_j equals the average amount of the GDP one person can contribute to city j per hour. And it differs from city to city. By using GDP, it is quite clear that the benefits to economy and social welfare from saved travel time were all included. However, travel time value in Equation (4) considered the value of work trip while failing to include the value of personal trip.

Researches concerns about the choice making of travelers would use average wage other than per capita GDP in calculations. Thus, travel time value of work trips is calculated as follows.

$$TV'' = W_{wage} / WH \qquad (5)$$

W_{wage} is the urban per capita disposable income. This indicator could well represent the wage according to the statistical indicator used in China.

Travel time value of personal trip is thought to have a linear relationship with those of work trip. And the ratio of time value between personal trip and work trip was recommended to choose from 0.2 to 0.6 in China [29].

This research considers the impact of HSR on cities, which means that the time value of work trip as well as the personal trip should be considered. Therefore, the following equation is used.

$$TV_j = r \times TV_j' + (1-r) \times \beta \times TV_j'' \qquad (6)$$

where TV_j is the travel time value calculated from Equation (4).

TV_j is the travel time value calculated from Equation (5).

r is the ratio of work trip.

β is the ratio of time value of personal trip to the one of work trip.

In Equation (6), the former part is to represent the travel time value of work trip and the latter is to represent the time value of personal trip. Travel time value of each city is calculated separately.

Because once travel fare is considered, passenger's choice among different transportation modes may differ

from the one with minimum travel time. In other words, HSR might be a preferable choice in some routes but no in all routes. Thus, to evaluate the accessibility of a city, the accessibility of each route should be decided first by Equation (3). In Equation (1), the minimum travel time is selected directly as it is assumed that the shorter time the more preferable the choice. The evaluation results of the two indicators will be contrasted to see how fare impacts on accessibility.

As can be seen from the Equation (3), travel fare is integrated in accessibility similarly to travel time in this research. Firstly, the data of travel fare of every mode between every two cities is collected. Then the accessibility between two cities is defined as the minimum one.

$$A_i = \sum_j a_{ij} \qquad (7)$$

where A_i is the accessibility of the city i evaluated by Generalized weighed travel time.

The Generalized weighed travel time considers all relationships within the area of study. But there is still no distance decay as weighed mean travel time.

To get a full view, the whole of any cities that might be influenced are seen as the population and the cities selected and calculated are seen as samples. One sample Kolmogorov-Smirnov (K-S) test will be carried out to identify what kind of distribution the population is. Then independent-samples t-test will be implemented to find the differences of mean accessibility between the two scenarios.

2.2. Transportation Equity Evaluation

Keep making investments to improve transportation system is important for the development of the society. However, the resource available for investing is limited. Which means that to overfulfil one city's transportation demand is unfair to the others. In other words, the distribution of investment among different cities would affect transportation equity.

The definition of transportation equity requires balance between transportation demand and transportation capability. In this research we define the Passenger Transportation Balance Index to indicate such a balance. The passenger transport is considered only here because HSR influence it most. And in the evaluation process how the freight transport is not included. The indicator, which is represented by y_i, is calculated by the following equation.

$$y_i = P_i / (A_i \times GDP_i) \qquad (8)$$

A_i is the accessibility of city i, which is derived from Equation (1) or Equation (3). P_i and GDP_i respectively represent the number of residents and gross domestic product of city i. $1/A_i$ here is to represent the passenger transportation capability of the city. A_i is in the position of denominator. Because the smaller its numerical value the better the accessibility. And ceteris paribus, the better the accessibility the larger value of the Passenger Transportation Balance Index is. P_i/GDP, which is the reciprocal of the per capita GDP, is the representative of Passenger transportation demand.

y_i is a relative indicator, as the accessibility indicator A_i. It would vary as the cities chosen for study changed. Thus, there is no certain meaning of its numerical value. It should only be used for comparisons among cities in the same evaluation process. By comparing value of y, it is easy to find out which city's transportation demand was more satisfied.

Basing on accessibility data, the transportation equity is discussed by Passenger Transportation Balance Gini coefficient, which is calculated by the data of accessibility evaluation. Gini coefficient (also known as the Gini index or Gini ratio) is an indicator for statistical dispersion. It is originally used to measure the inequality among values of a frequency distribution (for example levels of income). And it has been used in transportation research, too. Santos Bruno [3] used it to indicate the dispersion of accessibility values across all centers. Gini coefficient theoretically ranges from 0 to 1. A Gini coefficient of zero expresses perfect equality where all values are the same, corresponding to complete equality. A low value of Gini coefficient indicates a more equal distribution, while higher Gini coefficients indicate more unequal distribution. To be validly computed, no negative goods can be distributed, which the data of Transportation Balance could well satisfied.

If the probability function $f(y)$ is discrete, which is the case of Transportation Balance Index distribution, and where y_i (i=0 to n) is the points with nonzero probabilities, indexed in increasing order ($y_i < y_{i+1}$), then Transportation Balance Gini coefficient is calculated using Equation (9). Gini coefficient indicates the inequality state, which can be expressed with Lorenz curve as shown in **Figure 1**, and it is obtained by the ratio of

two areas: A/(A+B).

$$G = 1 - \left[\sum_{i=1}^{n} f(y_i)(S_{i-1} + S_i) \bigg/ S_n \right] \qquad (9)$$

Where $S_i = \sum_{j=1}^{i} f(y_j) y_j$ and $S_0 = 0$.

3. Case Study

3.1. Introduction to WuGuang HSR

WuGuang HSR is one of the most successful lines since it started commercial service at the end of the 2009. It is part of the unfinished 2100-km long Beijing–Guangzhou High-Speed Railway (JingGuang HSR). The main line of WuGuang HSR is 968-kilometre long, with Wuhan (capital of Hubei Province) and Guangzhou (capital of Guangdong Province) as its two ends. It costs three hours and thirty-three minutes in transferring passengers between the two ends by the fastest commercial service. With 15 stations in operation, WuGuang HSR crosses three provinces, Hubei, Hunan and Guangdong (as shown in **Figure 2**).

The corridor WuGuang HSR goes along is very busy since the accomplishment of the conventional JingGuang Railway in 1957. These two railways have an approximately same route. Wuhan, Guangzhou and Changsha are set as the three original stations, providing services with different combinations of intermediate stations.

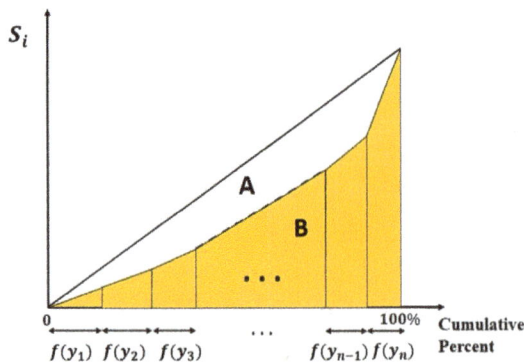

Figure 1. Transportation Balance Index distribution expressed in Lorenz curve and Gini coefficient.

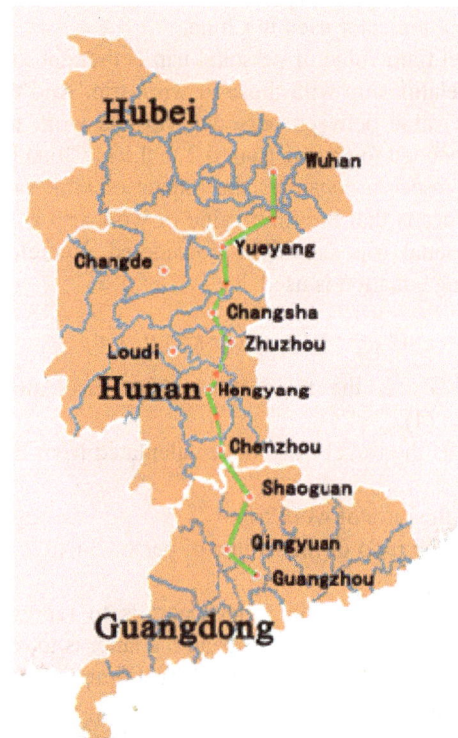

Figure 2. Location of WuGuang HSR.

The majority of seats are of the second class and the fare is about ¥ 0.46 per kilometer. A few seats are in the first class, and seats of the deluxe class are also available, which charge about 60% and 80% high respectively more comparing with those in the second class. The price of the second class is only like a half of the price of the flight between Wuhan and Guangzhou even considering the usual discount (about 50%). Each station's distance from the end of Wuhan and its administration in provincial level and city above the prefecture level is illustrated in the **Table 1**.

These stations are neither of the same size nor the same importance for the line. Wuhan, Changsha South, Guangzhou South and Guangzhou North are much bigger than the others not only because they are located in the capitals of the three provinces, but also because they are already serving as or planned to be important transport hubs. And what should be informed is that only the four cities, namely Wuhan, Changsha, Hengyang and Guangzhou, have an airport.

Among all these stations, Chibi North, Miluo East, Hengshan West, Leiyang West are smaller because they are newly built stations in small developing counties, where there is only small travel demand comparatively. The statistical survey conducted by the Statistic Bureau of Hunan province could well prove it. See **Table 2**.

HSR is now the most convenient when travel time as well as travel cost is considered in all public passenger transportation. From January to October, just the Hunan section of WuGuang HSR has already transferred 12.5 million passengers, which indicates that HSR has become an important transport mode.

3.2. Cities to Be Analyzed

The transportation equity is revealed by the data of accessibility in this research. To find out the changes of the accessibilities is more important than getting a precise numerical evaluation result of accessibility. Cities to be analyzed are firstly selected from all the cities which might be influenced by HSR. Because it is merely impossible to evaluate all the cities been influence at the same time, for the hardness of identifying them all. HSR is a huge investment which not only influences the cities which are along the line, but also the cities sharing the same transportation system. Thus, sampling is concerned to estimate characteristics of the whole population, which is all the cities be influenced.

The primary criterion in selecting cities is that the selected cities should be along WuGuang HSR or have direct train service to cities along WuGuang HSR. In other word, cities studied should be in the transport corridor. This criterion would help to get a clear result of the accessibility change by analyzing the impact of HSR. Secondly, the administration demarcations are considered to ensure the data is comparable in this study. That's because the statistic survey is done by different local administration, which usually has some differences in statistic method. To take a sample analysis, cities from one province would be helpful to make the results potent.

Table 1. Stations along the WuGuang HSR line.

Stations in Operation	Province	City	Distance from Wuhan (Km)
Wuhan		Wuhan	0
Xianning North	Hubei		85
Chibi North		Xianning	127
Yueyang East			209
Miluo East		Yueyang	279
Changsha South		Changsha	347
Zhuzhou West		Zhuzhou	387
Hengshan West	Hunan	Hengyang	455
Hengyang East			496
Leiyang West		Hengyang	552
Chenzhou West		Chenzhou	628
Shaoguan		Shaoguan	758
Qingyuan		Qingyuan	885
Guangzhou North	Guangdong	Guangzhou	922
Guangzhou South			968

Table 2. Passengers transferred in the first four months of 2010 (unit: person).

Station	Jan	Feb	Mar	April
Yueyang East	52,514	74,052	64,449	61,916
Miluo East	5780	12,989	9145	7893
Changsha South	260,697	233,697	329,016	344,025
Zhuzhou West	54,241	46,225	54,072	54,540
Hengshan West	28,830	30,744	34,606	27,539
Hengyang East	79,568	90,552	98,202	99,483
Leiyang West	19,705	21,268	23,052	25,516
Chenzhou West	96,055	68,077	103,072	107,246

Also, the level of the city is considered. There are three levels of cities which have an HSR station, decreasing in order of the sub-provincial level, the prefecture level and the county level. These levels do not only reveal cities' political status but also reflect their economic development. In this paper, cities of the county level are not considered for both their economic aggregate and the populations are small comparing with cities from higher levels. From **Table 2**. it can be found that stations in county level cities only share small part of the passenger flow. And in this specific case, county level cities are only short distances away from prefecture level cities which are taking charge of them in the politically hierarchy.

Considering criteria aforementioned, cities in Hunan province are chosen. Besides the five cities Yueyang, Changsha, Zhuzhou, Hengyang, and Chenzhou which have an HSR station, Loudi and Changde are selected. Loudi and Changde are prefecture level cities which are of importance to regional development. The nine cities are illustrated in **Table 3**.

The accessibility of a certain city is not only influenced by other cities within the region it belongs, but also cities beyond the region [30]. So Wuhan and Guangzhou are used in the evaluating process to ensure the result more attached to the reality. These seven cities are all at the prefecture level. Also, it is easy to find that Wuhan, Guangdong and Changsha, capitals of the respective province, are more developed than other cities from this table. Thus, Yueyang, Changsha, Zhuzhou, Hengyang, Chenzhou, Wuhan, Guangzhou, Loudi, Changde and Wuhan, Guangzhou are analyzed in the accessibility evaluation process.

4. Empirical Analysis

There are nine cities to be analyzed in the accessibility

Table 3. Cities selected for evaluations.

City	GDP (billion)	Resident Population (10^4)	Expendable income (￥1)
Yueyang	1.5394	547.61	19,558
Changsha	4.5471	704.41	26,451
Zhuzhou	1.2748	385.71	24,017
Hengyang	1.4203	714.84	17,866
Chenzhou	1.0818	458.36	17,606
Loudi	0.6807	378.46	16,937
Changde	1.4916	571.46	17,861
Wuhan	5.56593	978.54	23,738
Guangzhou	10.60448	1270.96	34,438

evaluation process. However, the evaluation results of Wuhan and Guangzhou are not discussed in this paper. Because as mega city of sub-province level, both Wuhan and Guangzhou have lots of transportation connections with cities allocated in a broad area. So only considering a few cities along HSR line is insufficient for the accuracy of their accessibility evaluation.

As aforementioned, there are only 4 cities which have airline services. Among those 4 cities, there are only direct flights between Wuhan and Guangzhou, which take 1 hour and 40 minutes on the plane. According to the administration provisions, passengers traveling by plane need to arrive at the airport, which usually located in the suburban area, to board the plane at least half an hour before the plane took off. Thus, minimum travel time between Wuhan and Guangzhou would add up to two and a half hours. Railway, buses are also considered in comparing with HSR while private vehicles are not considered in this paper.

4.1. Accessibility Evaluation by Weighed Mean Travel Time

The latest minimum travel time is collected in two scenarios, with or without the HSR taken into consideration. Accessibility represented by here is rather a comparative indicator, which reveals the change of accessibility other than showing the absolute value of accessibility.

According to Equations (1) and (2), the accessibility variation can be evaluated. And the evaluation results of are presented in **Table 4**.

According to the definition of accessibility measurement, the smaller the numerical value of is, the more accessible the city is. From evaluation results, it can be found that all the cities along HSR were exalted in accessibility. And these accessibilities have been promoted in similar ratios. Though no HSR station in, the accessibili-

Table 4. Accessibility evaluation results using weighed mean travel time.

	With HSR	Without HSR	Ratio[b]
Wuhan	5.44	11.19	51%
Yueyang	1.37	2.81	51%
Changsha	1.61	3.75	56%
Zhuzhou	0.681	1.48	54%
Hengyang	1.27	2.46	48%
Chenzhou	1.22	2.39	49%
Guangzhou	11.16	21.57	48%
Loudi	1.13	1.70	33%
Changde	2.68	3.37	25%

b. Ratio= (Without HSR-With HSR)/ Without HSR.

ties of Loudi and Changde have been improved too. That is because both direct trips and transfer are considered and passengers can move to a city with HSR and take it. Thus, Loudi and Changde also benefit from the advent of HSR.

4.2. Accessibility Evaluation by Generalized Weighed Indicator

The travel time and travel fare of each transport mode are collected in two scenarios, with or without the HSR taken into consideration. In the evaluation process, r in Equation (6) is 0.5 according to an investigation implemented on WuGuang HSR [31]. And β is valued 0.4 as recommended [32].

According to Equations (3), (4) and (6) the accessibility evaluation results of A_i calculated by Generalized weighed travel time are presented in **Table 5**.

After travel fare is taken into account, HSR is not attractive for it is more expensive than rail and bus. Passengers will choose the traditional railway or bus if they do not find the time saved by HSR worth the increase of fare. Thus, it is easy to interpret that the accessibility promotion illustrated is **Table 5** is not as much as in **Table 4**.

By contrasting the difference of evaluation results by weighed mean travel time and integrated indicator, it is found that the ticket price of HSR has great impact on accessibility. And the results in **Table 5** are closer to the observation that the accessibilities of cities with HSR do not have a promotion of about 50%.

Only find out the accessibility change of the seven cities is not sufficient to understand the impact of the HSR to the transportation corridor. In order to understand the differences between the two scenarios, the non-parameter analysis is carried out to find the distribution of accessibility. The results shown in **Table 5** are used in this procedure. The result is shown in **Table 6**.

Table 5. Accessibility Evaluation using Generalized Weighed Travel Time.

	With HSR	Without HSR	Ratio[a]
Wuhan	16.73	18.87	11.3%
Yueyang	7.45	7.87	5.3%
Changsha	6.90	7.42	7.0%
Zhuzhou	4.77	4.95	3.6%
Hengyang	5.98	7.33	18.4%
Chenzhou	7.39	8.47	12.7%
Guangzhou	25.65	36.67	30.0%
Loudi	5.41	5.87	7.8%
Changde	7.74	8.73	11.4%

a. Ratio = (Without HSR-With HSR)/ Without HSR.

Table 6. One-Sample kolmogorov-smirnov test.

		accessibility without HSR	accessibility with HSR
N		7	7
Normal Parameters (Test distribution is Normal.)	Mean	7.234	6.520
	Std. Deviation	1.372	1.143
Most Extreme Differences	Absolute	0.242	0.205
	Positive	0.138	0.143
	Negative	−0.242	−0.205
Kolmogorov-Smirnov Z		0.641	0.543
Asymp. Sig. (2-tailed)		0.807	0.930

Distribution of accessibilities in both scenarios can be seen as normal according to the results. The samples are independent of each other. To further understand the gap of standard deviations between the two scenarios, independent-samples t-test is carried out. The final results are shown in **Table 7**.

The purpose of independent-samples t-test is to test for significant differences between means. Levene's test checks for whether their variances are relatively similar or not. The significance for Levene's test is above 0.05. Thus the "Equal Variances Assumed" test is used.

The significance is 0.839 (0.839 > 0.05), which means there is no significant difference between values of mean accessibility in two scenarios.

In other words, despite the difference of mean accessibilities caused by the advent of WuGuang HSR, there is no significant accessibility change statistically. This conclusion dose not conflict with the evaluation result in **Table 4** that cities have HSR stations built in have an over 30% accessibility improvement. Because when one sample K-S test and independent T-Test are accomplished, all the cities been influenced are considered other than the samples chosen.

4.3. Transportation Equity Analysis

Cities are acting as centripetal as well as centrifugal forces in the whole system [33], which are both connected to the accessibility. The more accessible one city is the more benefit it would have in economic aspect. However, if decisions are made considering monetary aspect only, the disparities between developed regions and undeveloped ones will increase [2]. So the equity issue is an important one to be considered. And the equity among different regions is going to be considered. The following part is going to reveal the equity change through analyzing the change of accessibility. The

Table 7. Independent samples t-test of accessibility.

	Levene's Test for Equality of Variances		t-test for Equality of Means				
	F	Sig.	t	df	Sig. (2-tailed)	Mean Difference	Std. Error Difference
Equal variances assumed	0.043	0.839	−1.058	12	0.311	−0.714 29	0.674 92
Equal variances not assumed			−1.058	11.622	0.311	−0.714 29	0.674 92

Passenger Transportation Balance indexes of the two scenarios are calculated using Equations (3) and (7). Results are presented in **Table 8**.

By using the data in **Table 8**, Gini Coefficient can be calculated in **Table 9**.

The Gini coefficients calculated shows that after HSR is built, the cities influenced are slightly falling into a more unequal situation. However, due to the limited change, there is no potent evidence that WuGuang HSR has changed transportation equity along the transportation corridor.

There are some possible explanations for this slight influence. One is that the HSR would benefit cities in close extent, which resulted in similar change ration of accessibilities. Another is that because the network of HSR is still unfinished, the impacts brought in by a single line are limited.

In conclusion, the advent of HSR would improve the extent of satisfaction of cities' transportation demand. And it has also slightly worsened the transportation equity.

5. Discussion and Conclusions

This study analyzed various extents of transportation equity change and the accessibility change of cities along WuGuang High Speed Rail (HSR) line in China.

High speed is one of the most important features of HSR. According to accessibility evaluation, if travel time is considered only, cities with HSR stations attained a significant improvement in accessibility. The improvement is about 50% comparing with situation without HSR.

However, if the high price due to the cost of building and operating HSR is considered, it is found that there are only small improvements in accessibility, which for the majority of the cities the improvement rate is under 12%. In the process of using the generalized weighed travel time, the time value is slightly overvalued by using per capita GDP. If it is estimated by real income, which will result in a smaller time value, the evaluation would reveal a less impacted accessibility. This should have drawn the attention of HSR planners, that the ticket price of HSR is still unaffordable for a large part of citizens under the current income level in China. The high price determines that most of the WuGuang HSR users are on business trip and their companies will afford the cost. Thus, we suggest administers of HSR to supply discount tickets in low seasons and off-peak hours. Such measures could benefits more people on the one hand. On the other hand, it could help to improve the usage of HSR.

It is commonly believed that accessibility would improve greatly once HSR was built. In this research, the mean accessibility of the region WuGuang HSR lies in is evaluated to find out whether great improvement has occurred. It is found that WuGuang HSR did not statistical change accessibility of the region. The ticket price might be an important reason. Moreover, WuGuang HSR is still a single line other than part of a HSR network at present. And it is not well connected with conventional railway network. This could be possible explanations for the finding that no significant accessibility change is found in this region. That is to say how to take advantage of network effect should be considered in the planning process of HSR lines.

In this research, transportation resource versus transportation demand is defined as Passenger Transportation Balance index. And Transportation equity discussed here, which is a kind of equity among different areas, concerns about the distribution of Passenger Transportation Bal-

Table 8. The value of Passenger Transportation Balance indexes (Using the data of Generalized weighed travel time).

	With HSR	Without HSR	Ratio[a]
Yueyang	0.048	0.045	6%
Changsha	0.022	0.021	8%
Zhuzhou	0.063	0.061	4%
Hengyang	0.084	0.069	23%
Chenzhou	0.057	0.050	15%
Loudi	0.103	0.095	9%
Changde	0.049	0.044	13%

a. Ratio= (Without HSR-With HSR)/ Without HSR.

Table 9. Gini coefficient.

	With HSR	Without HSR
Gini Coefficient	0.652 6	0.649 1

ance index among different cities. Transportation equity is indicated by Gini coefficient with the results of Passenger Transportation Balance index. According to the evaluation, transportation equity has not been significantly affected. However, it is obvious that cities are not benefiting equally from HSR according to Passenger Transportation Balance index.

What have been achieved could well illustrate how cities would be influenced by HSR in aspects of accessibility and transportation equity. The degrees of the impacts are not only determined by the travel time shorten by HSR, but also determined by some other factors like income level. In future researches, more measurements could be applied to describe how cities would be influenced in more detailed way, like which social groups would benefit most from HSR. Further, how citizens of different group would be influenced as well as how transport equity among different groups would change will be studied by combining residents' travel choice model with this research in the future.

REFERENCES

[1] A. Monzón, E. Ortega and E. López, "Efficiency and Spatial Equity Impacts of High-speed Rail Extensions in Urban Areas," *Cities*, Vol. 30, 2013, pp. 18-30.

[2] J. Shi and N. Zhou, "A Quantitative Transportation Project Investment Evaluation Approach with both Equity and Efficiency Aspects," *Research in Transportation Economics*, Vol. 36, No. 1, 2012, pp. 93-100.

[3] B. Santos, A. Antunes and E. J. Miller, "Integrating Equity Objectives in a Road Network Design Model," *Transportation Research Record*, Vol. 2089, pp. 35-42.

[4] U. Blum, K.E. Haynes and C. Karlsson, "Introduction to the Special Issue: The Regional and Urban Effects of High-speed Trains," *The Annals of Regional Science*, Vol. 31, No. 1, 1997, pp. 1-20.

[5] K. T. Geurs and B. van Wee, "Accessibility Evaluation of Land-Use and Transport Strategies: Review and Research Directions," *Journal of Transport Geography*, Vol. 12, No. 2, 2004, pp. 127-140.

[6] M. Givoni and D. Banister, "Airline and Railway Integration," *Transport Policy*, Vol. 13, No. 5, 2006, pp. 386-397.

[7] C. Hsu, Y. Lee and Liao C, "Competition between High-Speed and Conventional Rail Systems: A Game Theoretical Approach," *Expert Systems with Applications*, Vol. 37, No. 4, 2010, pp. 3162-3170.

[8] C. Román, R. Espino and J. C. Martín, "Competition of High-Speed Train with Air Transport: The Case of Madrid-Barcelona," *Journal of Air Transport Management*, Vol. 13, No. 5, 2007, pp. 277-284.

[9] N. Adler, E. Pels and C. Nash, "High-speed Rail and Air Transport Competition: Game Engineering as Tool for Cost-Benefit Analysis," *Transportation Research Part B: Methodological Modelling Non-urban Transport Investment and Pricing*, Vol. 44, No. 7, 2010, pp. 812-833.

[10] E. Cascetta, A. Papola, F. Pagliara and V. Marzano, "Analy-sis of Mobility Impacts of the High Speed Rome-Naples Rail Link Using within Day Dynamic Mode Service Choice Models," *Journal of Transport Geography*, Vol. 19, No. 4, 2011, pp. 635-643.

[11] J. Gutiérrez and P. Urbano, "Accessibility in the European Union: the Impact of the Trans-European Road Network," *Journal of Transport Geography*, Vol. 4, No. 1, 1996, pp. 15-25.

[12] C. Chen and P. Hall, "The Impacts of High-speed Trains on British Economic Geography: A Study of the UK's InterCity 125/225 and Its Effects," *Journal of Transport Geography*, Vol. 19, No. 4, 2011, pp. 689-704.

[13] A. Reggiani, P. Bucci, G. Russo, A. Haas and P. Nijkamp, "Regional Llabour Markets and Job Accessibility in City Network Systems in Germany," *Journal of Transport Geography*, Vol. 19, No. 4, 2011, pp. 528-536.

[14] D. A. Smith and M. Timberlake, "Conceptualising and Mapping the Structure of the World System's City System," *Urban Studies*, Vol. 32, No. 2, 1995, pp. 287-302.

[15] P. Coto-Millán, V. Inglada and B. Rey, "Effects of Network Economies in High-Speed Rail: The Spanish Case," *The Annals of Regional Science*, Vol. 41, No. 4, 2007, pp. 911-925.

[16] J. Martín and G. Nombela, "Microeconomic Impacts of Investments in High Speed Trains in Spain," *The Annals of Regional Science*, Vol. 41, No. 3, 2007, pp. 715-733.

[17] K. E. Haynes, "Labor Markets and Regional Transportation Improvements: the Case of High-speed Trains: An introduction and Review," *The Annals of Regional Science*, Vol. 31, No. 1, 1997, pp. 57-76.

[18] K. S. Kim, "High-speed Rail Developments and Spatial Restructuring: A Case Study of the Capital Region in South Korea," *Cities*, Vol. 17, No. 4, 2000, pp. 251-262.

[19] M. Givoni, "Development and Impact of the Modern High-Speed Train," *Transport Reviews*, Vol. 26, No. 5, 2006, pp. 593-611.

[20] M. Garmendia, J. M. Urea and J. M. Coronado, "Long-distance Trips in A Sparsely Populated Region: The Impact of High-Speed Infrastructures," *Journal of Transport Geography*, Vol. 19, No. 4, 2011, pp. 537-551.

[21] D. Banister and M. Thurstain-Goodwin, "Quantification of the Non-Transport Benefits Resulting from Rail Investment," *Journal of Transport Geography*, Vol. 19, No. 2, pp. 212-223.

[22] J. M. Urea, P. Menerault and M. Garmendia, "The

High-Speed Rail Challenge for Big Intermediate Cities: A National, Regional and Local Perspective," *Cities*, Vol. 26, No. 5, 2009, pp. 266-279.

[23] L. Gou, "The Impact of Inter-City Express Rail Access to Major Metropolitan Area," Master Thesis, Southwest Jiaotong University, Chengdu, 2009.

[24] D. Meng, W. Chen and Y. Lu, "Impacts of High-Speed Railway on the Spatial Pattern of the Regional Accessibility in China (in Chinese)," *Areal Research and Development*, No. 4, 2011, pp. 6-10.

[25] B. Yang, "Study on the Influence of High-speed Railways on Regional Development," Master Thesis, East China Normal University, Shanghai, 2011.

[26] C. Zhang, "High-speed Rail Development Impact of the Urban System," Master Thesis, Nankai University, Tianjin, 2009.

[27] H. Zhu, T. You and J. Zhang, "High-Speed Railway Development Impact on Wuhan City (in Chinese)," *Comprehensive Transportation*, No. 3, 2011, pp. 43-48.

[28] G. Javier, "Location, Economic Potential and Daily Accessibility: An Analysis of the Accessibility Impact of the High-Speed Line Madrid-Barcelona-French Border,"

Journal of Transport Geography, Vol. 9, No. 4, 2001, pp. 229-242.

[29] H. Jiping, "The Significance and Methods of Research on Travel Time Value and Methods (in Chinese)," *Comprehensive Transportation*, No. 10, 2008, pp. 64-67.

[30] X. Ren, "Transportation Infrastructure, Factor Mobility and Manufacturing Location," Master Thesie, Chongqing University, Chongqing, 2010.

[31] L. Jianbin, "Survey and Analysis on Passenger Travel Characteristics and Distribution Features of Wuhan-Guangzhou High Speed Railway," *Railway Standard Design*, No. 11, 2011, pp. 1-4.

[32] Y. Tao, "Research on Economic Impacts of HSR," *Railway Transport and Economy*, Vol. 29, No. 1, 2007, pp. 4-6.

[33] J. Gutirrez, R. Gonzlez and G. Gmez, "The European High-Speed Train Network: Predicted Effects on Accessibility Patterns," *Journal of Transport Geography*, Vol. 4, No. 4, 1996, pp. 227-238.

Permissions

The contributors of this book come from diverse backgrounds, making this book a truly international effort. This book will bring forth new frontiers with its revolutionizing research information and detailed analysis of the nascent developments around the world.

We would like to thank all the contributing authors for lending their expertise to make the book truly unique. They have played a crucial role in the development of this book. Without their invaluable contributions this book wouldn't have been possible. They have made vital efforts to compile up to date information on the varied aspects of this subject to make this book a valuable addition to the collection of many professionals and students.

This book was conceptualized with the vision of imparting up-to-date information and advanced data in this field. To ensure the same, a matchless editorial board was set up. Every individual on the board went through rigorous rounds of assessment to prove their worth. After which they invested a large part of their time researching and compiling the most relevant data for our readers. Conferences and sessions were held from time to time between the editorial board and the contributing authors to present the data in the most comprehensible form. The editorial team has worked tirelessly to provide valuable and valid information to help people across the globe.

Every chapter published in this book has been scrutinized by our experts. Their significance has been extensively debated. The topics covered herein carry significant findings which will fuel the growth of the discipline. They may even be implemented as practical applications or may be referred to as a beginning point for another development. Chapters in this book were first published by Scientific Research Publishing Inc.; hereby published with permission under the Creative Commons Attribution License or equivalent.

The editorial board has been involved in producing this book since its inception. They have spent rigorous hours researching and exploring the diverse topics which have resulted in the successful publishing of this book. They have passed on their knowledge of decades through this book. To expedite this challenging task, the publisher supported the team at every step. A small team of assistant editors was also appointed to further simplify the editing procedure and attain best results for the readers.

Our editorial team has been hand-picked from every corner of the world. Their multi-ethnicity adds dynamic inputs to the discussions which result in innovative outcomes. These outcomes are then further discussed with the researchers and contributors who give their valuable feedback and opinion regarding the same. The feedback is then collaborated with the researches and they are edited in a comprehensive manner to aid the understanding of the subject.

Apart from the editorial board, the designing team has also invested a significant amount of their time in understanding the subject and creating the most relevant covers. They scrutinized every image to scout for the most suitable representation of the subject and create an appropriate cover for the book.

The publishing team has been involved in this book since its early stages. They were actively engaged in every process, be it collecting the data, connecting with the contributors or procuring relevant information. The team has been an ardent support to the editorial, designing and production team. Their endless efforts to recruit the best for this project, has resulted in the accomplishment of this book. They are a veteran in the field of academics and their pool of knowledge is as vast as their experience in printing. Their expertise and guidance has proved useful at every step. Their uncompromising quality standards have made this book an exceptional effort. Their encouragement from time to time has been an inspiration for everyone.

The publisher and the editorial board hope that this book will prove to be a valuable piece of knowledge for researchers, students, practitioners and scholars across the globe.

List of Contributors

Sheldon Waite and Erdal Oruklu
Department of Electrical and Computer Engineering, Illinois Institute of Technology, Chicago, USA

Luan Carlos Santos Silva, João Luiz Kovaleski, Silvia Gaia, Manon Garcia and Pedro Paulo de Andrade Júnior
Department of Production Engineering and Technology Transfer Research Group, Federal University of Technology—Paraná (UTFPR), Ponta Grossa, Brazil

Hyuk-Jae Roh
Saskatchewan Ministry of Highways and Infrastructure, Regina, Canada

Sandeep Datla
City of Edmonton, Edmonton, Canada

Satish Sharma
Faculty of Engineering, University of Regina, Regina, Canada

Michael D. Anderson and Jeffrey P. Wilson
Department of Civil and Environmental Engineering, University of Alabama in Huntsville, Huntsville, USA

Gregory A. Harris
Center for Management and Economic Research, University of Alabama in Huntsville, Huntsville, USA

Francesca Pagliara
Department of Civil, Architectural and Environmental Engineering, University of Naples Federico II, Naples, Italy

John Preston
School of Civil Engineering and the Environment, University of Southampton, Southampton, UK

Shauna L. Hallmark, Bo Wang and Yu Qiu and Robert Sperry
Iowa State University, Ames, USA

Christopher A. Bolin
Division of Sustainability, AquAeTer, Inc., Centennial, USA

Stephen T. Smith
Division of Sustainability, AquAeTer, Inc., Helena, USA

Shuangrui Fan, Tingyun Ji and Bergqvist Rickard
Logistics and Transport Research Group, Department of Business Administration, School of Business, Economics and Law at University of Gothenburg, Göteborg, Sweden

Wilmsmeier Gordon
Economic Commission for Latin America and the Caribbean (ECLAC), Santiago, Chile
Transport Research Institute (TRI), Edinburgh Napier University, Edinburgh, UK

Vaishali D. Khairnar and Srikhant N. Pradhan
Computer Department, Institute of Technology Nirma University, Ahmadabad, India

Tarek K. Refaat, Mai Hassan, Ramez M. Daoud and Hassanein H. Amer
Electronics Engineering Department, American University in Cairo, New Cairo, Egypt

Xueming Chen
L. Douglas Wilder School of Government and Public Affairs, Virginia Commonwealth University, Richmond, USA

Diala Jomaa, Siril Yella and Mark Dougherty
Department of Computer Engineering, Dalarna University, Borlänge, Sweden

Arshdeep Bahga and Vijay K. Madisetti
Electrical and Computer Engineering, Georgia Institute of Technology, Atlanta, USA

Fady W. Gendi, Tarek K. Refaat, Amir H. Sadek, Ramez M. Daoud, Hassanein H. Amer, Chahir S. Fahmy and Omar M. Kassem
Electronics Engineering Department, American University in Cairo, New Cairo, Egypt

Hany M. ElSayed
Electronics and Communication Department, Cairo University, Giza, Egypt

Christopher A. Bolin
AquAeTer, Inc., Division of Sustainability, Centennial, USA

Stephen T. Smith
AquAeTer, Inc., Division of Sustainability, Helena, USA

Michael F. Gorman
Department of MIS, Operations and Decision Sciences, University of Dayton, Dayton, USA

Yan Zhou, Anant Vyas and Danilo Santini
Argonne National Laboratory, Argonne, USA

Deshan Shan, Shengai Cui and Zhen Huang
Bridge Engineering Department, Southwest Jiaotong University, Chengdu, China

Jing Shi and Nian Zhou
Institute of Transportation Engineering, Tsinghua University, Beijing, China